The Conservation of Medicinal Plants

Proceedings of an International Consultation

Organized by

The World Health Organization (WHO)
IUCN – the World Conservation Union
and the World Wide Fund for Nature (WWF)

In association with

The Ministry of Health of the Royal Thai Government

Sponsored by

The United Nations Environment Programme

In technical collaboration with

The Food and Agriculture Organization of the United Nations (FAO)

World Health Organization, 1211 Geneva 27, Switzerland
IUCN – The World Conservation Union, 1196 Gland, Switzerland
The World Wide Fund for Nature, 1196 Gland, Switzerland

Programme Organisers

O Akerele (WHO)
N Farnsworth
O Hamann (IUCN)
V Heywood (IUCN
J A McNeely (IUCN)
G Stott (WHO)
H Synge (WWF)

The Conservation of Medicinal Plants

Proceedings of an International Consultation
21–27 March 1988 held at
Chiang Mai, Thailand

edited by
Olayiwola Akerele, Vernon Heywood and Hugh Synge

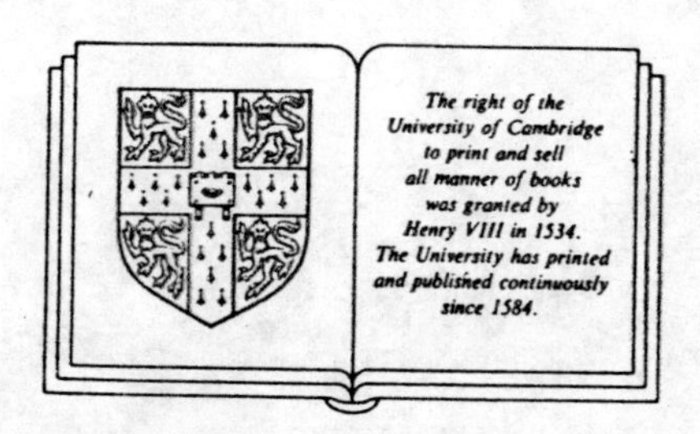
The right of the
University of Cambridge
to print and sell
all manner of books
was granted by
Henry VIII in 1534.
The University has printed
and published continuously
since 1584.

CAMBRIDGE UNIVERSITY PRESS
Cambridge
New York Port Chester Melbourne Sydney

Published by the Press Syndicate of the University of Cambridge
The Pitt Building, Trumpington Street, Cambridge CB2 1RP
40 West 20th Street, New York, NY 10011–4211, USA
10 Stamford Road, Oakleigh, Melbourne 3166, Australia

First published 1991

Printed in Great Britain at the University Press, Cambridge

British Library cataloguing in publication data available

Library of Congress cataloguing in publication data available

ISBN 0 521 39206 3 hardback

VW

Dedication

The book is dedicated to the fond memory of our late colleague, Dr Pricha Desawadi, an outstanding public health administrator and budding conservationist. May his soul attain Nirvana.

Contents

Techniques to Conserve Medicinal Plants

Policies to Conserve Medicinal Plants

Experiences from Programmes to Conserve Medicinal Plants

Contributors

Dr O Akerele
World Health Organisation, CH-1211 Geneva 27, Switzerland

Mr S K Alok
Joint Secretary, Ministry of Health and Family Welfare, 110011 New Delhi, India

Mr L de Alwis
30 Hotel Road, Mt Lavinia, Sri Lanka

Dr A Bonati
Inverni della Beffa, Via Ripamonti 99, 20141 Milano, Italy

Dr Cheng Zhong-ming
Nanjing Botanical Garden Mem. Sun Yat-Sen, Nanjing, Jiangsu, People's Republic of China

Dr A B Cunningham
Institute of Natural Resources, University of Natal, PO Box 375, Pietermaritzburg 3200, Republic of South Africa

Professor N R Farnsworth
University of Illinois at Chicago, College of Pharmacy, 833 South Wood Street, Chicago, Illinois 60612, USA

Professor O Hamann
Botanical Gardens, University of Copenhagen, O Farimagsgade 2 B, DK-1353 Copenhagen K, Denmark

Professor He Shan-an
Nanjing Botanical Garden Mem. Sun Yat-Sen, Nanjing, Jiangsu, People's Republic of China

Professor V H Heywood
IUCN Plants Office, Descanso House, 199 Kew Road, Richmond, Surrey, TW9 3BW, UK

Dr Akhtar Husain
Central Institute of Medicinal and Aromatic Plants, Post Bag No 1, PO Ram Sagar Misra Nagar, Lucknow - 226 016, India

Dr A S Islam
Professor of Botany, University of Dhaka, Dhaka, Bangladesh

Dr T C Jessup
c/o World Wide Fund for Nature, PO Box 133, Bogor 16001, Indonesia

Dr K Kartawinata
Unesco Regional Office for Science and Technology for Southeast Asia, J L Thamrin 14, Tromolpos 273/JKT, Jakarta, Indonesia

Hon. W J M Lokubandara
Minister of Education, Cultural Affairs and Information, Colombo, Sri Lanka

Mr C de Klemm
21 rue de Dantzig, 75015 Paris, France

Dr J O Kokwaro
Department of Botany, Faculty of Sciences, University of Nairobi, Nairobi, Kenya

Mr J A McNeely
IUCN, Avenue du Mont-Blanc, CH-1196 Gland, Switzerland

Dr C Padoch
Institute of Economic Botany, New York Botanical Garden, Bronx, New York 10458-5126, USA

Professor D Palevitch
Agricultural Research Organisation, Institute of Field and Garden Crops, The Volcani Center, 6 Bet-Dagan, 50250 Israel

Dr M J Plotkin
Conservation International, 1015 18th Street NW, Suite 1000, Washington DC 20036, USA

Dr P P Principe
ORD/OMMSQA (RD-680), US Environmental Protection Agency, Washington DC 20460, USA

The late Dr Pricha Desawadi
Formerly Deputy Director-General, Department of Medicinal Sciences, Ministry of Public Health, Yod-se, Bangkok 10100, Thailand

Dr M A Rifai
Herbarium Bogoriense, Puslitbang Biologi-LIPI, Bogor, Indonesia

Professor R E Schultes
Botanical Museum, Harvard University, Oxford Street, Cambridge, Mass. 02138, USA

Dr H M Schumacher
Deutsche Sammlung von Mikroorganismen und Zellkulturen GmbH, Mascheroder Weg 1B, D-3300 Braunschweig, Germany

Dr D D Soejarto
University of Illinois at Chicago, College of Pharmacy, 833 South Wood Street, Chicago, Illinois 60612, USA

Dr H Soedjito
Herbarium Bogoriense, Puslitbang Biologi-LIPI, Bogor, Indonesia

Mr H Synge
49 Kelvedon Close, Kingston-upon-Thames, KT2 5LF, UK

Mr J W Thorsell
IUCN, Avenue du Mont-Blanc, CH-1196 Gland, Switzerland

Mr P Wachtel
WWF International, Avenue du Mont-Blanc, CH-1196 Gland, Switzerland

Dr Xiao Pei-gen
Institute of Medicinal Plant Development (IMPLAD), Chinese Academy of Medicinal Sciences, Hai Dian District, Dong Bei Wang, Beijing, People's Republic of China

Preface

Nearly all cultures from ancient times to the present day have used plants as a source of medicines. A considerable percentage of the peoples in both developed and developing countries use medicinal plant remedies and, according to the World Health Organization (WHO), the number is on the increase, especially among younger people. In the industrialized countries, consumers are seeking visible alternatives to modern medicine with its dangers of over-medication. The escalating cost of sophisticated medical care is another factor.

Over the last decade or so, WHO's Health Assembly has passed a number of resolutions in response to a resurgence of interest in the study and use of traditional medicine in health care, and in recognition of the importance of medicinal plants to the health systems of many developing countries. In answer to WHO's call, health authorities and administrators in developing countries have decided to take traditional forms of medicine more seriously and to explore the possibility of utilizing them in primary health care.

This great surge of public interest in the use of plants as medicines has been based on the assumption that the plants will be available on a continuing basis. However, no concerted effort has been made to ensure this, in the face of the threats of increasing demand, a vastly increasing human population and extensive destruction of plant-rich habitats such as the tropical forests. In some developing countries, users have to boil their plant medicines using precious wild-collected fuelwood, and so add insult to ecological injury.

Today many medicinal plants face extinction or severe genetic loss, but detailed information is lacking. For most of the endangered medicinal plant species no conservation action has been taken. For example, there is very little material of them in genebanks. Also, too much emphasis has been put on the potential for discovering new wonder drugs, and too little on the many problems involved in the use of traditional medicines by local populations.

For most countries, there is not even a complete inventory of medicinal plants. Much of the knowledge on their use is held by traditional societies, whose very existence is now under threat. Little of this information has been recorded in a systematic manner. Besides the identification and selection of medicinal plants for use in health services, there is the potential that plants hold as an inexhaustible reservoir for the identification and isolation of useful chemical compounds for syndromes such as AIDS, for which there is yet no known cure.

In the light of this situation, WHO, IUCN and WWF decided that it would be timely to collaborate in convening an International Consultation on the conservation of medicinal plants, bringing together leading experts in different fields to exchange views on the problems, determine priorities and make recommendations for action. The experts at the meeting included adminstrators and policy-makers in health and conservation, and covered the disciplines of ethnomedicine, botany, education, pharmacology, nature conservation and economics. For IUCN and WWF, this meeting was an important part of their Joint Plants Programme, started in 1984.

The Consultation took place in Chiang Mai, Thailand, on 21-27 March 1988, with the Ministry of Public Health of the Royal Thai Government as host. A wide range of topics was covered, which included a review of medicinal plant policies (utilisation and conservation) in individual countries; the need for information systems, including databases; and the part that botanic gardens can play in the cultivation and conservation of endangered medicinal species.

A lively and stimulating exchange of views took place between the conservationists, scientists and health administrators, who were meeting for the first time in the same forum. The participants prepared and issued the "Chiang Mai Declaration - Saving Lives by Saving Plants" (page xix) – which affirms the importance of medicinal plants and calls on the United Nations, its agencies and Member States, as well as other international organisations, to take action for their conservation. The Consultation also outlined and reviewed a set of draft guidelines on the conservation of medicinal plants, which will be published separately and disseminated widely to governments and relevant institutions throughout the world for adaption to local situations.

The forty-first World Health Assembly (1988) drew attention to the Chiang Mai Declaration and endorsed the call for international cooperation and coordination to establish a basis for the conservation of medicinal plants, so as to ensure that adequate quantities are available for future generations. This places medicinal plants, their rational and sustainable use, and their conservation, firmly in the arena of public health policy and concern.

Olayiwola Akerele, WHO
Vernon Heywood, IUCN
Hugh Synge, WWF

Acknowledgements

WHO, IUCN and WWF would like to acknowledge the kind hospitality and cooperation of the Royal Thai Government. We would like to recognise the role of our colleagues in Thailand, especially the late Dr Pricha Desawadi, former Deputy Director-General, Department of Medical Sciences, Ministry of Public Health, and Mrs Wantana Mgamwat, Director, Division of Medicinal Research, Department of Medical Sciences, Ministry of Public Health, for their efforts in facilitating the organisation of the Consultation and for their active participation. We also wish to thank Professor Norman Farnsworth, Professor Ole Hamann and Dr Gordon Stott for their valued help in preparing the initial programme for the meeting, and Dr Liu Guo-Bin and Dr B B Gaitonde of WHO for their help.

WHO, IUCN and WWF wish to acknowledge the financial support provided by the Danish aid agency DANIDA and by the United Nations Environment Programme (UNEP), which enabled the participation of several delegates from developing countries; and a generous grant from the American Society of Pharmacognosy, which has helped with the follow-up to the Consultation.

The editors would like to thank most warmly Maryse Cestre, Penny Croucher, Erika Kiess and Nicky Powell for their secretarial support to the conference and in particular for their magnificent contribution to the arduous task of completing these Proceedings. They would also like to thank Sarah Stacey for the splendid work she did in promoting coverage of the meeting in the press; as a result of her work, the meeting received more publicity than any event ever before on the conservation of plants.

Olayiwola Akerele, WHO
Vernon Heywood, IUCN
Hugh Synge, WWF

The Chiang Mai Declaration

Saving Lives by Saving Plants

We, the health professionals and the plant conservation specialists who have come together for the first time at the WHO/IUCN/WWF International Consultation on Conservation of Medicinal Plants, held in Chiang Mai, 21-26 March 1988, do hereby reaffirm our commitment to the collective goal of "Health for All by the Year 2000" through the primary health care approach and to the principles of conservation and sustainable development outlined in the World Conservation Strategy. We:

- Recognise that medicinal plants are essential in primary health care, both in self-medication and in national health services;
- Are alarmed at the consequences of loss of plant diversity around the world;
- View with grave concern the fact that many of the plants that provide traditional and modern drugs are threatened;
- Draw the attention of the United Nations, its agencies and Member States, other international agencies and their members and non-governmental organisations to:
  - The vital importance of medicinal plants in health care;
  - The increasing and unacceptable loss of these medicinal plants due to habitat destruction and unsustainable harvesting practices;
  - The fact that plant resources in one country are often of critical importance to other countries;
  - The significant economic value of the medicinal plants used today and the great potential of the plant kingdom to provide new drugs;
  - The continuing disruption and loss of indigenous cultures, which often hold the key to finding new medicinal plant that may benefit the global community;
  - The urgent need for international cooperation and coordination to establish programmes for conservation of medicinal plants to ensure that adequate quantities are available for future generations.

We, the members of the Chiang Mai International Consultation, hereby call on all people to commit themselves to Save the Plants that Save Lives.

Chiang Mai, Thailand
26 March 1988

Introduction

Medicinal Plants: Policies and Priorities

Olayiwola Akerele
Programme Manager, Traditional Medicine, Division of Diagnostic, Therapeutic and Rehabilitative Technology, World Heath Organization, Geneva

Traditional medicine has been with the World Health Organization (WHO) for the last twelve years or so and for the rest of the world for the last several thousand years of recorded history. One might say that we are new at the game.

Traditional medicine is widespread throughout the world. It comprises those practices based on beliefs that were in existence, often for hundreds of years, before the development and spread of modern scientific medicine and which are still in use today. As its name implies, it is part of the tradition of each country and employs practices that have been handed down from generation to generation. Its acceptance by a population is largely conditioned by cultural factors and much of traditional medicine, therefore, may not be easily transferable from one culture to another. In dealing with traditional medicine, WHO aims at exploiting those aspects of it that provide safe and effective remedies for use in primary health care.

Acknowledging its potential value for the expansion of health services, the World Health Assembly has passed a number of resolutions. In 1976, it drew attention to the manpower reserve constituted by traditional practitioners (resolution WHA29.72). In 1977, it urged countries to utilize their traditional systems of medicine (resolution WHA30.49). In 1978, it called for a comprehensive approach to the subject of medicinal plants (resolution WHA 331.33.) This approach was to include:

- An inventory and therapeutic classification, periodically updated, of medicinal plants used in different countries;
- Scientific criteria and methods for assessing the safety of medicinal plant products and their efficacy in the treatment of specific conditions and diseases;
- International standards and specifications for identity, purity, strength and manufacturing practices;
- Methods for safe and effective use of medicinal plant products by

various levels of health worker;
- Dissemination of such information among Member States; and
- Designation of research and training centres for the study of medicinal plants.

In May 1987, the Fortieth World Health Assembly (resolution WHA 40.33) reaffirmed the main points of the earlier resolutions and the related recommendations made, in 1979, by the Alma-Ata Conference. This resolution provides a mandate for future action in this field and is summarized below. Member States were urged *inter alia*:

- To initiate comprehensive programmes for the identification, evaluation, preparation, cultivation and conservation of medicinal plants used in traditional medicine;
- To ensure quality control of drugs derived from traditional plant remedies by using modern techniques and applying suitable standards and good manufacturing practices.

Thus, the importance of conservation is recognized by WHO and its Member States and is considered to be an essential feature of national programmes on traditional medicine.

Several years ago the over-exploitation of wild-growing *Rauvolfia serpentina* in India for export exhausted the supply to a point where the Indian government placed an embargo on the export of this plant, which remains in place today. This has created a major problem in the United States of America, since the United States Pharmacopoeia requires that *Rauvolfia serpentina* be of Indian origin when used in a crude form. Another example of a plant that has been over-exploited in India for export to other Asian countries is *Coptis teeta*, which is now considered as endangered in India.

A general idea of WHO's priorities in this field for the next few years may be obtained from the medium-term programme for traditional medicine (WHO, 1987). I shall not attempt to describe all the activities listed but shall concentrate on those items that are directly relevant to the subject of this book.

In most developing countries, where coverage by health services is limited, it is to the traditional practitioner or to folk medicine that the majority of the population turns in sickness and the treatment is, in large part, based on the use of medicinal plants. Early in this century, the greater part of medical therapy in the industrialized countries was dependent on medicinal plants but, with the growth of the pharmaceutical industry, their use fell out of favour. Even so, 25% of all prescriptions dispensed from community pharmacies from 1959 to 1980 in the U.S.A. contained plant extracts or active principles prepared from higher plants (Farnsworth, 1984).

According to a survey by the International Trade Centre (1982), trade in medicinal plants and their derivatives in pharmacy has declined in many industrialized countries, owing to the volume of competitive synthetic products currently marketed. Now the pendulum is swinging back and the value of medicinal plants in treatment is receiving increasing attention worldwide.

Overall, the trade in botanicals has increased, through their use in the health food and cosmetic industries. Imports, however, represent only a minute percentage of the value of internal trade in medicinal plants. For example, in the United States, imports in 1980 were valued at US$44.6 million compared to an internal trade in medicinals and botanicals in 1981 estimated at US$3.9 billion.

In a recent report (WHO, 1986), the production of traditional plant remedies in China was valued at US$571 million and the country-wide sales of crude plant drugs at US$1.4 billion annually (Li Chaojin, 1987).

It is evident that, even if imports by industrialized countries are declining (which is doubtful), the potential for internal trade in medicinal plants and their derivatives in many developing countries is tremendous.

The attention paid by health authorities and administrations to the use of medicinal plants has increased considerably, although for different reasons. In developing countries, this has largely resulted from a decision to take traditional forms of medicine more seriously and to explore the possibility of utilizing them in primary health care. In other countries, health authorities have been compelled to exercise closer surveillance because of the growth in the popular use of herbs and plants in self-treatment and in the health food industry, developments not necessarily to their liking.

It is often claimed and is widely believed that remedies of natural origin are harmless and carry no risk to the consumer. We should keep in mind that nothing could be further from the truth, particularly where there is a risk of toxic plants being used by mistake or where "herbal preparations" are marketed with the addition of undeclared potent synthetic substances.

That many traditional remedies are of therapeutic value is no longer open to doubt. However, the use of manufactured products should be governed by the same standards of safety and efficacy as are required for modern pharmaceutical products. Proof of safety should take precedence over establishing efficacy, and accuracy in labelling the constituents of medicinal plant remedies is critical for safety evaluation and drug control.

In developed countries, a knowledge of medicinal plants is no longer required in the training of health staff; pharmacognosy, the study of drugs of plant or animal origin, has virtually disappeared from the curriculum and the subject is passed over in silence. By contrast, in recent years, there has been a great surge of public interest in the use of herbs and plants, a subject which has received extensive coverage in the press and lay publications, much of it uncritical and unverified and some even dangerous.

Ensuring safety, therefore, in the use of medicinal plants and remedies derived from them requires not only measures for control but also a substantial effort in public information and professional education and in making readily available up-to-date and authoritative data on their beneficial properties and possible harmful effects.

Where safe and simple medicinal plant remedies have been employed traditionally for a long time in the treatment of minor self-limiting conditions, establishing efficacy may not be so important, provided their composition is known.

However, where a medicinal plant remedy, traditional or otherwise, is to be marketed for the treatment of more serious conditions, the manufacturer should demonstrate both safety and efficacy before the product is licensed. A pre-market review is indispensable if the dangers of exaggerated claims and ineffective or unduly toxic medicaments are to be avoided.

With this as an introduction, let me describe some of the activities undertaken by WHO.

A WHO/DANIDA Inter-Regional Workshop on Appropriate Methodology for Selection and Use of Traditional Remedies in National Primary Health Care Programmes was held in Bangkok, Thailand, in November/December 1985, with participants from Indonesia, Malaysia, Nepal, the Philippines and Thailand. The Workshop was the first in a series intended to address problems of safety and efficacy of traditional remedies, including related issues of standards, stability and dosage formulation. The overall objective of this first Workshop was for participants to acquire the methodology needed for the introduction and utilization of natural substances in health services.

Subsequently, a visit was made to three countries (Indonesia, Nepal and Thailand) to assess the impact of the course in improving methodology and the application of newly acquired knowledge to the use of medicinal plants in primary health care. All three countries have conducted satellite workshops for their nationals and have proceeded to select a number of single plant remedies for use in their health services.

Another workshop, to follow up the Bangkok meeting, is planned and will address some of the problems identified by participants. WHO is now planning, with the support of DANIDA, two similar workshops for countries of East, Central and Southern Africa and for the West African countries.

The use of medicinal plants in traditional medicine finds its natural expansion and further development in primary health care. It is at this level that the transition from traditional practice to medical care can most easily be made.

In China, traditional medicine is an integral part of the formal health system and is utilized in about 40% of cases at the primary care level. Supplies are assured by the state-owned Chinese Crude Drugs Company which has branches in all provinces, autonomous regions, municipalities and counties. Formerly, crude drugs were mostly collected in the wild state but, with more brought under cultivation, natural sources are becoming depleted. Special encouragement has, therefore, been given for the cultivation of medicinal plants. Agricultural departments at all levels take part in formulating policy and establishing plantations which now cover some 330,000 hectares (Li Chaojin, 1986).

It was with a view to sharing experience in this area that, in 1985, WHO sponsored an Inter-regional Seminar on the Role of Traditional Medicine in Primary Health Care in China (WHO, 1986). It gave senior administrators, responsible for national health policy in 19 countries, an opportunity of studying the utilization of traditional medicine in primary health care and of examining the possibility of adopting comparable approaches in their own health services. The response to the Seminar was prompt, several of the countries represented (Bangladesh, India, Nepal, Philippines, Sri Lanka and Sudan) taking action to make better use of their traditional systems of medicine.

In India, the Ministry of Health and Family Welfare took the valuable step of holding, in 1986, four regional seminars on medicinal plants, their collection, cultivation, exploitation, conservation and rational use. Copies of these reports may be obtained from the Ministry of Health and Family Welfare, New Delhi-110011, India. Participants were representative of the various disciplines, occupations and institutions concerned. This important and interesting initiative could very well be taken by other countries. It would create greater awareness of the extent of their wealth of medicinal plants and would stimulate the different departments involved to work together.

At the Fourth International Conference of Drug Regulatory Authorities, co-sponsored by the Ministry of Health and Welfare of Japan and WHO and held in Tokyo in July 1986, a workshop was held on traditional herbal medicines. It was acknowledged that these medicines play an important part in health care in many countries, developed as well as developing. It was noted that truly traditional practices are more amenable to influence through education and training than to statutory control. The workshop also concentrated on the exploitation of traditional medicine through over-the-counter sales of labelled products on a commercial basis and addressed the need for legislation, quality, standards, and information.

Countries wishing to make full use of their heritage of traditional medicine and the wealth of medicinal plants which most of them possess, have a special interest in sponsoring ethno-medical studies, bringing together botanists, clinicians, pharmacologists and others for the purpose and in making adequate resources available.

A first step would be to review on a national basis the utilization of medicinal plants in general and of medicaments derived from them. Such an examination might reveal opportunities for making greater use of safe and effective galenical preparations which might stimulate local cultivation and production and at the same time permit economies to be made, saving scarce foreign currency.

In any event, experience in the preparation and utilization of galenicals is a necessary prerequisite for clinical evaluation of traditional remedies, when these have been identified as meriting further study. Such evaluation may present many problems, particularly with compound medicines containing several ingredients.

An important feature of traditional Chinese therapy is the preference of practitioners for compound prescriptions over single substances, it being held that some constituents are effective only in the presence of others. This renders assessment of efficacy and, eventually, identification of active principles much more difficult than for simple preparations. The whole subject of the rationale for compound prescriptions of medicinal plants offers a vast field for research. The addition of modern synthetic drugs to traditional remedies complicates the matter of evaluation even further.

National inventories of medicinal plants are essential if sound programmes for their rational use and exploitation are to be developed. Such inventories, still to be made in many countries, need to describe the geographic and climatic distribution of medicinal plants, their source (collection from the wild, cultivation in *in situ* or *ex situ* in botanic gardens, commercial plantations) and an indication of their relative abundance or scarcity.

For each plant there would be an account of its utilization (e.g. folk medicine, traditional healers, pharmaceutical or food industries) and its place in commerce (e.g. local use, internal trade, export). There would also be a description of its constituents, pharmacological properties, and therapeutic indications.

Logically, the investigation, utilization and exploitation of medicinal plants by a country should also include measures for their conservation. This is the subject of our Consultation and it is expected that a number of national programmes for the conservation of medicinal plants will result from this initiative. Conservation and inventories of medicinal plants go hand in hand, the latter being essential for the identification of endangered species, for setting priorities and for monitoring the situation.

Most developing countries have an abundance of medicinal plants which are used in their traditional forms of medicine. The planning of pharmacological and clinical studies to assess their safety and therapeutic efficacy, the decision to cultivate them commercially, and the development of policies for their conservation all require some form of ranking or assessment of their relative values and importance. There is, however, no simple way of making such comparisons; thus:

- Certain plants may be common but are widely and safely used in folk medicine which, presumably, makes them important;
- Other plants may be used in treating serious conditions and are important because of their therapeutic value; while
- Still others may be of great economic importance as items of internal trade or export, and so on.

Furthermore, within the same geographical area, a medicinal plant may be considered to have very important properties by some communities but not by others.

Obviously, any conclusion about the relative importance of a particular medicinal plant must depend on the criteria applied and the context in which it is considered.

It is necessary, therefore, to take a comprehensive approach and to bring together the main disciplines and interests concerned – health, agriculture, industry, trade, universities – under some form of coordinating mechanism. Such a body would assess needs and priorities, formulate national policy, help to mobilize resources, and ensure the orderly development of work and research in this field.

Medicinal plants, their study, evaluation, utilization and conservation, form an important part of the proposed medium-term programme in traditional medicine for WHO and Member States over the next few years (WHO, 1987).

Situation analyses in countries will include inventories of medicinal plants, assessment of the safety and effectiveness of traditional remedies, and comparisons of the relative advantages of western and traditional methods in the treatment of specific conditions.

Countries will lay special emphasis on making safety a foremost consideration, not only through the training of professional and technical staff and the application of standards, specifications and good manufacturing practices but also by seeing that the public is kept well-informed on the subject.

The activities described will be mostly implemented by countries themselves, with relatively little need for external financing. WHO's involvement will have to be largely catalytic in nature and this is as it should be. Fortunately, the Traditional Medicine Programme has many linkages – with other programmes in WHO, with many international agen- cies and organizations, and with numerous government and university departments.

The challenge will be to find ways of working together. The 1987 World Health Assembly resolution (WHA40.33) serves to remind all concerned that much remains to be done. Strong official support in their own countries is probably what workers in this field need most.

Of almost equal importance is ready access to technical expertise at home and abroad, the ability to keep in touch with developments in other countries, and the opportunity to exchange ideas and experience. This traditional role of international organizations and universities is one that has considerable potential for expansion, at least so far as medicinal plants are concerned.

Collaboration with United Nations agencies, multinational and non-governmental organizations active in fields related to traditional medicine has been increasing recently and will continue. Linkages with and between WHO Collaborating Centres for Traditional Medicine, with their wider responsibilities, will form an important part of the programme.

I have tried to give you an idea of WHO's policies and activities in relation to medicinal plants and their rational and sustainable utilization. Significant progress over the next few years will depend on the imagination and determination which can be brought to bear on the subject.

References

Akerele, O., Stott, G. & Lu Welbo (Eds) (1987). The role of traditional medicine in Primary Health Care in China. *American Journal of Chinese Medicine*, Suppl. No.1.

Farnsworth, N.R. (1984). How can the well be dry when it is filled with water? *Economic Botany*, 38, 4-13.

International Trade Centre UNCTAD/GATT (1982). *Markets for Selected Medicinal Plants and their derivatives*. Geneva.

Li Chaojin (1987). Management of Chinese Traditional Drugs. In *The Role of Traditional Medicine in Primary Health Care in China*, eds. O. Akerele, G. Stott and Lu Welbo. *American Journal of Chinese Medicine*, Suppl. N°. 1, 39-41.

Proceedings of the Fourth International Conference of Drug Regulatory Authorities, 1986. Tokyo.

WHO (1987). *Global Medium-Term Programme (Traditional Medicine) covering a specific period 1990-1995*. (WHO document TRM/MTP/87.1)

The Joint IUCN-WWF Plants Conservation Programme and its Interest in Medicinal Plants

Ole Hamann
Botanical Gardens, University of Copenhagen, Denmark

Introduction

Living resource conservation for sustainable development is the main goal of the world's largest, independent conservation organizations, the International Union for Conservation of Nature and Natural Resources (IUCN) and the World Wide Fund for Nature (WWF).

Around the world, over-exploitation of living resources takes place in order to meet short term needs, but more often than not the process destroys exactly those resources on which the welfare of millions of people depend in the long term. This lack of sustainable utilization is, in fact, widening the gap between rich and poor countries in the world, because there is a clear relation between conservation and development: they operate in the same global context, and the underlying problems that must be overcome if either is to be successful are identical (IUCN, UNEP & WWF, 1980).

Most tropical countries are developing countries with grave economic, social and perhaps political problems. In these countries, an accelerating loss of biological diversity – plants, animals and microorganisms – leads to a breakdown of ecological processes and life-support systems. These problems affect developing and developed countries alike, and international action and collaboration are needed to find and implement solutions.

The Case of Tropical Forests

The Food and Agriculture Organization of the United Nations (FAO) has estimated that about 56% closed tropical forest remained on earth in 1980; 44% closed tropical forest, or rain forest, had already disappeared during the period 1940 to 1980 (Lanly, 1982). The annual deforestation was estimated to be about 75,000 km^2, but this figure included rain forests only. An additional 38,000 km^2 of other forest types are being lost annually, i.e. drier forests, monsoon forests, etc. If this continues, almost all rain forests, which are not protected as reserves or national parks, will be seriously altered or destroyed in just 30 years, and all rain forests of the world will be negatively affected in just 80 years. Presumably only few areas in Amazonia and Central Africa will remain intact, because of the vast size and remoteness of some parts of these areas.

However, such projections for the fate of the tropical forests rarely take into account the population growth in the world. The effect of the enormous population growth on the global ecosystems will be very great. Recently it has been estimated that we already use about 40% of the biological primary production of the globe, directly or indirectly (Vitousek *et al.*, 1986). Considering that the world population may reach 7.7 billion by the year 2020, one may wonder whether the biosphere will have the capacity to support such a number of people.

Population growth is one important factor to consider in relation to conservation and sustainable utilization of living resources, particularly in tropical countries where the growth rate is highest (WRI, 1985). Another important factor is poverty. The World Bank has estimated that more than 1 billion people in tropical countries live in absolute poverty. This leads, naturally and easily understood, to a tremendous pressure on the natural resources, which in many cases far exceeds the level of sustainable utilization. A vicious circle is created, where environmental degradation, population growth and poverty interlock and prevent sound development.

What is being Done?

The complex problems inherent in development and sustainable utilization are more and more recognized politically. Recently they were analyzed by the World Commission on Environment and Development, otherwise known as the Brundtland Commission, after the Norwegian Prime Minister Gro Harlem Brundtland who chaired the commission. The report of the commission, *Our Common Future,* represents a broad north-south, east-west consensus on the major conservation and development issues facing the world.

The report addresses the United Nations family in a major effort to build the political will required to turn the development process into a new direction which is both sustainable and equitable. Its basic concept is sustainable development, and it points to the need for a new international policy of solidarity, which aims at conserving natural resources as a necessary factor for economic growth. The report thus emphasizes the importance of establishing a healthy and productive link between the environment and economic growth.

It is probably still too early to see new international initiatives as a result of the Brundtland Commission's report, but it is already having profound influence in some countries. In Denmark, for example, the Danish International Development Agency (Danida) has formulated a new, overall strategy for incorporating environmental considerations in the development aid programmes (Danida, 1988). Following this sectoral strategies have been formulated, and thereafter specific country strategies will be made for the work of Danida in its collaboration with its main partners, viz. Kenya, Tanzania, India, Bangladesh and China.

An Ongoing Process

These recent developments may be regarded as the culmination of a process which began in the 1970s, and which to a very large extent has been driven by non-governmental conservation organizations such as IUCN and WWF. The major principles and objectives of conservation, which had been brought forward by the conservation movement during the previous decades, were presented in the *World Conservation Strategy* (IUCN, UNEP & WWF, 1980). This introduced the concept of sustainable utilization as a central element in conservation, and in 1987, the Brundtland report then added the political dimension to the issue in the effort to mobilize international action.

The *World Conservation Strategy* broke the idea of conservation being for the few, being elitist, and linked instead conservation with the utilization and management of natural resources. It outlined the three main objectives for living resource conservation:

1. To maintain essential ecological processes and life-support systems (such as soil regeneration and protection, the recycling of nutrients, and the cleansing of waters), on which human survival and development depend;
2. To preserve genetic diversity (the range of genetic material found in the world's organisms) on which depend the functioning of many of the above processes and life-support systems, the breeding programmes necessary for the protection and improvement of cultivated plants, domesticated animals and microorganisms, as well as much scientific and medical advance, technical innovation, and the security of the many industries that use living resources; and

3. To ensure the sustainable utilization of species and ecosystems (notably fish and other wildlife, forests and grazing lands), which support millions of rural communities as well as major industries.

The *World Conservation Strategy* is exerting great influence on the policies of many countries. Quite a number have taken the next step forward in preparing a *National Conservation Strategy* which is a national plan building on the principles of the *World Conservation Strategy.* Zambia is one tropical country which has completed its own *National Conservation Strategy* with the assistance of IUCN, and several other countries are in the process of doing so.

The Joint IUCN-WWF Plants Conservation Programme

The first phase, 1984-1987. The Plants Conservation Programme was launched in 1984, because it was realized that plants had been neglected by the conservation organizations. During the first 3 years of the programme, about 2 million US dollars were used on some 50 different projects and activities, in what may be called the first phase of the programme. Compared to the magnitude of the problems this investment is small, but nevertheless it is considerable for an effort especially targeted at plants. In order to maximize the effect, a major tactic from the start was to design the programme to be catalytic, so as to provide other, larger bodies with the arguments and methods to work for plant conservation. This first phase of the programme has been a success, because it has placed plants on the international conservation agenda. Three main factors lie behind this preliminary success; one, the programme was realistically formulated from the beginning, building on the principles of the *World Conservation Strategy* and using the lessons learned from previous programmes of IUCN and WWF, such as the one on tropical forest conservation; two, a central coordination was ensured from the headquarters of IUCN and WWF; and three, some of the leading botanists of the world agreed to become involved by joining a Plant Advisory Group, which has contributed much in intellectual and scientific guidance.

The projects and activities in the Plants Conservation Programme can roughly be divided into two types: 1), the strategic activities, in which the basic principles and methods for plant conservation are developed, and 2), on-the-ground model projects where the principles and methods are tried out in the field. The activities and projects are then grouped according to the following 6 major objectives of the programme:

1. Spreading the message. To build public awareness on how plants contribute to our lives, and so stimulate action by pressure on governments and industry and by help to conservation bodies;
2. Building the capacity to conserve. To encourage and help other organizations to conserve plants, in particular the members of IUCN and the National Organizations of WWF, by providing the basic services they need;
3. Conserving plant genetic resources. To conserve sufficient diversity within crops and their wild relatives to ensure that all their potential is available for use in the future. Working closely with the International Board for Plant Genetic Resources (IBPGR) and FAO, the programme concentrates on the relatively new subject of conserving plant genetic resources *in situ* as a complement to *ex situ* techniques;
4. Conserving wild plants of economic value. To emphasize the conservation of wild plants that are important, particularly for rural people in developing countries. Within this large and neglected group of plants, the programme is giving special attention to the conservation of medicinal plants used in primary health care;
5. Strengthening the capacity of botanic gardens to achieve conservation. To build the c. 1500 botanic gardens and arboreta of the world into an effective, data-rich network of institutions with the primary aim of plant conservation; to do this the programme concentrates on education, awareness, training and networking; and
6. Promoting plant conservation in selected countries. To identify, develop and implement on-the-ground model projects in priority countries.

Through this set of objectives and associated activities the programme is approaching the problems from several directions. The achievements from 1984 to 1987, and the activities planned for 1988-1990 are described in a booklet, which is available from IUCN (Synge, 1988).

The second phase, 1988-1990. Several of the projects initiated during the first phase will have profound influence on the way the programme will proceed: One is the project to produce the book *Centres of Plant Diversity: A Guide and Strategy for their Conservation*. This will identify those plant sites and vegetation types around the world whose conservation, in whole or in part, will ensure the survival of the most species. This book will serve as the guide to the most cost-effective and practical way of saving as many plant species as possible, and will provide the basis for future planning of on-the-ground projects.

A second project group of major importance has involved botanic gardens. Following an international conference in 1985, a Botanic Gardens Conservation Secretariat was established by 1987, and a *Botanic Gardens Conservation Strategy* has been completed (Anon, 1989). Through the Secretariat the ways and means for focussing the work of botanic gardens on plant conservation are provided, and a whole new constituency for the conservation movement is being activated.

A third project with implications for future activities concerns wild plants of economic value. Professor Jack Hawkes, one of the fathers of the crop genetic resources conservation system, has made a preliminary study for the programme on the feasibility of ranking economic plants in approximate order of value to man, to serve as the basis for assessing conservation priorities. Eventually this ranking system will come to function as the cornerstone of the Plants Conservation Programme's activities on economic plants.

Medicinal Plants

Until 1984, medicinal plants had not been one of the main concerns of conservation organizations, despite the fact that the world's plant resources rapidly are diminishing. The IUCN database on plants currently holds information on 43,000 plant species, of which more than 18,000 are threatened; most of these data, however, originate from temperate and subtropical countries, and from islands. When it comes to the plant world of the tropics, which comprises about two thirds of all higher plants, the amount of information available at the species level is inadequate for assessing the status of these plants. But the trend is very clear: by combining data on the rate of destruction of tropical forests with the theory of island biogeography, Dr. Peter Raven of the Missouri Botanical Garden has estimated that no less than 60,000 plants, nearly 1 in 4 of the world total, could become extinct by the middle of the next century, if present trends continue. The areas that will suffer the greatest losses are exactly those of the tropics, where medicinal plants are most important to daily health, and where medicinal plants constitute an important component of the diversity of the plant kingdom. Not only do the tropics contain the largest number of species per unit area, but the tropical plants also display an amazing variability in their genetic makeup, e.g. in the biochemical compounds which they produce. The multitude of ecological interactions among the organisms in the tropical rain forest, for example, produce a vast array of adaptations, many of which are displayed in plants by production of an immense number of biochemical compounds.

About half the world's medicinal compounds are still derived or obtained from plants. The medicinal products from plants and other biota are, in general, more important in developing countries than in industrialized nations. But even in these, where the focus very much is on chemical discovery and synthesis of pharmaceuticals, drug products from biota are major contributors to the human health services sector of the economy, and plant-derived drugs contribute billions of dollars to the economy each year.

It is also noteworthy that some of the most important drugs of the past 50 years or so, which have revolutionized modern medical practices, have almost all first been isolated from plants, and often from plants which for one purpose or another have been employed in primitive or ancient societies (Schultes, 1986). These "wonder drugs" include the curare alkaloids; penicillin and other antibiotics; cortisone; reserpine; vincoleucoblastine; the Veratrum alkaloids; podophyllotoxin; strophantine and other new therapeutical agents (Schultes & Swain, 1976).

The level of demand for biotic drug products has remained more or less stable for the last two decades (Oldfield, 1984). Often the cultivation and harvesting of medicinal plants is less costly than artificial drug synthesis; well known examples of products are atropine, digoxin, and morphine. Natural products may also be used as building blocks for the synthesis of "semisynthetic" drugs; this is the case for plant saponins, which can be extracted and easily altered chemically to produce sapogenins for the manufacture of steroidal drugs. Furthermore, the medicinal plants may give the chemical blueprints for the development of related synthetic drugs, as for example cocaine from *Erythroxylon coca* which provided the chemical structure for the synthesis of procaine and other related local anesthetics. Natural extraction and artificial drug synthesis represent complementary research areas, although they are not often seen as such (Oldfield, 1984).

Medicinal Plants in the Tropics and their Conservation

It has been estimated that as many as 75 to 90 % of the worlds rural people rely on herbal traditional medicine as their primary health care. It appears neither possible nor desirable to replace this herbal medicine with western medicine, at least during this century. Consequently there is a growing interest in medicinal plants and traditional medicine: within the last decade, the governing bodies of the World Health Organization of the United Nations (WHO) have called for the intensification of efforts in the development of national traditional medicine activities. Many member states have responded by placing greater emphasis on the utilization of traditional remedies including medicinal plants, which constitute the greatest part of traditional medicine. However, with the increasing use of medicinal plants in many countries, and with the accelerating destruction of natural resources in the tropics, it has become clear that the exploitation of medicinal plants must be accompanied by conservation measures. Otherwise these plants become depleted as resources or may even face extinction.

Some countries have instituted measures to protect endangered species of medicinal plants but the majority have yet to take such action. Through the Plants Conservation Programme, in which medicinal plants represent a priority in the conservation of plants of economic value, IUCN and WWF are collaborating with the government of Sri Lanka on a broad project for conserving the medicinal plants of this country. Minor projects in other countries have also been initiated. However, there is a long way from realizing the existence and urgency of the problem to actually starting actions for solving it: First of all the necessary basic data is often lacking or is inadequate. Just to establish the correct identity of the plant material presents a vast job, for which the resources may be small or absent, particularly in those developing countries where conservation of medicinal plants is most urgently required. More knowledge is also needed about the distribution, ecology and status of these plants, as a prerequisite for the study and assessment of their pharmaceutical and medicinal properties and potentials.

As suggested by Professor Richard Schultes of Harvard University, the research area of "ethnobotany" may provide a shortcut for identification of those plants which are of the greatest interest: "It is (at once) obvious that a vast reservoir of still virgin information on plant properties remains to be tapped and salvaged. This ethnopharmacological information has not only its academic interest but can be put to practical use for the benefit of all mankind." (Schultes, 1986).

But once the basic data are available or at least adequate, the actual conservation problems and requirements need to be addressed, such as whether emphasis should be placed on *in situ* or *ex situ* techniques, whether there is basis for cultivation schemes, etc.

Thus there is a need to focus international attention on the conservation and sustainable utilization of medicinal plants, and a need to outline the principles and methods for implementing suitable conservation policies. This is recognized by many countries. However, as mentioned above, the scientific and economic resources to do the job are often lacking where they are most needed; but those industrialized countries, who are gaining economically from the use and exploitation of these resources have an obligation to assist and collaborate in conserving the living resources and have also a self-interest in doing so. Much more emphasis ought to be placed on assisting those rural people in developing countries who have the most direct relation to medicinal plant use. They ought to benefit from some of the economic gains deriving from the exploitation of medicinal plants.

Conclusion

The Joint IUCN-WWF Plants Conservation Programme is emphasizing medicinal plants as a major, but neglected group of plants for which conservation is a priority. However, so far there has been little co-ordinated action on an international scale to ensure the sustainable utilization and conservation of medicinal plants. The present International Consultation is therefore a historic event, in bringing together the conservation and the health care professionals and in seeking to combine the many different interests in the subject, with the intention of producing a set of guidelines for countries wishing to make the best possible use of their medicinal plant resources and to conserve them for the future. The formulation of such guidelines and actions following therefrom will represent an important contribution towards sustainable development, as called for by the World Commission on Environment and Development.

References

Anon. (1989) *The Botanic Gardens Conservation Strategy*. WWF, IUCN, BGCS.

Danida (1988). *Environment and Development: A Plan of Action*. Copenhagen: Danish International Development Agency. Ministry of Foreign Affairs.

IUCN, UNEP & WWF (1980). *World Conservation Strategy. Living Resource Conservation for Sustainable Development*. Gland: IUCN.

Lanly, J.-P. (1982). Tropical Forest Resources. *Forestry Paper No. 30*. Rome: FAO.

Oldfield, M. L. (1984). *The value of conserving genetic resources*. Washington, D.C.: US Department of the Interior, National Park Service.

Schultes, R. E. (1986). Ethnopharmacological conservation: a key to progress in medicine. *Opera Botanica*, 92, 217-224.

Schultes, R. E. & Swain, T. (1976). The Plant Kingdom: a virgin field for new biodynamic constituents. In *The Recent Chemistry of Natural Products, including Tobacco*. New York: Proceedings Phillip Morris Science Symposium.

Synge, H. (1988). *The Joint IUCN-WWF Plants Conservation Programme*. Kew: IUCN.

Vitousek, P. M., Ehrlich, P. R., Ehrlich, A. H. & Matson, P. A. (1986). Human appropriation of the products of Photosynthesis. *Bioscience*, 36, 368-373.

World Commission on Environment and Development (1987). *Our Common Future.* Oxford: Oxford: University Press.

WRI (1985). *Tropical Forests: A Call for Action.* Washington, D.C.: World Resources Institute.

The Issue of Medicinal Plants

Global Importance of Medicinal Plants

Norman R. Farnsworth and **Djaja D. Soejarto**
Program for Collaborative Research in the Pharmaceutical Sciences, College of Pharmacy, University of Illinois at Chicago, Chicago, Illinois, U.S.A.

Introduction

Ancient Man is known to have utilized plants as drugs for millennia. Based on current knowledge, at least in the West, we know that extracts of some of these plants are useful in a crude form, i.e. *Atropa belladonna* Tincture as an antispasmodic, *Rauvolfia serpentina* roots for hypertension and as a tranquilizer, *Papaver somniferum* extract or tincture as an analgesic, etc. Further, we know that at least 121 chemical substances of known structure are still extracted from plants that are useful as drugs throughout the world (Anon, 1982a). A large number of plants are used in traditional medical practices, and have been for more than 3000 years, such as in Chinese Traditional Medicine, Ayurvedic Medicine, Unani Medicine, etc., most of which probably exert therapeutic effects and would be proven as such if they were properly evaluated by Western standards. Still further, plants have been employed for centuries by primitive cultures; most of these are less likely to pass the test of modern experimental verification of efficacy. Finally, there are a large number of so-called herbal remedies, mainly sold in health food stores in developed countries, many of which remain to be verified for their real therapeutic effects.

Several years ago the World Health Organization made an attempt to identify all medicinal plants that exist in the world. It was admitted that the compilation of names of medicinal plants undoubtedly contained many replicates since botanical verification was not attempted. Further, the list only provided Latin binomials and the countries where the plants were used, but excluded data indicating what the plants were used for. More than 20,000 species were included on this list. In our own NAPRALERT database we document ethnomedical uses alone for *ca.* 9,200 of 33,000 species of monocots, dicots, gymnosperms, pteridophytes, bryophytes and lichens, which would suggest that *ca.* 28% of plants on earth have been used ethnomedically. It is generally reported that 5,000 of 35,000 species of plants growing in the People's Republic of China are used as drugs in Chinese Traditional Medicine. This would suggest that about 14% of plants have some drug application in that country. Thus if 14-28% of these plants are used as drugs, and we accept the conservative number of higher plant species on earth to be 250,000, one can estimate that 35-70,000 species have at one time or another been used in some culture for medicinal purposes.

If we are to discuss the global importance of medicinal plants, it is necessary to define the term medicinal plant. We have chosen to select the broadest definition that includes all higher plants that have been alleged to have medicinal properties, i.e. effects that relate to health, or which have been proven to be useful as drugs by Western standards, or which contain constituents that are used as drugs. Complicating the matter of identifying all such plants is that there are no readily available sources of information, that is, the information is found scattered throughout the world in difficult to obtain books on medical botany, in surveys of the medicinal uses of plants for specific geographic regions and/or for different ethnic groups, in anthropological writings, as notes on voucher specimens in various herbaria throughout the world, entered by botanists who collected the specimens, in a variety of review articles and in other sources. Secrets locked in the minds of indigenous peoples, shamans, traditional healers and the like, are less likely to be discovered, for reasons to be pointed out below.

Finally, how does one measure "importance". It is possible to document, in many cases, that medicinal plants and their active principles are able to alleviate or cure human suffering and illness. Another way is to measure the monetary value of the cost of plants entering into global commerce, but data are available only for plants that are entering into commerce on a major scale, i.e. in hundreds or thousands of tons per year. Most likely the largest numbers that could be obtained by the assessment of "value" in these terms could not be backed up by hard data. Still, a further way is to measure the gross tonnage of plants that enter into global commerce.

On a global basis it appears that the only consolidated source of information on the monetary value of medicinal plants entering into global commerce is that provided by the International Trade Centre (UNCTAD/GATT). The most recent compilations of data from this agency involving medicinal plants and their derivatives was published in 1982 (Anon., 1982a). However, this information must be considered as grossly incomplete. Major data on export/import of medicinal plants involve West Germany, Singapore and Hong Kong. However, these countries produce little, if any, medicinal plants. Other producing countries often do not report details of specific plants but only report figures for "seeds" or "roots" or "medicinal plants". Further, the UNCTAD/GATT also compiles data on "Spices, A Survey of World Markets" (1982) (Anon., 1982b) and "Markets for Selected Essential Oils and Oleoresins" (1974) (Anon., 1974). Since many spices and essential oils have, in addition to food uses, medicinal applications, the problem of analysis of the global importance of "medicinal plants" becomes further complicated.

As it will become obvious from this essay, we are placing major emphasis on actual or potential medicinal plants from the tropical rain forests, since these plants are generally agreed to be most susceptible to possible loss.

It is an accepted fact that the tropical rain forests are the richest biome on earth. It is also an accepted fact that the tropical rain forests still represent one of the last and true great frontiers of wilderness, which still evoke awe and wonderment. In view of the fact that our institution is currently involved with the United States National Cancer Institute's recently resurrected plant program (Booth, 1987), and mindful of warnings from conservationists that the tropical rain forests or tropical moist forests or simply rain forests (Myers, 1980) may be decimated within the next decade or so, in addition to a concern brought by personal knowledge of the situation during a number of years of field as well as laboratory experience, we consider it timely to put into perspective the prospect that this tropical biome may still contribute to the alleviation of human suffering, before it vanishes from the face of the earth.

In terms of human existence, the tropical rain forests represent a store of renewable natural resources, which have for eons, by virtue of their richness in both animal and plant species, contributed a myriad of items for the survival and well-being of man. These include basic food supplies, clothing, shelter, fuel, spices, industrial raw materials, and medicine. It is to this last item, medicinal or ethical drugs, that this paper is addressed.

Definition and Extent of the Tropical Rain Forests

Alexander von Humboldt and many others referred to the tropical rain forests as the great "Hylaea". The Germans called them *Regenwald*, the British jungle, and in Latin the word is *pluviisylva*. To American tourists in South America, they are "green Hell". Following the classical definition of Schimper, later adopted by Richards (Richards, 1952), the tropical rain forests are defined as an evergreen plant community, at least thirty meters in height, rich in great woody lianas and in arborescent and herbaceous epiphytes, but the woody forms predominate. Mature forests have a closed canopy and contain several more or less distinct strata, thus creating a habitat complexity characteristic of the tropical rain forests. Although other definitions, variants of the above wording, have been offered, this definition remains valid to this day and reflects the true nature of this forest ecosystem. Tropical rain forests are a climax vegetation that has developed only in the tropical belt, where the climatic conditions are characterized by constant high temperature and humidity, with abundant rains. This translates into a mean temperature of 24°C or higher and essentially frost-free, with an annual precipitation of 2,000 mm or more, and not less than 100 mm in any month, for two out of three years (Richards, 1952, Myers, 1984, Myers, 1987). Such areas lie within the boundary of the tropic of Cancer in the north and the tropic of Capricorn in the south, and normally occurring below 1,300 m above sea level, and include lowland rain forest, montane (and submontane) rain forest, cloud forest, riverine forest, swamp and bog forest, and wetter forms of lowland seasonal forest (IUCN, 1979). Under this definition, the tropical rain forests presently cover an area of about 9 million sq. km, about 7% of the earth's land surface. Of this extension, 5.1 million sq. km. are in tropical America, 1.9-2.1 million in Asia (largely in Southeast Asia), and 1.8 million in Africa, with patches in a few Indian Ocean and Pacific Islands (Sommer, 1976, Myers, 1980, Myers, 1984).

Biotic Richness

Estimates on the biotic richness of the tropical rain forests place the number of species (both animal and plants) between 3 and 30 million (Raven, 1980, Erwin, 1983, Myers, 1984, Myers, 1987). The great bulk of these comprise the insects, which live in the forest canopies (Erwin, 1983). The number of species of seed plants occurring in the tropics is estimated at 155,000 (Prance, 1977), of which approximately 120,000 (including 30,000 undescribed species) occur in the tropical moist forests alone (Myers, 1984, Myers, 1987). Of these, roughly three-fifths occur in tropical America and one-fifth each in tropical Asia and Africa.

The biotic richness of the tropical rain forests is due to the high biological diversity of this ecosystem, compared with that of the temperate regions. In a sample plot of 1 hectare of tropical rain forest, up to 100 tree species may be found, compared to only about 10-15, rarely to 35, at the most, in a temperate forest (Leigh, 1982). To better appreciate the plant species richness of the tropical rain forests, a comparison is often made between angiosperm floras of Brazil, possessing about 40,000 plant species (mostly in an area of 3 million sq. km comprising the

Amazonian rain forest), Colombia, with about 35,000 species (also, most of them occurring in the rain forest that occupies one-third of the 1.1 million sq. km of the territory), and Southeast Asia, with about 25,000 species (in the rain forest covering 2.1 million sq. km) *vs.* the United States, possessing about 20,000 species in an area of 9 million sq. km. (including Hawaii and Puerto Rico), or between Costa Rica, with about 8,000 plant species in an area of 52,000 sq. km *vs.* Great Britain, with 1,443 species in an area of 244,000 sq. km, or New Zealand, with 1,996 plant species in an area of 268,000 sq. km. (Myers, 1984, Gentry, 1986).

Drugs from the Tropical Rain Forests

No one would seriously challenge the fact that man is still largely dependent on plants in treating his ailments. According to an estimate of the World Health Organization (WHO), approximately 80% of the people in developing countries rely chiefly on traditional medicines for their primary health care needs, of which a major portion involves the use of plant extracts or their active principles (Farnsworth *et al.*, 1985). In a country like the People's Republic of China, with a population of more than 1 billion, the main type of drug therapy is still in the form of plant extracts (Anon., 1975 and see chapters by Xiao Pei-gen and He Shen-an and Cheng Zhong-Ming in this volume). Even in the United States, where synthetics dominate the drug market scene, plant products still represent an important source of prescription drugs. Approximately one-fourth of all prescriptions dispensed from community pharmacies in the United States contain one or more ingredients derived from higher plants (Farnsworth & Morris, 1976), which in 1980 was valued at $8.112 billion (Farnsworth & Soejarto, 1985), at the consumer level.

How much do the tropical rain forests contribute to the primary health care of man? With such a tremendous store of biotic richness, certainly one expects them to be a storehouse, and a source, of potentially important medicinal drugs. Unfortunately, the answer to this question is impossible to assess. At present, there is no straightforward explanation that can be given in terms of numbers, though it must be considerable. The following may serve to illustrate as examples in support of this assertion.

One-half of the forty species of flowering plants that yield prescription drugs valued at $8.112 billion *in the United States* in 1980 (Farnsworth & Soejarto, 1985) consist of plants that originate from the tropics. For these plants a direct monetary value may be calculated. In addition to these, there are many other plants (at least 58 species) whose products also entered into the prescription market in 1980, but which the United States Food and Drug Administration (FDA) does not recognize as drugs. One-third of these plants also originate from tropical rain forest areas, but no direct monetary value can be given. There are a large number of plants, probably in the order of 400 species, whose products constitute the herbs marketed by the herbal industry in "health-food" stores throughout the United States and which were valued at $360 million dollars in 1981 (Tyler, 1986). Only a small fraction of these consist of tropical plant species, whose monetary value is impossible to estimate. Outside of the United States, many of the plants and plant products just mentioned are also used in European countries, which add considerably to the monetary value of drugs derived from tropical plants.

However, human population in the developed countries represents only one-third of the total world population of 4.4 billion in 1980 (Peterson, 1984). The other two-thirds are located in the developing countries (including the People's Republic of China), for which a monetary value of plant drugs, least of all those derived from tropical plants, is next to impossible to estimate. We know, for example, that at least 6,500 species of plants are used locally in Asia as home remedies (Perry & Metzger, 1980), and at least 1,300 species are employed in the northwest Amazon alone by the native Indians for medicine, poisons, and narcotics (Schultes, 1979). Our own NAPRALERT database contains 1,900 species of flowering plants from tropical America which have also been used locally as home remedies. Certainly, there is a value that may be assigned to these plants, when one weighs the importance of tropical plants in the maintenance of man's primary health care.

In terms of modern pharmaceutical products, namely, prescription drugs, the value of tropical plants in the developing countries is small, since the population that inhabits the third world accounts for only about 15% of the worldwide prescription market, mainly Asia (7%) and Latin America (6%) (Griffin, 1982).

Overall, the *monetary* value of tropical plants in the primary health care of man does not seem to be overwhelmingly great. Their real value, however, must be measured in terms of the alleviation of human suffering and in terms of the number of lives that have been saved by plant drugs through the millenia.

What is, then, the potential value that the tropical rain forests may have to contribute to modern medicine?

Based on data which we have computerized in our NAPRALERT database system, at present, there are about 121 clinically useful prescription drugs worldwide that are derived from higher plants, of which 45 (37%) are currently used in the United States of America (Farnsworth *et al.*, 1985). The other 76 (63%) are being used elsewhere, and still need further investigations to demonstrate their safety and efficacy, in compliance with the regulations set forth by the FDA, before they can be used and marketed in the United States. Only 95 plant species are involved in the production of these 121 drugs (Farnsworth *et al.*, 1985). Of the 95 species, 39 are plants originating in and around the tropical rain forests. These are listed in Table 1. Other useful drugs produced from *non-* tropical plants are listed in Table 2. Of the 45 drugs derived from the 38 tropical rain forest species (Table 1), only 23 (50%) are presently in use and marketed in the United States. Therefore, 23 of 121 drugs (20%) that are currently used in the United States have been derived from plants originated from the tropics, while the remaining 80% derive from plants from other than the tropics, mainly the north temperate and subtropical regions. Indeed, these figures do not provide a convincing demonstration on the importance of the tropical rain forests as a source of therapeutic chemical compounds. Does this mean that tropical rain forests contain a lesser number of medicinally important plants? Far from it!

The main explanation that may be given for the higher proportion of prescription drugs from temperate and subtropical regions is the fact that phytochemical studies and medical advances in the temperate areas, involving temperate and subtropical plants, have taken place for a much longer period than those for the tropical regions. Coupled with this has been the problem of accessibility, namely, logistical and political problems in the collection of plant samples from the tropical rain forests, for study. As a result, our knowledge on the medicinal plants of the temperate regions, with a poorer flora – only one-third that of the tropics – is better.

Data presented in Table 3 were identified based on the import-export data of medicinal plants entering world commerce as compiled by the International Trade Centre for 1982 (Anon., 1982a). It can be seen that there are 74 species of plants that are listed in this compendium. Some of the plants are used *per se* as drugs, some are used in advanced (extractive) form, some are used as a source of purified drugs, some as a source of chemicals that are used for the synthesis of prescription drugs and some are used as herbal remedies. For the most part, these major medicinal plants entering into global commerce are required in ton or hundreds of tons amounts and are cultivated.

When unpublished data provided by the pharmaceutical firms throughout the world are considered, we estimate that about 5,000 species of higher plants worldwide have been exhaustively studied as a source of new drugs for human use (Farnsworth & Soejarto, 1985). Many of these plants originated from the temperate regions. Similarly, the National Cancer Institute's Anticancer Screening Program screened approximately 35,000 species of higher plants for the period of 1960-1982. The greater percentage of these species were collected in regions outside the range of the moist tropical forests. At least 85% of the world flora of higher plants, the greater part of which from the tropical rain forests, remain to be screened for anticancer activities (Spjut, 1985).

Drugs for the Twenty-First Century

Among the 45 plant drugs of known structure derived from the tropical rain forest species (Table 1), which include those that are of major importance used in therapy, none is currently produced through synthesis. Consider, for example, the anticancer vinblastine and vincristine from *Catharanthus roseus*, the tranquilizers rescinnamine and reserpine from *Rauvolfia serpentina*, the antimalarial quinine from *Cinchona* species, the antiarrhythmic quinidine from the same source, the antiglaucoma pilocarpine from *Pilocarpus jaborandi*, and the topical anaesthetic cocaine from *Erythroxylum coca*. The other 76 plant drugs from the non-tropical species (Table 3) include those of major importance, such as the painkiller morphine and the anticough codeine from *Papaver somniferum*, the spasmolytic and cold medicine atropine from *Atropa belladonna* and *Hyoscyamus niger*, and the cardiac glycosides for the treat- ment of congestive heart failure from *Digitalis* species. Given these and other synthetically produced drugs (Anon., 1982c), one may ask if all these drugs have now satisfied the therapeutic needs of the human population. Far from it!

Looking ahead to the 21st century, we see that for a number of diseases or symptoms, improved and satisfactory cures still remain to be sought and developed (Brodie & Smith, 1985, Tyler, 1986). These are:

1. Viral diseases, such as herpes (genitalis, simplex, and zoster), AIDS, and certain cancers.
2. Diseases of unknown etiology, including arthritis, some cancers, muscular dystrophy, and Parkinsonism.
3. Self-inflicted diseases, namely, alcoholism, liver disease, drug-dependency, obesity, smoking, and the like.
4. Genetic diseases, ranging from cystic fibrosis and hemophilia to sickle cell disease.
5. The control of symptoms, such as pain, elevated cholesterol levels, hypertension, and the general susceptibility to diseases of various kinds.

Obviously, much still remains to be done. The search for the cures and/or alleviation of these diseases, therefore, should continue. In view of the fact that only a small proportion of the tropical rain forest plants have been thoroughly investigated for their medicinal potential, our attention is naturally directed to this rich biome, as a possible source of new plant drugs.

This leads us to the question of what is the prospect of finding new drugs from the tropics?

Before answering this question, let us first examine the factors that may adversely affect the discovery and development of new plant drugs.

Factors that Affect Plant Drug Discovery and Development

The discovery of new drugs from plants is a complex process, which has been the subject of previous papers (Farnsworth & Bingel, 1977, Farnsworth, 1982, 1984). Putting aside the technical aspects of drug discovery, there remain in the United States two major stumbling blocks for the discovery and development of new plant drugs.

First, the cost of drug development is very high, in the vicinity of $50-100 million for each new drug (Tyler, 1986). To aggravate this situation, no drugs can be patented should there be prior disclosure about them in the scientific literature. Even if patent protection is granted, the date of the protection becomes effective from the beginning of the drug development, and not when the drug begins to be marketed, thus reducing the patent life considerably, with a much diminished return on investment. For these reasons, even large U.S. pharmaceutical industries shy away from plant drug development. This leaves the search and development of plant drugs to government agencies, such as the National Cancer Institute, and international organizations, such as the World Health Organization, for whom the public interest is served above all else.

Second, the present government's regulations pertaining to the marketing of new drugs are restrictive to drug development. In order for a new drug to be approved by the FDA, namely, the U.S. government regulatory agency, it has to be shown that it is safe *and* effective. Obviously, it takes a long period of time (more than 10 years) and a large investment to comply with these requirements. When the drug in question is derived from a plant long used in folk medicine, there is an added problem with patent protection, which altogether stifles any incentive for innovation and development.

Problem of New Drug Discovery from the Tropical Rain Forests

Aside from the high cost and rigid requirements for the development of new drugs from plants, there is a special problem that has to be considered with plants from the tropical rain forests. This concerns rain forest conversion and depletion, through commercial logging, fuelwood consumption, cattle ranching, and forest farming, which are estimated at 8-11 million hectares (about 1%) annually (Myers, 1980, FAO/UNEP, 1982, Myers, 1984, Burley, 1986). Because of the rapidly increasing encroachment of the human population, which is projected to double from the 1980 figure of 4.4 billion by the year 2025, a great portion of which will take place in the developing countries (Peterson, 1984, Myers, 1987), where the tropical rain forests are situated, the rain forest depletion rate is expected to increase at an even faster rate in the years ahead.

In addition to direct human pressures just mentioned, another major indirect factor in the depletion of the rain forest that must be taken into account is the environmental threat posed by genetic engineering, which will produce organisms that will be capable of making agriculture or animal husbandry, or both, profitable on virtually any tropical land surface, within one to three decades. One distinguished tropical ecologist has even stated that once this happens in the various tropical rain forest habitats, it is "goodbye, rain forest" (Janzen, 1987).

The consequence of interest from a pharmaceutical point of view is two-fold. First, the extinction of medicinal genetic resources, second, the disappearance of the human cultures which have developed in and around it and depended on it for its existence (Poore, 1976). Both these consequences have a direct bearing on the discovery of new drugs from the tropical rain forests.

Tropical medicinal genetic resources may contribute to pharmaceutical and health services in three ways (Oldfield, 1981): (i) they may be used directly as pharmaceuticals (plant extracts and products), (ii) they may serve as templates for chemical synthesis of related medicinal compounds, and (iii) they may be used as investigative, evaluative or other research tools in the drug development and testing process (namely, chemical compounds as well as animal species). Based on the conservative figure of 2.5 million species occurring in the tropical rain forests, up to 825,000 species may become extinct in a low deforestation case, and up to 1,250,000 in a high deforestation case, by the year 2000 (Lovejoy, 1980). A greater proportion of the species that will become extinct comprise the insects, which live on the forest canopies. With the extinction of these species, gone forever will be whatever kinds of organism that may be useful in medicine and medical research.

Estimates on the number of plant species that will become extinct by the years 2000-2050 vary from 5% (Irwin, 1978) to 10% (Lucas, 1978) and even 25% (IUCN, 1986). Whatever the correct figure should be, within a decade or so, a large number of plant species from the tropical rain forests will be gone forever, lost to humanity. Based on the estimate that the monetary value in terms of pharmaceutical potential of one angiosperm species that is destined to become extinct by the year 2000 in the United States alone is $203 million dollars (1980 buying power) (Farnsworth & Soejarto, 1985), the monetary value of the plant species that are destined to become extinct from the tropical rain forests by the year 2000 is incalculable. The true value of these species, however, may be even more than what the monetary value may indicate. Because of the high species diversity, but *low species density* characteristic of the tropical rain forests, most of the species that are destined to become extinct are the true rain forest species, of which we know very little of their chemistry or pharmacology. Thus, the greater portion of the plants that will be lost forever are those that may be of greatest potential value.

Plants have been used by man since the beginning of human culture for various purposes, for his survival, including medicine. At present, we know thousands of such plants which have at one time or another been used for medicinal purposes. Such folk or ethnomedical uses represent *leads* that may short-cut the discovery of modern therapeutic drugs, either directly from the plants or from their synthetic analogs. Although some skeptics regard such uses as mere "old-wife tales", the fact remains that many important modern plant drugs (digitoxin, reserpine, tubocurarine, ephedrine, to name a few) have been discovered by following leads from folk uses. In fact, 74% of the 121 biologically active plant-derived compounds presently in use worldwide, have been discovered through follow-up research work to verify the authenticity of information concerning the folk/ethnomedical uses of the plants (Farnsworth *et al.*, 1985). With the disappearance of the human cultures that have developed in and around the tropical rain forests, gone also will be the traditions and knowledge concerning medicinally useful plants from the tropical rain forest regions, before we have a chance to study and document them. Evidence is already abundant to show that such cultures are already replaced by a more "modern" one, such that asked of what kind of preparation should be taken for headaches, a native Indian of the Amazon rain forests once said "Take aspirin". The question is "how much medical value has been and will be lost with the disappearance of the rain forest biome?"

Prospects

Time is running short before the entire tropical rain forest biome disappears, and with it a large proportion (10%-50%) of the species which live within it. Opinions differ as to when this will happen, but everybody agrees that sooner or later it will happen. Once considered a renewable resource, the tropical moist forests are now a non-renewable resource, through advancing human pressures (Gómez-Pompa *et al.*, 1972). One way to stem or postpone the inevitable is certainly the conservation of this tropical biome, of which much has been written (Poore, 1976, Myers, 1980, Murphy, 1980, Oldfield, 1981, Myers, 1987).

What is more important from the drug discovery and development perspective is the urgency for major efforts in the exploration of the rain forest biome for the collection of plants to be submitted to massive screening programs. Because of the high cost of drug discovery and development, such an effort must be of necessity selective, with a focus on only one or two biological activities, giving priorities to the cure and/or alleviation of major disease conditions. It seems timely, therefore, that the NCI re-established its plant program in 1986 for the search for anti-cancer and anti-AIDS therapeutic agents from the tropical rain forests. Yet, while this program is going on, at least two rain forest angiosperm species become extinct every day. This does not seem to matter, except for the fact that perhaps it is among those species that have gone forever that major anti-cancer/anti-AIDS compounds might have been found. When one considers this possibility, and the fact that the chances of finding anti-cancer agents from plants that could enter into clinical trials is very low – in the order of one in 1,600 (Suffness, personal communication) – the prospect does not look as rosy any more. One has to persevere, however, especially in light of the fact that the only anti-cancer agents of plant origin in use today, vinblastine and vincristine from *Catharanthus roseus*, are alkaloids, a chemical class of compounds widely distributed and concentrated in the rain forest species (Levin, 1976, Moody, 1978).

If one looks at other types of biological activities, the prospect of finding new drugs remains optimistic. For example, the recent discoveries of anti-implantation agents (Kong *et al.*, 1985, Lal *et al.*, 1987), antimalarial agents (O'Neill *et al.*, 1987), immunostimulants (Wagner *et al.*, 1985), cardioactive agents (Aladesanmi & Ilesanmi, 1987), and antiviral agents (Corthout *et al.*, 1987, Joshi *et al.*, 1986, Wachsman *et al.*, 1987), to mention only a few, from tropical rain forest species, attests to this assertion. Then, there are also pleasant surprises that may await the persistent researcher, such as the discovery of thiarubrine-A, a potent antibiotic, from an East African rain forest species of the genus *Aspilia* (Compositae), through careful field observations of chimpanzees' dietary habit (Ghitelman, 1986).

As pointed out above, information on ethnomedical uses of plants has provided an important lead in the discovery of the greater number of clinically useful plant drugs worldwide. This means that it is as urgent to carry out extensive explorations in the tropical rain forest areas, within whatever time still remains, to study and document the ethnomedical uses of rain forest plants, followed by laboratory investigations. Again, because of the high cost of plant drug development, field and laboratory searches must, of necessity, be selective, limiting the efforts and resources to a number of high priority disease conditions. Although a massive amount of data is already available on the medicinal uses of plants from various tropical countries, a large percentage of these concern common, weedy, cultivated, and easily accessible secondary forest plants, which are genetically hardy and often with a wide geographical distribution, and thus are least affected by the extinction process. Certainly, in a flora as rich as the tropical rain forests, many plants that have been used for millennia by the rain forest people still remain unknown to the world, and whatever value such plants may have to offer to modern medicine will remain hidden, and what is worse, will even be lost forever, if nothing is done about them.

As a result of field explorations and laboratory research outlined above, plant species of high interest will be identified. For purposes of chemical isolation and structure elucidation, followed by bioassays using laboratory animals, and eventual clinical trials, larger amounts of plant materials will still be needed, before any attempts at synthesis can be made. At the end, when the synthetic process is difficult or cost-prohibitive, or simply impossible, the final source of plant materials for isolation will still be the rain forest species, which may be rare, threatened, or in danger of extinction. Although cultivation of species of interest will be the logical route to pursue, in practice, this may not always be easy to do. Tropical rain forest species have a very narrow genetic tolerance, so that they may be difficult to propagate outside of their own habitat, not the least to say the long time it takes for tree species to mature before harvesting. Such a scenario calls for a different approach in the propagation of the plant species of interest. This concerns *ex situ* medicinal genetic resource conservation, such as is being done with our crop plants (Wilkes, 1985). Botanical gardens, tropical as well as extratropical, should play a key role in this direction, as exemplified by the conservation of *Melastoma tetramerum*, an endangered species of the Bonin Islands (Shimozono & Iwatsuki, 1986). A particular emphasis should be given to *in vitro* preservation, a form of *ex situ* conservation. The idea is to conserve medicinal genetic resources in the form of tissue or shoot cultures, which should be kept in existence *ad infinitum*. The state of our technology is such that the art of keeping *in vitro* plant cultures alive to perpetuity (Seitz, 1987), and still maintaining the genetic variability and capability of producing medicinally useful chemical compounds, and if necessary, of reproducing into an entire plant, is within a possibility, not a mere whim, and must be explored.

Summary and Conclusions

Based on data presented herein there is no doubt that plants have served mankind over the millenia as sources of potent drugs and have been used *per se* in the crude form as drugs. Currently one can document that at least 137 species of plants are collected from the wild or are cultivated in ton quantities. Most of these are produced from cultivated plants; little is known about the natural abundance of those few major species that are collected from wild growing plants.

Documenting information on plants that do not have well recognized medical uses is much more difficult. Many plants are used in traditional medical practices at the village, country or regional levels and data on the extent of use for most of these plants is difficult to document accurately since many of these are not exported. It is safe to assume that from 35-70,000 species of higher plants are used as medicines in various cultures of the world. Specific data on the plants and their uses could be documented through an examination of books, review articles, scientific reports and other publications on the uses of plants as medicines, as well as through an examination of notes affixed to voucher specimens by botanists in the various herbaria of the world.

It is clear that plants can be useful in their crude or advanced forms as drugs, that plants offer a source of drugs in their pure state and that biologically active secondary metabolites from plants can serve as templates for the synthesis of modern drugs. It is perhaps the latter role that is most important for consideration of conservation of genetic material in future drug development programs.

Despite its alarming rapid depletion rate, the tropical rain forests still represent a great storehouse of medicinal genetic resources which may yield important drugs to treat a number of diseases or symptoms, for which improved or satisfactory cures still remain unavailable at present. The true potential value of the tropical forest plants has not been fully realized for this purpose. Only about 15% or so of all angiosperm species on earth have been examined chemically and pharmacologically, in detail or superficially, for their medical potential. A great many of these comprise plants from the temperate and subtropical regions. The greater percentage of the tropical moist forest plants which represent at least 40% of all the angiosperm species, have not been studied for their medicinal potential.

Because of the large number of angiosperm species, which are projected to become extinct within a decade or so, due to the mass extinction process as a consequence of the rain forest disappearance, and many of which may include plants of major medicinal value, urgent measures must be taken to uncover and exploit fully the potential medicinal value of the rain forest species. Such measures should include major drug discovery and development programs, field explorations in the tropical rain forests to collect and document ethnomedical data and to collect and identify plants for preliminary screening for high priority biological activities, as well as plant genetic conservation. This last measure should be directed primarily toward rain forest medicinal plant species which have been shown to possess highly active, high-priority biological activities. Both *in situ* and *ex situ* conservation measures should be adopted. Similarly, an evaluation should be made to assess the conservation status of all currently important medicinal plants, with emphasis on those species originating from the tropical rain forest areas, and measures be taken to conserve the genetic pool of species which have been identified as threatened or endangered.

Only with efforts and persistence, can the potential value of the tropical rain forests to modern medicine and for the welfare of man be fully realized.

References

Aladesanmi, A.J. & Ilesanmi, O.R. (1987). Phytochemical and pharmacological investigation of the cardioactive constituents of the leaf of *Dysoxylum lenticellare*. *J. Nat. Prod.*, 50, 1041-1044.

Anonymous (1974). *Markets for selected essential oils and oleoresins*. Geneva: International Trade Centre. 297 pp.

Anonymous (1975). *Herbal Pharmacology in the People's Republic of China*. Washington, D.C.: National Academy of Sciences.

Anonymous (1982a). *Markets for selected medicinal plants and their derivatives*. Geneva: International Trade Centre. 206 pp.

Anonymous (1982b). *Spices: A survey of the world market*. Volume II. Geneva: International Trade Centre. 258 pp.

Anonymous (1982c). WHO's model list of essential drugs. *Am. Pharm.*, NS22, 22-23.

Booth, W. (1987). Combing the earth for cures of cancer, AIDS. *Science*, 237, 969-970.

Brodie, D.C. & Smith, W.E. (1985). Implications of new technology for pharmacy education and practice. *Am. J. Hosp. Pharm.*, 42, 81-95.

Budowski, G. (1978). A strategy for saving wild plants: experience from Central America. In *Extinction is Forever*, 3rd printing, ed. G.T. Prance & T.S. Elias. Bronx: New York Botanical Garden.

Burley, W. (1986). Testimony of William Burley – World Resources Institute – on S.1747 and S.1748 for the United States Senate Committee on Foreign Relations, March 19.

Corthout, J., Pieters, L., Claeys, M., *et al.* (1987). Antiviraly active constituents of *Spondias mombin* L. *Pharmac. Weekbl. Sci. Ed.*, 9, 222.

Erwin, T.L. (1983). Tropical forest canopies: the last biotic frontier. *Bull. Entomol. Soc. Amer.* (Spring), 14-19.

Farnsworth, N.R. (1982). Rational approaches applicable to the search for and discovery of new drugs from plants. In *Proceedings of the First Latinamerican and Caribbean Symposium on Pharmacologically Active Natural Products*, 21-28 June 1980, Havana, Cuba. UNESCO.

Farnsworth, N.R. (1984). How can the well be dry when it is filled with water? *Econ. Bot.*, 38, 4-13.

Farnsworth, N.R., Akerele, O., Bingel, A.S. *et al.* (1985). Medicinal plants in therapy. *Bull. World Health Organiz.*, 63, 965-981.

Farnsworth, N.R. & Bingel, A.S. (1977). Problems and prospects of discovering new drugs from higher plants by pharmacological screening. In *New Natural Products and Plant Drugs with Pharmacological, Biological or Therapeutic Activity*, ed. by H. Wagner & P. Wolff. New York: Springer-Verlag.

Farnsworth, N.R. & Morris, R.W. (1976). Higher plants: the sleeping giants of drug development. *Am. J. Pharm.*, 148, 46-52.

Farnsworth, N.R. & Soejarto, D.D. (1985). Potential consequence of plant extinction in the United States on the current and future availability of prescription drugs. *Econ. Bot.*, 39, 231-240.

Food and Agriculture Organization/United Nations Environment Program (1982). *Tropical Forest Resources*. Rome: FAO, and Nairobi: UNEP.

Gentry, A.H. (1986). Endemism in tropical versus temperate plant communities. In *Conservation Biology*, ed. M.E. Soulé. Sunderland, Massachusetts: Sinauer Associates, Inc.

Ghitelman, D. (1986). An antibiotic discovered by chimps. *MD* (August) 1986, 39-49.

Gómez-Pompa, A., Vasquez-Yanes, C. & Guevara, S. (1972). The tropical rain forest: a nonrenewable resource. *Science* (Washington, D.C.), 177, 762-765.

Griffin, Jr., R.J. (1982). Pharmaceuticals in third world markets. *Am. Pharm.*, NS22, 18-21.

IUCN (1979). Categories, objectives and criteria for protected areas. Annex to General Assembly Paper G.A. 78/24. In *Proceedings of IUCN Fourteenth Technical Meeting.* Morges: IUCN.

IUCN (1986). 60,000 Plants under threat. *Threatened Plants Newsletter*, No. 16, 2-3.

Irwin, H.S. (1978). Preface. In *Extinction is Forever*, 3rd printing, ed. G.T. Prance & T.S. Elias. Bronx: New York Botanical Garden.

Janzen, D.H. (1987). Letters: Conservation and agricultural economics. *Science* (Washington, D.C.), 236, 1159.

Joshi, M.M., Chowdhury, B.L., Vishnoi, S.P. *et al.* (1986). Antiviral activity of (+)-odorinol. *Planta Med.*, 53, 254-255.

Kane, C.J.M. (1986). Studies on purification and characterization of a novel herpesvirus inhibitor from natural product. *Dissert. Abstr. Intl.*, 47(104), 1530B.

Kong, Y.C., Ng, K.-H., Wat, K.-H. *et al.* (1985). Yuehchukene – a novel anti-implantation indole alkaloid from *Murraya paniculata*. *Planta Med.*, 1985, 304-307.

Lal, R., Gandhi, M., Sankaranarayan, A. *et al.* (1987). Antifertility effect of *Azadirachta indica* oil administered *per os* to female albino rats on selected days of pregnancy. *Fitoterapia*, 58, 239-242.

Leigh, Jr., E.G. (1982). Introduction: why are there so many kinds of tropical trees? In *The Ecology of a Tropical Forest*, ed. E.G. Leigh, Jr., A.S. Rand & D.M. Windsor. Washington, D.C.: Smithsonian Institution.

Levin, D.A. (1976). Alkaloid-bearing plants: an ecogeographic perspective. *Amer. Nat.*, 110, 261-284.

Lovejoy, T.E. (1980). A projection of species extinctions. In *The Global 2000 Report to the President: Entering the Twenty-First Century*, Vol. 2. Washington, D.C.: Government Printing Office.

Lucas, G.L. (1978). Conservation: recent developments in international cooperation. In *Extinction is Forever*, 3rd printing, ed. G.T. Prance & T.S. Elias. Bronx: New York Botanical Garden.

Moody, S. (1978). Latitude, continental drift, and the percentage of alkaloid-bearing plants in floras. *Amer. Nat.*, 111, 965-968.

Murphy, J. (1980). Environment: the quiet apocalypse. *Time Magazine*, October 13, p. 80.

Myers, N. (1980a). *Conversion of Tropical Moist Forests*. Washington, D.C.: National Academy of Sciences.

Myers, N. (1980b). The problem of disappearing species: what can be done? *Ambio*, 9, 229-235.

Myers, N. (1984). *The Primary Source*. New York: W.W. Norton & Co.

Myers, N. (1987). Tropical deforestation and a mega-extinction spasm. In *Conservation Biology*, ed. M.E. Soulé. Sunderland, Massachusetts: Sinauer Associates, Inc.

Oldfield, M.L. (1981). Tropical deforestation and genetic resources conservation. In *Blowing in the Wind: Deforestation and Long-range Implications*, ed. V.H. Sutlive, N. Altshuler & M.D. Zamora. Williamsburg, Virginia: Department of Anthropology, College of William and Mary.

O'Neill, M.J., Bray, D.H., Boardman, P. *et al.* (1987). Plants as sources of antimalarial drugs, part 4: activity of *Brucea javanica* fruits against chloroquinone-resistant *Plasmodium falciparum* in vitro and against *Plasmodium berghei* in vivo. *J. Nat. Prod.*, 50, 41-48.

Perry, L.M. & Metzger, J. (1980). *Medicinal Plants of East and Southeast Asia*. Cambridge, Massachusetts: MIT.

Peterson, J. (1984). Global population projections through the 21st century: a scenario for this issue. *Ambio*, 13, 134-141.

Poore, D. (1976). The values of tropical moist forest ecosystems. *Unasylva*, 28 (Nos. 112-113), 127-146.

Prance, G.T. (1977). Floristic inventory of the tropics: where do we stand? *Ann. Missouri Bot. Gard.*, 64, 659-684.

Raven, P. (1980). *Research Priorities In Tropical Biology*. Washington, D.C.: National Academy of Sciences.

Richards, P.W. (1952). *Tropical Rain Forest*. Cambridge, England: Cambridge University Press.

Schultes, R.E. (1979). The Amazonia as a source of new economic plants. *Econ. Bot.*, 33, 259-266.

Seitz, U. (1987). Cryopreservation of plant cell cultures. *Planta Med.*, 53, 311-314.

Shimozono, F. & Iwatsuki, K. (1986). Botanical gardens and the conservation of an endangered species in the Bonin island. *Ambio*, 15, 1921.

Sommer, A. (1976). Attempt at an assessment of the world's tropical forests. *Unasylva*, 28(112/113), 5-27.

Spjut, R.W. (1985). Limitations of a random screen: search for new anticancer drugs in higher plants. *Econ. Bot.*, 39, 266-288.

Tyler, V.E. (1979). Plight of plant drug research in the United States today. *Econ. Bot.*, 33, 377-383.

Tyler, V.E. (1986). Plant drugs in the twenty-first century. *Econ. Bot.*, 40, 279-288.

Wagner, H., Kreutzkamp, B. & Jurcic, K. (1985). Inhaltsstoffe und Pharmakologie der *Okoubaka aubrevillei*-Rinde. *Planta Med.*, 1985, 404-407.

Wachsman, M.B., Damonte, E.B., Coto, C.E. *et al.* (1987). Antiviral effects of *Melia azedarach* L. leaf extracts on Sindbis virus-infected cells. *Antiviral Res.*, 8, 1-12.

Wilkes, G. (1985). Germplasm conservation toward the year 2000: potential for new crops and enhancement of present crops. In *Plant Genetic Resources: A Conservation Imperative*, ed. C.W. Yeatman, D. Kafton & G. Wilkes. Selected AAAS Symposium 87.

TABLE 1. TROPICAL PLANTS YIELDING CLINICALLY USEFUL DRUGS [a]

Species (Family)	Tropical Origin Af	Am	As[c]	Drug	Action/Clinical Use
Adhatoda vasica Nees (Rubiaceae)			*	Vasicine (peganine)	Oxytocic
Anamirta cocculus (L.) W.& A. (Menispermaceae)			*	Picrotoxin	Analeptic
Ananas comosus (L.) Merr. (Bromeliaceae)		*		Bromelain	Antiinflammatory, Proteolytic agent
Andrographis paniculata Nees (Acanthaceae)			*	Andrographolide Neoandrographolide	Bacillary dysentery Bacillary dysentery
Ardisia japonica Bl. (Myrsinaceae)		*	*	Bergenin	Antitussive
Areca catechu L. (Palmae)			*	Arecoline	Anthelminthic
Azadirachta indica Juss. (Meliaceae)			*	Azadirachtin	Insecticide (on non-food plants)
Carica papaya L. (Caricaceae)		*		Chymopapain	Proteolytic; mucolytic
Catharanthus roseus (L.) G.Don (Apocynaceae)	*			Vinblastine (VLB) Vincristine (VCR)	Antitumor Antitumor
Centella asiatica (L.) Urban (Umbelliferae)	*	*	*	Asiaticoside	Vulnerary
Cephaelis ipecacuanha (Brot.) A.Richard (Rubiaceae)		*		Emetine	Amoebicide, emetic
Chondodendron tomentosum R. & P. (Menispermaceae)		*		Tubocurarine	Skeletal muscle relaxant
Cinchona ledgeriana Moens ex Trimen (Rubiaceae)		*		Quinidine Quinine	Antiarrhythmic Antimalarial; antipyretic
Cinnamomum camphora (L.) Presl (Lauraceae)			*	Camphor	Rubefacient
Cissampelos pareira L. (Menispermaceae)			*	Cissampeline	Skeletal muscle relaxant
Crotalaria sessiliflora L. (Leguminosae)			*	Monocrotaline	Antitumor (topical)
Curcuma longa L. (Zingiberaceae)			*	Curcumin	Choleretic
Datura metel L. (Solanaceae)		*		Scopolamine	Sedative
Dioscorea spp. (Dioscoreaceae)		*	*	Diosgenin	Contraceptive
Erythroxylum coca Lamk. (Erythroxylaceae)		*		Cocaine	Local anesthetic
Gossypium species (Malvaceae)	*	*	*	Gossypol	Male contraceptive

TABLE 1. TROPICAL PLANTS YIELDING CLINICALLY USEFUL DRUGS[a] (cont.)

Species (Family)	Tropical Origin			Drug	Action/Clinical Use
	Af	Am	As[c]		
Lonchocarpus nicou (Aubl.) DC. (Leguminosae)		*		Rotenone	Piscicide
Mucuna deeringiana (Bort.) Merr. (Leguminosae)			?	L-Dopa [b]	Antiparkinsonism
Nicotiana tabacum L. (Solanaceae)		*		Nicotine	Insecticide
Ocotea glaziovii Mez (Lauraceae)		*		Glaziovine	Antidepressant
Pausinystalia yohimbe (K Schum.) Pierre ex Beille (Rubiaceae)	*			Yohimbine	Aphrodisiac
Physostigma venenosum Balf. (Leguminosae)	*			Physostigmine	Cholinesterase inhibitor
Pilocarpus jaborandi Holmes (Rutaceae)		*		Pilocarpine	Parasympathomimetic
Piper methysticum Forst.f. (Piperaceae)			*	Kawain [b]	Tranquilizer
Quisqualis indica L. (Combretaceae)			*	Quisqualic acid	Anthelminthic
Rauvolfia canescens L. (Apocynaceae)		*		Deserpidine	Antihypertensive; tranquilizer
Rauvolfia serpentina (L). Benth. ex Kurz (Apocynaceae)			*	Ajmalicine	Circulatory disorders
				Rescinnamine	Antihypertensive; tranquilizer
				Reserpine	Antihypertensive; tranquilizer
Ricinus communis L. (Euphorbiaceae)	*			Castor oil	Laxative
Rorippa indica (L.) Hochreut. (Cruciferae)			*	Rorifone	Antitussive
Simarouba glauca DC. (Simaroubaceae)		*		Glaucarubin	Amoebicide
Stevia rebaudiana Bertoni (Compositae)		*		Stevioside	Sweetener
				Rebaudioside A	Sweetener
Strophanthus gratus Baill. (Apocynaceae)	*			Ouabain	Cardiotonic
Strychnos nux-vomica L. (Loganiaceae)			*	Strychnine	CNS stimulant
Theobroma cacao L. (Sterculiaceae)		*		Theobromine	Diuretic

a. Based on data from Farnsworth *et al.* (1985) with some updated material.
b. Also produced by total synthesis.
c. Af = Africa; Am = America; As = Asia.

TABLE 2. NON-TROPICAL PLANTS YIELDING CLINICALLY USEFUL DRUGS [a]

Species	Drug	Action/Clinical Use
Adonis vernalis L.	Adoniside	Cardiotonic
Aesculus hippocastanum L.	Aescin	Antiinflammatory
Agrimonia eupatoria L.	Agrimophol	Anthelmintic
Ammi majus L.	Xanthotoxin	Leukoderma, vitiligo
Ammi visnaga (L.) Lamk.	Khellin	Bronchodilator
Anabasis aphylla L.	Anabasine	Skeletal Muscle Relaxant
Anisodus tanguticus (Maxim.) Pascher	Anisodamine; anisodine	Anticholinergic
Artemisia annua L.	Artemisinin	Antimalarial
Artemisia maritima L.	Santonin	Anthelmintic
Atropa belladonna L.	Atropine; hyoscyamine	Anticholinergic
Berberis vulgaris L.	Berberine	Antibacterial
Brassica nigra (L.)Koch	Allyl isothiocyanate	Rubefacient
Camellia sinensis (L.) Kuntze	Caffeine	CNS Stimulant
Cannabis sativa L.	Δ9-Tetrahydro-cannabinol	Antiemetic, decrease ocular tension
Cassia acutifolia Delile	Sennosides A & B	Laxative
Cassia senna L. var. *senna* (= *C. angustifolia* Vahl)	Sennosides A & B	Laxative
Cassia species	Danthron [b]	Laxative
Citrus species	Hesperidin, rutin	Capillary antihem-orrhagic
Colchicum autumnale L.	Colchiceine amide; demecolcine	Cancer
	Colchicine	Gout
Convallaria majalis L.	Convallatoxin	Cardiotonic
Coptis japonica Makino	Palmatine	Antipyretic; detoxicant
Corydalis ambigua (Pallas) Cham. & Schlechtdl.	(±)-Tetrahydro-palmatine	Analgesic; sedative; tranquilizer
Cynara scolymus L.	Cynarin	Choleretic
Cytisus scoparius (L.) Link	Sparteine	Oxytocic
Daphne genkwa Sieb. & Zucc.	Yuanhuacine; yuanhuadine	Abortifacient
Digenia simplex (Wulf.) Agardh [d]	Kainic acid	Ascaricide
Digitalis lanata Ehrh.	Acetyldigitoxin, deslanoside, digitoxin, lanatosides A,B,C	Cardiotonic

TABLE 2. NON-TROPICAL PLANTS YIELDING CLINICALLY USEFUL DRUGS [a] (cont.)

Species	Drug	Action/Clinical Use
Digitalis purpurea L.	Digitalin, digitoxin, gitalin	Cardiotonic
Ephedra sinica Stapf	Ephedrine[b] pseudoephedrine[b], nor-pseudoephedrine[b]	Bronchodilator
Fraxinus rhynchophylla Hance	Aesculetin	Antidysentery
Gaultheria procumbens L.	Methyl salicylate [b]	Rubifacient
Glaucium flavum Crantz	Glaucine	Antitussive
Glycyrrhiza glabra L.	Glycyrrhizin	Antiinflammatory; sweetener
Hemsleya amabilis Diels	Hemsleyadin	Bacillary dysentery, antipyretic
Hydrangea macrophylla (Thunb.) Seringe var. *thunbergii* (Siebold) Makino	Phyllodulcin	Sweetener
Hydrastis canadensis L.	Hydrastine	Hemostatic; astringent
Hyoscyamus niger L.	Hyoscyamine; atropine	Anticholinergic
Larrea divaricata Cav.	Nordihydroguaia-retic acid	Antioxidant
Lobelia inflata L.	α-Lobeline	Tobacco deterrant; respiratory stimulant
Lycoris squamigera Maxim.	Galanthamine	Cholinesterase inhibitor
Mentha species	Menthol	Rubefacient
Papaver somniferum L.	Morphine, codeine; noscapine papaverine	Analgesic, antitussive Antitussive Smooth muscle relaxant
Podophyllum peltatum L.	Etoposide, [c] teniposide [c] podophyllotoxin	Antitumor agent
Potentilla fragarioides L.	(+)-Catechin	Hemostatic
Rhododendron molle G. Don	Rhomitoxin	Antihypertensive
Salix alba L.	Salicin	Analgesic
Sanguinaria canadensis L.	Sanguinarine	Dental plaque inhibitor
Silybum marianum (L.) Gaertn.	Silymarin	Antihepatotoxic
Sophora pachycarpa Schrenk ex C.A. Meyer	Pachycarpine	Ecbolic
Stephania sinica Diels	Rotundine	Analgesic; sedative; tranquilizer
Thymus vulgaris L.	Thymol	Antifungal (topical)
Trichosanthes kirilowii Maxim.	Trichosanthin	Abortifacient

TABLE 2. NON-TROPICAL PLANTS YIELDING CLINICALLY USEFUL DRUGS [a] (cont.)

Species	Drug	Action/Clinical Use
Urginea maritima (L.) Baker	Scillarin A & B	Cardiotonic
Valeriana officinalis L.	Valepotriates	Sedative
Veratrum album L.	Protoveratrines A & B	Antihypertensive
Vinca minor L.	Vincamine	Cerebral stimulant
Several other plants	Allantoin [b]	Vulnerary
	Benzyl benzoate [b]	Scabicide
	Borneol	Antipyretic; analgesic; antiinflammatory
	Pinitol	Expectorant

a. Based on data from Farnsworth *et al.* (1985) with some updated material.
b. Also produced by total synthesis.
c. Semi-synthetic.
d. A marine plant.

TABLE 3 MAJOR MEDICINAL PLANTS ENTERING GLOBAL COMMERCE [1]

Plant Name	Used as a Source of	Country of Production or Cultivation
Acorus calamus L.	Crude drug, essential oil	Cultivated in India
Agave sisalana Perrine	Hecogenin	East Africa, Brazil
Aloe ferox Mill.	Dried leaf extract	South Africa, probably cultivated
Aloe vera (L.) Burm.f.	Fresh leaf juice	Cultivated in USA
Aloe species (others)	Dried leaf extract	
Ammi majus L.	Khellin	Cultivated in Asia and Mediterranean region
Ammi visnaga (L.) Lamk.	Xanthotoxin	Same as above
Anthemis nobilis L.	Crude drug	Same as above
Atropa acuminata Royle ex Lindl.	Same as above	Cultivated in India
Atropa belladonna L.	Atropine, total alkaloids	Central and southern Europe, cultivated in USA, UK, Eastern Europe, India, China
Berberis vulgaris L.	Berberine and crude drug	Europe, Asia
Carica papaya L.	Papain, chymopapain	Cultivated in Sri Lanka, Zaire, Uganda, Mozambique, Tanzania, South Africa, India
Cassia acutifolia Delile	Crude drug; sennosides	Cultivated in India
Cassia senna L. var. *senna* (= *C. angustifolia Vahl*)	Crude drug; sennosides	Cultivated in Egypt
Catharanthus roseus (L.) G.Don	Vinblastine, vincristine, ajmalicine	Pantropical; cultivated in USA, India and other countries
Cephaelis acuminata Karst.	Crude drug	Nicaragua, Colombia, Costa Rica, Panama
Cephaelis ipecacuanha (Brot) A.Richard	Crude drug	Brazil
Cinchona calisaya Wedd.	Quinine, quinidine, cinchonine, cinchonidine, total alkaloids	Cultivated in Indonesia, Zaire, Tanzania, Burundi, India, Kenya, Guatemala, Peru, Ecuador, Bolivia, Rwanda, Sri Lanka, Colombia, Costa Rica
Cinchona ledgeriana Moens	Same as above	Same as above
Cinchona officinalis L.	Same as above	Same as above

TABLE 3 MAJOR MEDICINAL PLANTS ENTERING GLOBAL COMMERCE [1] (Cont.)

Plant Name	Used as a Source of	Country of Production or Cultivation
Cinchona pubescens Vahl (= *C. succirubra* Pav. ex Klotsch)	Same as above	Same as above
Datura metel L.	Atropine, scopolamine, total alkaloids	Cultivated in Asia
Datura metel L. var. *fastuosa*	Same as above	Same as above
Datura stramonium L.	Same as above	Cultivated in southern Europe and Asia
Digitalis lanata Ehrh.	Lanatoside A, purpurea glycoside A, digitoxin, digoxin, acetyl digitoxin	Same as above
Digitalis purpurea L.	Lanatosides A,B,C (converted also to acetyldigitoxin and digitoxin)	Cultivated in India and temperate zones
Dioscorea composita Hemsl.	Diosgenin	Mexico, Guatemala
Dioscorea deltoidea Wallich	Diosgenin	Nepal, India, China
Dioscorea floribunda Mart. & Gal.	Diosgenin	Cultivated in India (introduced)
Duboisia leichhardtii F.v. Muell.	Atropine	Cultivated in Australia
Eleutherococcus senticosus (Rupr. & Maxim.) Maxim.	Crude drug	USSR, China
Ephedra sinica Stapf	Crude drug; ephedrine	China
Glycyrrhiza glabra L.	Glycyrrhizin, crude extract	Cultivated in Spain, Turkey, Iraq, China, Mongolia, USSR, South Africa, USA, France, Italy, Iran, Afghanistan, Syria, Lebanon, Israel, UK, China
Glycyrrhiza glabra L. var *typica* Regel & Herd.	Same as above	Same as above, especially in Spain
Glycyrrhiza glabra L. var *glandulifera* Waldst. & Kit.	Same as above	Same as above, especially in USSR
Glycyrrhiza glabra L. var *beta-violacea*	Same as above	Same as above, especially in Iran
Glycyrrhiza uralensis Fisch.	Same as above	Same as above, especially in China

TABLE 3 MAJOR MEDICINAL PLANTS ENTERING GLOBAL COMMERCE [1] (Cont.)

Plant Name	Used as a Source of	Country of Production or Cultivation
Harpagophytum procumbens DC.	Crude drug	South Africa, Namibia, Lesotho, Botswana
Hibiscus sabdariffa L.	Crude drug	Sudan, China, Thailand, Egypt
Hyoscyamus muticus L.	Atropine	
Hyoscayamus niger L.	Atropine, total alkaloids	Cultivated in temperate zones
Matricaria chamomilla L.	Crude drug	Cultivated in Eastern Europe
Panax ginseng C.A.Meyer	Crude drug and extracts	Cultivated in South Korea, China
Panax quinquefolius L.	Crude drug and extracts	Cultivated in USA; some wild collections
Panax pseudoginseng Wallich	Crude drug and extracts	Japan
Panax pseudoginseng Wallich var. *japonicus*	Crude drug and extracts	Japan
Papaver somniferum L.	Crude; morphine, codeine, thebaine	Cultivated in Turkey, India, Burma, Thailand
Pausinystalia yohimbe (K. Schum.) Pierre ex Beille	Crude drug; yohimbine	Cameroon, Nigeria, Rwanda
Peumus boldus Molina	Crude drug	Peru, Chile, Zaire
Physostigma venenosum Balf.	Physostigmine (eserine)	Sierra Leone, Cameroon, introduced to India and Brazil
Pilocarpus jaborandi Holmes	Pilocarpine	Tropical America, especially Brazil
Plantago indica L.	Seed husks	Cultivated in India
Plantago ovata Forsk.	Seed husks	Cultivated in India, Rwanda
Plantago psyllium L.	Seed husks	Cultivated in India
Rauvolfia tetraphylla L. (= *R. canescens* L.)	Reserpine	Not stated
Rauvolfia sellowii Muell.-Arg.	Reserpine	Not stated
Rauvolfia serpentina L. Benth. ex Kurz	Reserpine, rescinnamine, ajmaline, ajmalicine, serpentine, crude drug	Thailand, Zaire, India, Bangladesh, Sri Lanka, Burma, Malaysia, Indonesia, Nepal
Rauvolfia vomitoria Afz.	Reserpine, ajmaline	Zaire, Mozambique
Rhamnus frangula L.	Crude drug	Europe

TABLE 3 MAJOR MEDICINAL PLANTS ENTERING GLOBAL COMMERCE[1] (Cont.)

Plant Name	Used as a Source of	Country of Production or Cultivation
Rhamnus purshiana DC.	Crude drug; casanthrosides	USA
Rheum emodi Wallich	Crude drug	India
Rheum officinale Baillon	Crude drug	China
Rheum palmatum L.	Crude drug	China
Rheum webbianum Royle	Crude drug	India
Silybum marianum Gartn.	Silybin	Mediterranean region
Smilax species	Crude drug	Asia, Far East, India, Central and South America
Swertia chirata Roxb.	Crude drug	India
Urginea indica Kunth	Same as above	India
Urginea maritima L.	Crude drug, scillarin A & B	Mediterranean region, Egypt, Turkey
Valeriana officinalis L.	Crude drug	Cultivated in Eastern Europe and Netherlands
Valeriana officinalis L. var. *latifolia* Miq.	Crude drug	Cultivated in Japan
Valeriana wallichii DC. (= *V. jatamansi* Jones)	Crude drug	Cultivated in India

1. According to data found in Anon. (1974, 1982a, 1982b).

Traditional Knowledge of Medicinal Plants – The Search for New Jungle Medicines

Mark J. Plotkin Ph.D*
Director, Plant Conservation, World Wildlife Fund, and Associate in Ethnobotanical Conservation, Harvard Botanical Museum

> *"Through most of man's history, botany and medicine were, for all practical purposes, synonymous fields of knowledge, and the shaman, or witch-doctor – usually an accomplished botanist – represents probably the oldest professional man in the evolution of human culture."*
>
> R.E. Schultes, 1972

Conservation

In 1980, *Harvard* magazine asked several of the University's most prominent faculty members what they considered to be the single most serious problem facing mankind. Dr E. O. Wilson, noted entomologist and sociobiologist, wrote:

> "What event likely to occur in the 1980s will our descendants most regret, even those living a thousand years from now? My opinion is not conventional, although I wish it were. The worst thing that can happen – will happen – is not energy depletion, economic collapse, limited nuclear war, or conquest by a totalitarian government. As terrible as these catastrophes would be for us, they can be repaired within a few generations. The one process ongoing in the 1980s that will take millions of years to correct is the loss of genetic and species diversity by the destruction of natural habitats. This is the folly our descendants are least likely to forgive us."

There exists widespread agreement among biologists that the rate of species extinction is increasing at an alarming rate. Although many temperate life forms like the California condor (*Gymnogyps californicus*) and the black-footed ferret (*Mustela nigripes*) are on the verge of extinct-

---

* *Now* Vice President, Conservation International, Washington, D.C.

ion and may disappear during the next decade, the majority of the world's threatened species inhabit the tropical forests. Though these forests cover less than 10% of the earth's surface, they are believed to contain over 50% of the world's species (Wilson, 1985).

The staggering amount of species diversity concentrated in tropical regions is best demonstrated with several comparative examples. The Rio Negro in central Brazil contains more species of fish than are found in all of the rivers of the U.S. combined (M. Goulding, pers. comm.). Manu National Park in southeastern Peru is home to more species of birds than are found in the entire United States (J. O'Neill, pers. comm.). A hectare of forest in the northeastern region of our country typically contains about 20 species of trees, whereas a similar plot in western Amazonia may contain more than 300 (A. Gentry, pers. comm.).

One problem faced by biologists working to prevent diminution of biological diversity is the lack of basic knowledge about most of the world's species. Carolus Linnaeus inaugurated the Latin binomial system in 1753, and approximately 1.7 million species of plants and animals have been described since then (Wilson, 1985). In a 1964 study, it was estimated that the number of insect species alone was 3 million (Williams, 1964), and more recent work has caused scientists to revise estimates sharply upwards. In the early 1980's, Terry Erwin of the Smithsonian Institution developed a technique for sampling the invertebrates of the rainforest canopy, a poorly-known fauna due to the difficulties of access and the prevalence of stinging insects. The results of these studies led scientists to re-think the magnitude of the diversity of life on earth. When 19 specimens of a single tree species were investigated using Erwin's technique, more than 1200 different species of beetles were collected (Eckholm, 1986). Currently, Erwin estimates that there are over 30 million species of insects in the world (T. Erwin, pers. comm.).

Compared to the insects, the angiosperms (flowering plants) are much better known. It is postulated that there are some 250,000 species worldwide, 90% of which are already known to science (Eckholm, 1986). Nevertheless, major expeditions to the tropics, particularly the Amazon region, continue to bring back new species. Yet the flora of the tropics, like the fauna, is faced with serious threats to its very survival. For example, as much as 95% of the Atlantic Coastal Forest of eastern Brazil has already been destroyed (Mittermeier, 1982). On the island of Madagascar, where 80% of the flowering plants are believed to be endemic, well over half of the original forest cover has been removed or seriously disturbed (Mittermeier, pers. comm.). In the Hawaiian archipelago, where the rate of endemism is higher than 90%, 14% of the flora is already believed to be extinct (Frame *et al.*, 1989).

Extinction *is* a natural process – since the origin of life 10 million years ago, many species have disappeared. Yet, to view these recent extinctions as natural is to misinterpret the geologic record. It has been postulated that the present rate of global species extinctions is 400 times faster than in the recent geologic past, and that this rate is rapidly accelerating (Wilson, 1985). The only similar examples in the history of this planet were the massive species die-out at the end of the Paleozoic and Mesozoic Eras. A striking feature of these historic natural disasters, however, is that the extinctions were primarily faunal. Indeed, there is little evidence in the fossil record of mass extinctions of vascular plants (Knoll, 1984). In the past, plants were presumably more resistant to extinction than the dominant animal life forms like dinosaurs. Consequently, plant diversity has increased through time, with especially high rates of speciation occurring during the Carboniferous and Cretaceous periods (Knoll, 1984). The unpleasant conclusion is that man is causing the first major reduction of global plant diversity since the origin of life.

Why should the medical community in the United States be concerned about the destruction of distant tropical rain forests? Because rainforest plants are complex chemical storehouses that contain many undiscovered biodynamic compounds with unrealized potential for use in modern medicine. We can gain access to these materials *only* if we study and conserve the species that contain them.

Yet, in a region like the Amazon, home to tens of thousands of plant species, how do we decide where to begin? Random screening of plant materials for new biodynamic compounds has proved to be very expensive. A more cost-effect method to find new and useful plant compounds is the science of ethnobotany.

Ethnobotany

The term "ethnobotany" was first used by Harshberger (1896), who defined it as the study of "plants used by primitive and aboriginal people". The term was broadened by Robbins, Harrington & Freire-Marreco (1916), who suggested that the science of ethnobotany should include the investigation and evaluation of the knowledge of all phases of life amongst primitive societies and of the effects of the vegetal environment upon the life customs, beliefs and history of these tribal peoples. Twenty-five years later, Jones (1941) put forward a more concise definition: "The study of the interrelationships of primitive men and plants". Schultes (1967) expanded this to include "the relationships between man and his ambient vegetation". Ford (1980) regarded ethnobotany as "the totality of the people with the plants in a culture and the *direct* interaction by the people with the plants". Ethnobotany therefore is considered to be a subset of the science of economic botany which "emphasizes the uses of plants, their potential for incorporation into another (usually Western) culture, and [suggests that these people] have *indirect* contact with the plants through their by-products" (Ford, 1980).

For the purposes of this paper, I prefer the concept of ethnobotany promulgated by Jones. Nevertheless, I would like to restate his definition of ethnobotany as "the study of tribal peoples and their utilization of tropical plants."

Some of the earliest evidence of plant use by Amerindians are fossil remains of peyote (*Lophophora williamsii*) excavated from caves in Texas and dated at about 7000 B.C. In lowland South America, artifacts from coastal Ecuador indicate that coca (*Erythroxylum* spp.) was in use in 2100 B.C. (Plowman, 1984). The Sinu culture (1200-1600) of northwestern Colombia produced numerous gold pectorals with mushroom-like representations. Schultes & Bright (1979) concluded that these were hallucinogenic mushrooms of the genus *Psilocybe*. That no South American tribe is known to consume hallucinogenic mushrooms today makes one wonder about the quantity of ethnobotanical information which has already been lost.

Compared with those of Mexico, written ethnobotanical records from lowland South America at the time of the Conquest are relatively scant. R.E. Schultes, who has conducted extensive research on the data recorded by the Spaniards both in Mexico and in South America, has concluded (pers. comm.) that the ecclesiastical class who helped conquer and settle Mexico were generally better educated and less exploitative than the group who subjugated the native peoples of western South America. While Mexico was still being "pacified", the King of Spain sent his personal physician Francisco Hernandez to the New World for five years to study the medicinal plants of the Aztecs (Schultes & Hofmann, 1979). Furthermore, the codices of both the Mayans and the Aztecs have been deciphered and have yielded a wealth of information on the useful plants of both tribes. Although some of the pottery from highland cultures like the Mochica of central Peru depict ritual uses of hallucinogenic fungi (*Psilocybe*) and hallucinogenic cacti (*Trichocereus*), our knowledge of South American ethnobotany is much less detailed than it is for Mexico.

A major barrier to know where to focus our research efforts in the search for new plant medicines is the haphazard way in which most of the early ethnobotanical data was collected. Most of the major figures who collected important ethnobotanical data up until this century originally went to the New World for reasons other than documentation of ethnobotanical lore. Amerigo Vespucci was carrying out geographic exploration when he made the first discovery of coca chewing on the Guajira Peninsula of Colombia. Charles-Marie de la Condamine, during an expedition to Ecuador to determine the shape of the earth, made the first major scientific voyage down the Amazon and stumbled on to such ethnobotanical treasures as rubber (*Hevea*), ipecac (*Cephaelis*), quinine (*Cinchona*) and curare (possibly *Strychnos*). The great German explorer Alexander von Humboldt, together with his French colleague Aime Bonpland, traveled extensively in the New World tropics. They were making general collections of the biota when they observed the manufacture of both curare (from *Strychnos*) and the hallucinogenic snuff yopo (*Anadenanthera peregrina*). While travelling in the Amazon making general collections of the flora, the British botanist Richard Spruce observed the native use of other hallucinogenic plants.

A turning point in the history of South American ethnobotany was the arrival of R.E. Schultes in the Colombian Amazon in 1941. After completing his Ph.D. dissertation on the ethnobotany of the Indians in Oaxaca, Mexico, Schultes decided to initiate a study of plants employed in the manufacture of arrow poisons. He remained in the northwest Amazon until 1954, living with the Indians, participating in their native rituals, and conducting ethnobotanical research. Although he eventually went to work on the USDA project to harvest natural rubber from the Amazon during the Second World War, he continued collecting, eventually sending home over 24,000 plant specimens. Nearly 2,000 of these species had been employed medicinally by the natives, while others had been used for everything from clothing to contraceptives. Most of the ethnobotanists working in South America today are following in his footsteps.

Tropical forest peoples represent the key to understanding, utilizing and protecting tropical plant diversity. The degree to which they understand and are able to sustainably use this diversity is astounding. The Barasana Indians of Amazonian Colombia can identify all of the tree species in their territory without having to refer to the fruit or flowers (S.H. Jones, pers. comm. to E.W. Davis), a feat that no university-trained botanist is able to accomplish, and a single Amazonian tribe of Indians may use over 100 species of plants for medicinal purposes alone. Nevertheless, to this day, very few tribes have been subjected to a complete ethnobotanical analysis. Robert Goodland of the World Bank wrote (1981):

> "Indigenous knowledge is essential for the use, identification and cataloguing of the [tropical] biota. As tribal groups disappear their knowledge vanishes with them ... The preservation of these groups is a significant economic opportunity for the [developing] nation, not a luxury."

Since Amazonian Indians are often the only ones who know both the properties of these plants and how they can best be utilized, their knowledge must be considered as an essential component of all efforts to conserve and develop the Amazon. Failure to document this ethnobotanical lore would represent a tremendous economic and scientific loss to mankind.

Ethnobotany and Medicine

Plants have traditionally served as man's most important weapon against pathogens – in fact, it seems that even Neanderthals knew and made use of medicinal plants (Solecki, 1975). As early as 2000 B.C., the Chinese were using moulds to treat festering ulcers, and the ancient Egyptians are known to have applied mouldy bread to open wounds (Oldfield, 1984). It is only relatively recently, however, with the advent of modern technology and synthetic chemistry that we have been able to reduce our almost total dependence on the Plant Kingdom as a source of medicines. We nonetheless continue to rely on plants to a much greater degree than is commonly realized. Almost half of all prescriptions dispensed in this country contain substances of natural origin – and over 50% of these medications contain a plant-derived active principle (Farnsworth, 1977). In 1974 alone, the U.S. imported $24.4 million of medicinal plants (Caufield, 1982).

There are four basic ways in which plants that are used by tribal peoples are valuable for modern medicine. First, plants from the tropics are sometimes used as sources of direct therapeutic agents. The alkaloid D-tubocurarine is extracted from the South American jungle liana *Chondrodendron tomentosum* and is widely used as a muscle relaxant in surgery (Lewis & Elvin-Lewis, 1977). Chemists have so far been unable to produce this drug synthetically in a form which has all the attributes of the natural product and we therefore continue to rely on collection of this plant from the wild (R.E. Schultes, pers. comm.). Surprisingly, harvesting of medicinal plants is often less costly than artificial drug synthesis. In 1973, less than 10% of the 76 drug compounds from higher plants employed in U.S. prescription drugs were produced commercially by total chemical synthesis. Reserpine, an important hypotensive agent extracted from *Rauwolfia*, is a good example. The synthesis of the drug in the mid-1970's cost approximately $1.25 per gram whereas commercial extraction from the plant cost only about $0.75 per gram (Oldfield, 1984).

Tropical plants are also used as sources of starting points for the elaboration of more complex semi-synthetic compounds. An example of this would be saponin extracts that are chemically altered to produce sapogenins necessary for the manufacture of steroidal drugs. Until relatively recently 95% of all steroids were obtained from extracts of neotropical yams of the genus *Dioscorea* (Oldfield, 1984).

Flora from the tropics can serve as sources of substances that can be used as models for new synthetic compounds. Cocaine from the coca plant, *Erythroxylum coca*, has served as a model for the synthesis of a number of local anesthetics such as procaine. New and unusual chemical substances found in plants will continue to serve as "blueprints" for novel synthetic substances and will prove to be increasingly important in the future (Oldfield, 1984).

Plants can also be used as taxonomic markers for the discovery of new compounds. From a phytochemical standpoint, the Plant Kingdom has been investigated in a very haphazard manner; some families have been relatively well-studied while others have been almost completely overlooked. For example, many uses have been documented for the Liliaceae, and the family is known to be rich in alkaloids. Although little was known of the chemistry of the Orchidaceae, plants of this family were investigated because of its close relationship to the Liliaceae. The research demonstrated that not only was the Orchidaceae rich in alkaloids, but that many of these alkaloids were unique and thought to be of extreme interest for the future (Schultes, 1973).

What then is the value of ethnobotany to the search for new biodynamic compounds? Of the hundreds of thousands of species of living plants, only a fraction have been investigated in the laboratory (Farnsworth, 1988). This poor understanding of plants is particularly acute in the tropics; the noted Brazilian phytochemist Otto Gottlieb (1981) recently wrote: "Nothing at all is known about the chemical composition of 99.6% of our flora". It is worth noting that Brazil probably has more species of flowering plants (approximately 55,000) than any other country on earth (G. Prance, pers. comm.).

The importance of ethnobotanical inquiry as a cost-effective means of locating new and useful tropical plant compounds cannot be overemphasised. Most of the secondary plant compounds employed in modern medicine were first "discovered" through ethnobotanical investigation. There are some 119 pure chemical substances extracted from higher plants that are used in medicine throughout the world and 74% of these compounds have the same or related use as the plants from which they were derived (Farnsworth, 1988).

The rosy periwinkle (*Catharanthus roseus*) represents a classic example of the importance of plants used by local peoples This herbaceous plant, native to southeastern Madagascar, is the source of over 75 alkaloids, two of which are used to treat childhood leukemia and Hodgkin's disease with a very high success rate. Annual sales of these alkaloids worldwide in 1980 were estimated to reach $50 million wholesale prior to 100% markup for the retail market (International Marketing Statistics, 1980, quoted in Myers, 1984 [unpublished report]). This species was first investigated in the laboratory because of its use by local people as an oral hypoglycemic agent (Cordell & Farnsworth, 1976). Thus, we can see that investigation of plants used for medicinal purposes by "unsophisticated" peoples can provide us with new biodynamic compounds that may have very important applications in our own society.

We must also consider the importance of medicinal plants in the developing countries themselves. The World Health Organization has estimated that 80% of the people in the world rely on traditional medicine for primary health care needs (Farnsworth, 1988). In many cases, developing countries simply cannot afford to spend millions of dollars on imported medicines that they could produce or extract from tropical forest plants. Several African and Asian nations have begun to encourage traditional medicine as an integral component of their public health care programs. Indigenous medicines are relatively inexpensive, are locally available, and are usually readily accepted by the local populace (Prescott-Allen, 1982). The ideal situation would be the establishment of local pharmaceutical firms that would create jobs, reduce unemployment, reduce import expenditures, generate foreign exchange, encourage documentation of traditional ethnomedical lore and be based on the conservation and sustainable use of the tropical forest.

The Search for New Jungle Medicines

What can modern medicine learn from the witchdoctor? Certainly, much more than one might think. To most of us, the witchdoctor or medicine man or shaman is an object of some derision. In countless films, he has often been portrayed in a grass skirt with a bone through his nose, waving rubber snakes and shouting gibberish. Those of us, however, who have been fortunate enough to live in the tropical forest and work with these people have a very different impression. These medicine men usually have a profound knowledge of tropical plants and the healing properties for which they may be employed. A single shaman of the Wayana tribe in the northeast Amazon, for example, may use over one hundred different species for medicinal purposes alone. Furthermore, a great many of the remedies *are* effective. Fungal infections of the skin are a common affliction in the humid tropics and modern medicine can only suppress – not cure – serious cases. On more than one occasion, I have had serious infections successfully treated by shamans using jungle plants.

Unfortunately, however, the oral traditions of these medicine men is not being passed onto the next generation. With the advent of Western medicine in many of these remote areas, the young members of the tribe demonstrate little interest in learning the traditional ethnomedical lore. Of all the shamans with whom I have lived and worked in the northeast Amazon, not a single one had an apprentice. We are, in my opinion, facing a critical situation – unless we act now, thousands and thousands of years of accumulated knowledge of how to use rainforest plants is going to disappear before the turn of the century.

What can the medical community do to aid both the struggle to conserve tropical forests and the search for new jungle medicines? Certainly, a much more prominent role needs to be played if both of these efforts are to prove successful. Many reasons for species conservation have been presented to the general public: aesthetic, ethical, etc., but the most relevant for the medical profession is the utilitarian: that species are of direct benefit to us. The few examples given in this paper – reserpine, d-tubocurarine, vincristine – are indicative of the kinds of undiscovered compounds which are undoubtedly out there. We now know that synthetics are not the only answers to our medical needs – and the European pharmaceutical firms are showing renewed interest in the potential of the tropical flora. There is heightened awareness in the United States as well. The National Cancer Institute recently awarded over $2.5 million in contracts to the New York Botanical Garden, the Missouri Botanical Garden and the University of Illinois to collect and test tropical plant species for anti-tumor activity.

With the moral and financial support of the U.S. medical community, the conservation movement can help protect and utilise tropical species for human welfare. For most of us, the days of believing that we have already found everything the Plant Kingdom had to offer are over. This attitude was addressed in a 1954 paper by DeRopp:

> "The situation results, at least in part, from the rather contemptuous attitude which certain chemists and pharmacologists in the West have developed toward both folk remedies and drugs of

plant origin ... They further fell into the error of supposing that because they had learned the trick of synthesizing certain substances, they were better chemists than Mother Nature who, besides creating compounds too numerous to mention, also synthesized the aforesaid chemists and pharmacologists ..."

Acknowledgments

I would like to thank the following people for providing useful information: Dr R.E. Schultes (Harvard Botanical Museum); Dr N. Nickerson (Tufts University); Dr Michael Goulding (Goeldi Museum); Dr John O'Neill (Louisiana State University); Dr Alwyn Gentry (Missouri Botanical Garden); Dr Terry Erwin (Smithsonian Institution); Professor G. Prance (New York Botanical Garden, now Royal Botanic Gardens, Kew); Dr. S.H. Jones (Cambridge University); Dr E. Wade Davis (Harvard University); and Dr R.A. Mittermeier, J. MacKnight, T. McShane and J. Trent (World Wildlife Fund).

References

Caufield, C. (1982). *Tropical Moist Forests*. London: International Institute for Environment and Development.

Cordell, G. & Farnsworth, N. (1976). A review of selected potential anti-cancer plant principles. *Heterocycles*, 4(2), 393-427.

Eckholm, E. (1986). Species are lost before they are found. *New York Times* (September 14, 1986), pp. 1-3.

Farnsworth, N. (1977). Foreword in *Major Medicinal Plants* by J. Morton. Springfield, Illinois: Charles C. Thomas.

Farnsworth, N. (1988). Screening plants for new medicines. In *Biodiversity*, ed. E.O. Wilson, pp. 83-97. Washington, D.C.: National Academy Press.

Ford, R. (1980). Ethnobotany: historical diversity and synthesis. In *The Nature and Status of Ethnobotany* ed. R. Ford, pp. 33-49. Ann Arbor: Museum of Anthropology, University of Michigan.

Frame, D., Wagner, W.L., Herbst, D.R. & Sohmer, S.H. (1989). The Hawaiian Islands. In *Floristic Inventory of Tropical Countries: The Status of Plant Systematics, Collections, and Vegetation, plus Recommendations for the Future* eds. D.G. Campbell & H.D. Hammond, pp. 181-186. New York: The New York Botanical Garden.

Goodland, R. (1981). *Economic Development and Tribal Peoples*. Washington, D.C.: World Bank.

Gottlieb, O. (1981). New and underutilized plants in the Americas: solution to problems of inventory through systematics. *Interciencia*, 6(1), 22-29.

Harshberger, J. (1896). Purposes of ethnobotany. *Bot. Gaz.*, 21(3), 146-154.

International Marketing Statistics (1980). *National Prescriptions Audit.* Ambler, Pennsylvania.

Jones, V. (1941). The nature and scope of ethnobotany. *Chron. Bot.*, 6(10), 219-221.

Knoll, A. (1984). Patterns of extinction in the fossil record of vascular plants. In *Extinctions*, ed. M. Nitecki, pp. 21-68. Chicago: University of Chicago Press.

Lewis, W. & Elvin-Lewis, M. (1977). *Medical Botany.* New York: Wiley-Interscience.

Mittermeier, R.A. (1982). The world's endangered primates: an introduction and a case study – the monkeys of Brazil's Atlantic Forest. In *Primates and the Tropical Forest* eds. R.A. Mittermeier & M.J. Plotkin, pp. 11-22. Washington, D.C.: World Wildlife Fund-U.S.; Pasadena, California: The L.S.B. Leakey Foundation.

Oldfield, M. (1984). *The Value of Conserving Genetic Resources.* Washington, D.C.: U.S. Dept. of the Interior.

Plowman, T. (1984). The origin, evolution, and diffusion of coca, *Erythroxylum* spp., in South and Central America. In *Pre-Columbian Plant Migration*, ed. D. Stone, pp. 125-163. Cambridge, Mass.: Peabody Museum.

Prescott-Allen, R. & P. (1982). *What's Wildlife Worth?* London: International Institute for Environment and Development.

Robbins, W., Harrington, J. & Freire-Marreco, B. (1916). Ethnobotany of the Tewa Indians. *Bureau of American Ethnology Bulletin*, 55.

Schultes, R.E. (1967). The place of ethnobotany in the ethnopharmacologic search for psychotomimetic drugs. In *Ethnopharmacologic Search for Psychoactive Drugs*, ed. D. Efron, Washington, D.C.: U.S. Govt. Printing Office.

Schultes, R.E. (1973). Orchids and human affairs: What of the future? *Bull. Am. Orchid Soc.*, (42), 785-789.

Schultes, R.E. & Bright, A. (1979). Ancient gold pectorals from Colombia: mushroom effigies? *Leafl. Bot. Mus. Harvard Univ.*, 27, 113-141.

Schultes, R.E. & Hofmann, A. (1979). *Plants of the Gods*. New York: McGraw-Hill.

Solecki, R. (1975). Shanidar IV, a Neanderthal flower burial in northern Iraq. *Science*, 190, 880-881.

Williams, C. (1964). *Patterns in the Balance of Nature.* New York: Academic Press.

Wilson, E.O. (1985). The biological diversity crisis. *Bioscience*, 55(11), 700-706.

The Reason For Ethnobotanical Conservation

Richard Evans Schultes, F.M.L.S.,
Jeffrey Professor of Biology and Director Emeritus, Botanical Museum, Harvard University, Cambridge, Mass.

Nunc vos potentes omnes herbas deprecor,
exoro maiestatem vestrum, quas parens
tellus generavit et cunctis dono dedit.
A Roman prayer to all herbs

In the few parts of the world still not affected by fast encroaching civilization, there exists a wealth of information on the properties of plants that is still available to us. It will not long be there for us to salvage. It has been built up by peoples in primitive societies over millennia by trial and error, for they have had to rely on their ambient vegetation for foods, medicines and all the other necessities of life.

When our civilization arrives with roads, missionary activities, commercial interests, tourism or otherwise, the products of our culture are rapidly adopted and, often even in one generation, replace what has for hundreds of years been part of their culture. This erosion of native ethnobotanical knowledge and use is nowhere more rapid than in the realm of biodynamic plants – medicinal, narcotic and toxic species (Schultes & Hofmann, 1979, 1980)

There have long been two strongly divergent poles in our evaluation of the worth of ethnobotanical studies. Many investigators have been carried away with enthusiasm that native peoples have had some special intuition that permitted them to seek out "nature's secrets". Others have cast aside or denigrated all native folk lore as not worthy of serious scientific attention. Naturally, both points of view are extreme and are unwarranted.

Recently, the realization that the aboriginal knowledge of plant properties is of both academic and practical value has matured. And various investigators in sundry fields have recognized the need to save native plant lore before it is entombed with the culture that gave it birth (Bruhn & Holmstedt, 1982; Gottlieb, 1979).

The Brazilian chemist, Otto Gottlieb, for example, wrote: "Since Indians in the Amazon are often the only ones who know both the properties of the forest species and how they can best be utilized, their knowledge must be considered an essential component of all efforts to conserve and develop the Amazon" (Gottlieb, 1979).

Davies wrote similarly: "The tragedy is that the Indian is one of the main keys to the successful occupation of the Amazon and as he disappears his vast wealth of knowledge goes with him" (Davies, 1977).

Interest in ethnobotany goes beyond these and other similar statements. The Society for Economic Botany and its journal *Economic Botany* have become increasingly more dedicated to ethnobotany. The Society for Ethnobiology has recently been established and is publishing its own journal. The highly successful *Journal of Ethnopharmacology*, now only in its 15th volume, enjoys worldwide circulation. Europe has several publications specializing in the study of medicinal plants used in native pharmacopoeias. Some governments – especially Mexico, China and India – are fostering the scientific study of native remedies. The newly organized Society of Ethnobotanists of India has just published a most useful *World Directory of Ethnobotanists*, an up-dating of a *Directory of Ethnobotanists* produced in 1976 by the Ethnobotanical Laboratory of the University at Ann Arbor; the number listed has increased from 150 to 490. Various congresses, meetings and symposia are including sections devoted to ethnobotany. Under the sponsorship of the Species Survival Commission of the prestigious International Union for Conservation of Nature and Natural Resources and the World Wildlife Fund U.S., the Botanical Museum of Harvard University has set up a subgroup on Ethnobotanical Conservation, the purposes of which are to knit together specialists in the many disciplines associated with ethnobotany on a worldwide basis and especially to encourage the conservation of ethnobotanical information in imminent danger of extinction.

Any denigration of native folk medicine is not supported by the contents of western pharmacopeias nor by the history of some of the most recently discovered drugs from the Plant Kingdom that have revolutionized the practice of modern medicine: the curare alkaloids; penicillin and the many other antibiotics; cortisone and the steroids; reserpine; vincoleucoblastine; the *Veratrum*-alkaloids; podophyllotoxin; strophanthine; physostigmine; and other new therapeutic or research agents (Schultes & Farnsworth, 1980).

Nor is this denigration supported by chemical and pharmacological investigations of native medicinal plants under current study (Schultes & Swain, 1976). One excellent example is the statistical analysis of 25 medicinal plants employed by the Aztecs. Twenty-five have been shown pharmcologically to produce the effects that the Aztecs claimed; four possibly could have the properties reported by the Indians; five (only

20%) seem unable to induce the physiological activities attributed to them in the native medicine (Ortíz de Montellano, 1975). A further example is the recent ethnopharmacological analysis of the loganiaceous genus *Buddleja*: a high degree of correlation exists between the wide spectrum of native medicinal uses and what is known of the chemical composition of this genus of 100 species of the tropics and subtropics of both hemispheres (Houghton, 1984). Similar correlations in the few groups of plants that have been ethnobotanically studied can be cited.

It is clear that modern science can no longer afford to ignore reports of any aboriginal uses of plants simply because they seem to fall beyond the limits of our credence. On the contrary, their uses should stimulate examination under the impartial searchlight of modern scientific analysis (Schultes, 1983).

Several plant explorers of the past century – von Martius and Spruce, for example – were of the opinion that the Indians of the Amazon utilized a limited vegetal pharmacopoeia. In his field notes on *paricá* – the hallucinogenic snuff prepared from *Anadenanthera peregrina* – Spruce wrote: "Throughout the Uaupés, this is almost the sole medicinal agent employed.... I have never known any other remedies applied except occasionally the milk of some tree (and they are not particular as to the species) by way of plasters in the case of some wound or internal pain." In another context, Spruce reported: "Among the native tribes of the Uaupés and of the upper tributaries of the Orinoco, niopo or paricá is the chief curative agent" (Spruce, 1873).

Von Martius had the same opinion as Spruce. He wrote: "...of external applications, I have seen only the following. For a wound or bruise or swelling, the milky juice of some tree is spread thick on the skin where it hardens into a sort of plaster and is allowed to remain on until it falls of itself. Almost any milky tree may serve, if the juice be not acrid; but the Heveas (India-rubbers), Sapotads and some Clusias are preferred. Such a plaster has sometimes an excellent effect in protecting the injured part from the external air".

It is not easy to reconcile this opinion of some of the greatest of field botanists who spent many years in the northwest Amazon with my own observations over the past 40 years amongst numerous tribes of the same region. There are, however, several possible reasons. The Indians recognize in general two "kinds" of medicines: those with purely physical effects and these are known and used by members of the tribe who might be called regular practioners or the ethnopharmacologists of the tribe who do not resort to magic or superstition but who possess a wide knowledge of the flora and the properties of a great number of species; and the plants with psychic effects – usually the hallucinogens, which are considered to be sacred and which the medicine man or payé manipulates. Spruce himself stated: "I have never been so fortunate as to see a genuine payé at work.. With the native and still unchristianized tribes I have for the most part held only passing intercourse during some of my voyages. Once I lived for seven months at a time among them on the river Uaupés, but even there I failed to catch a payé. When I was exploring the Jauarité cataracts ... news came that a famous payé .. 'would arrive that night and remain until next day' and I congratulated myself on so fine a chance of getting to know some of the secrets of his 'medicine'. He did nor reach the port until 10 p.m., and when he learned that there was a white payé (meaning myself) in the village, he and his attendants immediately threw back into the canoe his goods ... and resumed their dangerous voyage... in the night-time. I was told he had with him several palm-leaf boxes containing his apparatus... I could only regret that his dread of a supposed rival had prevented the interview which to me would have been full of interest; the more so since I was prepared to barter with him for the whole of his materia medica, if my stock-in-trade would have sufficed". Spruce, like most botanists engaged in making general collections of the flora, probably could not devote much attention to time-consuming ethnobotanical research, and in view of his incomparably complete collections of that strongly endemic flora, science is the richer because he spent most of his efforts in floristic botanizing (Schultes, 1873).

For fourteen years – 1941 to 1954 – I was able to live permanently in the northwest Amazon, and I have briefly visited the same region almost annually since 1954, making a total of 24,000 plant collecions.

Of these, there are notes on the aboriginal uses as medicines, narcotics and poisons of nearly 2000 species. Certainly many uses escaped my attention and will be discovered by future investigators, if they can in some way beat the rapidly advancing acculturation and loss of native plant lore. I have incorporated reports of plant uses from this region collected by some of my students and colleagues and from herbarium reports.

The flora of the Amazon is extensive – with probably some 80,000 species. One expert has estimated that nothing is known about the chemical composition of 99.6% of the Amazon's flora. Certainly almost all of the biodynamic species for which I have notes have never been chemically analyzed. Some of the uses may be of little or no practical value, but for many it is possible to see or appreciate their physiological effectiveness. Few may actually be curative, but an appreciable number probably are alleviative. Whatever the case may be, if a plant has any physical activity it indicates the presence of a least one active chemical constituent. We should know what these constituents are: they may not be of any value in our western pharmacopoeias; they may find wholly different uses in our technology; a few may yield drugs for modern medicine to treat the same conditions for which they are applied in primitive societies. And many species hold promise of the discovery of new chemical compounds, for it is now realized that unstudied tropical flora as rich as that of the Amazon represents a vast emporium of unknown chemical compounds awaiting discovery.

What can we say about some of the interesting and promising biodynamic plants of the northwest Amazon? Included now in my ethnopharmacological notes and those of several of my students and colleagues are at least 32 species the uses of which suggest possible cardiovascular activity; 90 are involved as probable major ingredients of arrow poisons; 27 seem to be insecticidal or insect-repellent; 57 are employed for their ichthyotoxic properties; four are valued as presumed oral contraceptives; more than 85 are taken as vermifuges; over 100 are believed to be febrifuges; a few are styptics; four dozen are applied to clean or hasten the healing of infected sores and wounds; five or six, it is claimed, relieve conjuntivitis; six are said to be stimulants; 11 are esteemed as hallucinogens or narcotics; and so the list could continue.

It is probable that there are few regions in the world where the indigenous populations possess a fuller acquaintance with the properties of their plants. The region is sparsely populated by numerous tribes of very diverse origins, cultures, languages and methods of using bioactive plants. Until recently the area has been by nature isolated and protected from external penetration, since rapids and waterfalls have rendered navigation impossible. Furthermore, the region is floristically the most variable and the richest in the Amazon Valley. All of these factors have tended to contribute towards the extreme ethnopharmacological wealth of the northwestern sector of the hylea.

The appreciation and utilization of plants for medicinal purposes, however, varies significantly from tribe to tribe. Some – the Colombian Sionas, Kofáns, Witotos, Boras, Yukunas, Tanimukas, Kubeos, Tukanos, Barasanas, Makunas, Kuripakos, Puinaves and certain tribes of Makus, for example, have rich pharmacopoeias. Other groups – the Waoranis of Ecuador – living in the same species-rich forests – have a dearth of plants therapeutically employed: intensive research indicates that they use only 35 species, 30 of which are valued in treating only six conditions (Davis & Yost, 1983); while their neighbours, the Kofáns, have at least 80 species for 27 different ailments.

An interesting and possibly significant aspect of the medicinally used plants in the northwest Amazon is their employment most commonly as simples: only rarely are two or more species mixed for therapeutic use – quite in contrast to the usual situation in the preparation of curares.

Peoples on every continent have learned to tip their arrows or darts with poisonous preparations, derived mainly from plants. In number of species so employed, South America is the centre for the use of arrow poisons or curares. And in diversity of ingredients, it appears that the northwestern part of the Amazon represens the epicentre.

Although curares have been carefully studied as sources of medicinally valuable compounds, there remains much to do from this point of view. It would hardly be an exaggeration to state that every tribe (and often every payé) dedicated to the preparation of curare has a different formula.

While most curare formulae call for a number of vegetal ingredients, the most widely prepared Amazon arrow poisons have as their basically active plants members of the Menispermaceae (species of *Abuta, Curarea, Chondrodendron* and other genera) or of the loganiaceous genus *Strychnos*. Alkaloids from the menispermaceous genera have, of course, become extremely important in western medicine during the past 50 years. Recent ethnopharmacological studies have, however, uncovered new sources of curares – even curares prepared of one species.

More than 150 years ago, the German botanist von Martius discovered that Indians on the Rio Japurá in Brazil were preparing a curare based on the annonaceous *Unonopsis veneficiorum*. Recently it has been learned that the distant Kofán Indians of Colombia and Ecuador prepare an arrow poison from the fruits of this plant (Schultes, 1969). This species contains bisbenzylquinolic compounds, but it is not known that these constituents can be responsible for the curariform activity. The Kofáns use another annonaceous plant in making a curare: the bark of a small tree of the genus *Anaxagorea*. The Kofán Indians likewise use the fruits of the lauraceous *Ocotea venenosa*, which also contain bisbenzylquinolic constituents. Another interesting toxic plant utilized by the Kofáns is the thymaeliaceous *Schoenobiblos peruvianus*, the fruits of which are the sole ingredients of an arrow poison; they are also a favourite fish poison of this tribe (Schultes, 1969). The Thymaeliaceae is rich in coumarin derivatives, but these are not believed to have curare activity.

In 1954, a new hallucinogenic snuff was identified in the northwest Amazon, prepared from the resin-like bark exudate of several species of the myristicaceous genus *Virola*. During the study of this interesting use it was discovered that the Waika Indians of Brazil tip their darts with the exudate and with no other ingredient as a commonly employed curare. While the hallucinogenic principles – tryptamines – have been identified, there is no suspicion of the constitutuent that can act as an arrow poison (Schultes & Holmstedt, 1968).

Amongst the other species reportedly used as active components of arrow poisons are *Vochysia columbiensis*, utilized by the Makú Indians of the Río Piraparaná who enjoy the reputation of making the best curare in the region; and the caryocaraceous *Anthodiscus obovatus* from which the Tukanos of the Vaupés prepare with a species of *Strychnos* a relatively strong arrow poison.

While many Indian tribes cultivate species of the euphorbiaceous *Phyllanthus* and the composite *Clibadium* and utilize the bark of wild species of the legume *Lonchocarpus* as fish poisons, many other species of lesser ichthyotoxic value have recently been identified as members of the euphorbiaceous *Nealchornia*, the myrsinaceous *Conomorpha*, the acanthaceous *Mendoncia* and the connaraceous *Connarus* (Schultes, 1969). A curious fish poison is prepared from the leaves of the araceous *Philodendron crasspedodromum* amongst the Indians of the Vaupés: the leaves, tied up and left to ferment for several days, are then crushed and thrown into still water. The Waoranis of Ecuador esteem the bark of the bignoniaceous *Minquartia guianensis* as an ichthyotoxic agent (Davis & Yost, 1983). A very unusual disovery has been the use by the Tikunas of the Colombian Amazonas of the dried pulp of the fruit of the bombacaceous *Patinoa ichthyotoxica*; no chemical constituent known from this family is known to have toxic properties (Schultes & Cuatrecasas, 1972).

There are many plants that the natives classify as poisonous and for which they have no use. Especially interesting are cucurbitaceous species in the genus *Gurania*, a family that deserves closer phytochemical study.

Several species of the marcgraviaceous genus *Souroubea* are valued in the Vaupés as the source of calmative teas administered to elderly natives suffering from "susto" (psychological fear) or to induce sleep. Two or three plants – especially the cultivated cucurbitaceous *Cayaponia ophthalmica* – are employed, apparently with some success, in treating conjunctivitis, a very frequent condition in the region (Schultes, 1964).

One of the medicinal uses most worthy of scientific evaluation is the application of the reddish resin-like bark exudate of several species of *Virola* and of the guttiferous *Vismia* to fungal infections of the skin, an extremely common affliction in the wet tropics. The condition often seems to clear up with this treatment in 10 or 15 days, but whether it represents a cure or merely suppression cannot be known at the present state of our technical understanding. Recent preliminary chemical studies have yielded several chemical constituents from *Virola* – lignans and neolignans – that may possibly account for antifungal activity. Other plants are employed in treating infections of the skin or of the mucous membrane of the mouth: the gum extracted from the pseudobulbs of the orchid *Eriopsis sceptrum*, a decoction of the bark of several species of

Vochysia, an infusion of the leaves of *Souroubea crassipetala* and the powdered bark of the rubiaceous *Calycophyllum acreanum* and *C. spruceanum*. A warm decoction of the leaves of *Anthurium crassinervium* var. *caatingae* is used by the Kubeos as an ear-wash to relieve a painful condition due probably to fungal infection.

One of the commonest medicinal plants of the Makunas is the malpighiaceous *Mezia includens*: the root is considered to be strongly laxative when crushed and soaked in water in which a flour (from *Manihot esculenta*) has been sitting for several hours. The boiled leaves make an emetic tea and, when they are applied warm as a cataplasm on the abdomen, are said to help a condition that seems to be hepatitis (Schultes, 1975).

Despite its toxicity, *Aristolochia medicinalis* is adminstered in the Vaupés to calm what appear to be epileptic seizures. The treatment is reported sometimes to be worse than the disease, since use of this tea, it is alleged, can cause insanity if not used with caution.

Only several species were encountered in use as presumed oral contraceptives: *Philodendron dyscarpium, Urospatha antisyleptica* and *Anthurium tessmannii* – all members of the aroid family. The Bara-Makú of the Río Piraparaná, who call the moraceous *Pourouma cecropiaefolia we-wit-kat-tu* ("no children medicine") scrape the bark from the root, rub the scrapings in water and give the drink to women; according to the natives, the drink can cause permanent sterility. These same Indians report that the leaves of *Vochysia lomatophylla* in warm chicha (a slightly fermented drink made from *Manihot esculenta*) has aborifacient properties. It is perhaps significant that the distant Campa Indians of Peru value this *Vochysia* as a possible contraceptive.

Several plants are widely employed as styptics to staunch the flow of blood from wounds: *Helosis guianensis* of the Balanophoraceae; *Costus erythrocoryne* and *Quiina leptoclada* of the Zingiberaceae and Quiinaceae, respectively.

A recently published ethnobotanical study of the bignoniaceous genus *Martinella* may be extremely significant. An extract of the root of *M. obovata* is widely employed throughout northern South America as an "eye medicine." References to this use over such an extensive area is "compelling evidence that *Martinella* contains medically useful properties" (Gentry & Cook, 1984). In the Vaupés, another bignoniaceous liana – *Arrabidea xanthophylla* – is valued in treating conjunctivitis.

The number of species for which vermifugal and febrifugal properties are claimed is naturally very high in view of the prevalence of intestinal parasites and various fevers, especially malaria. Very few of the plants so employed have been chemically or pharmacologically examined, although numerous species belong to genera recognized as having astringent properties. It will be sufficient to mention several members of the solanaceous genus *Brunfelsia* in connection with febrifugal activity: *B. chiricaspi, B. grandiflora* and *B. grandiflora* subsp. *schultesii*. The ingestion of a decoction of the leaves rapidly induces a sensation of chills: in fact, the vernacular name *chiricaspi* means "chill plant." In addition to this febrifugal use, these plants have additional applications in local medicine: in treating rheumatic pains and arthritis as well as snakebites. They may occacionally be added to the hallucinogenic drink *ayahuasca* (prepared from the malpighiaceous *Banisteriopsis caapi*) in order to lengthen and intensify the psychoactivity of the narcotic. And *Brunfelsia*, taken alone, can itself have hallucinogenic effects (Plowman, 1977). Yet little is known of the chemistry of such a well known and highly esteemed native medicine. Recent chemical studies of *B. grandiflora* subsp. *schultesii* disclosed a novel convulsant – pyrrole-3-carboxyamidine – which has been named brunfelsamidine, but it is not clear that this represents the physiologically active constitutent (Lloyd *et al.*, 1985).

Ethnopharmacology has recently been defined as "the observation, identifiction, description and experimental investigation of the ingredients and the effects of indigenous drugs." It is clearly a highly interdisciplinary field. It is "... not just a science of the past utilizing an outmoded approach. It still constitutes a scientific backbone in the development of active therapeutics based upon traditional medicine of various ethnic groups. Although not highly esteemed at the moment, it is a challenge to modern pharmacologists" (Bruhn & Holmstedt, 1982).

In addition, then, to its academic interest to anthropology and botany, a major reason for ethnopharmacological conservation is the search for potential new therapeutic agents for western medicine. The pharmaceutical industry in the United States has attained in the prescription market alone annual sales in excess of $3,000,000,000 from medicinal agents first discovered in plants, many of them found in use amongst unlettered peoples in aboriginal societies the world around (Schultes & Farnsworth, 1980). Can we afford any longer to neglect this prolific and promising treasure-trove of ethnopharmacological knowledge that may not long be available to us for the benefit of mankind?

Selected References

Altschul, S. von R. (1970). Ethnogynocological notes in the Harvard University Herbaria. *Bot. Mus. Leafl., Harvard Univ.*, 22 (1970), 333-343.

Bruhn, J.C. & Holmstedt, B. (1982). Ethnopharmacology: objectives, principles and perspectives. In *Natural Products as Medical Agents*, eds E. Reinhard & J.L. Beal. Hippocrates, Stuttgart, 405-430.

Davies, S.H. (1977). *Victims of the Miracle – Development of the Indians of Brazil.*

Davis, E.W. & Yost, J.A. (1983). The ethnobotany of the Waorani of eastern Ecuador. *Bot. Mus. Leafl., Harvard Univ.*, 29, 159-218.

Gentry, A.H. & Cook, K. (1984). *Martinella* (Bignoniaceae): widely used eye medicine of South America. *Journ. Ethnopharm.*, 11, 337-343.

Gottlieb, O. (1979). Chemical studies on medicinal Myristicaceae from Amazonia. *Journ. Ethnopharm.*, 1, 309-343; personal comm.

Holmstedt, B. & Bruhn, J.G. (1983). Ethnopharmacology – a challenge. *Journ. Ethnopharm.*, 8, 251-256.

Houghton, P.H. (1984). Ethnopharmacology of some *Buddleja* species. *Journ. Ethnopharm.*, 11, 293-308.

Lloyd, H.A., Fales, H.M., Goldman, M.E., Jerina, D.M., Plowman, T. & Schultes, R.E. (1985). Brunfelsamine: a novel convalsant from the medicinal plant *Brunfelsia grandiflora* in *Tetrahed. Letters*, 26, 2623-2634.

von Martius, K.F.P. *Systema Materiae Medicae Vegetabilis Brasiliensis*.

Ortíz de Montellano, B. (1975). Empirical Aztec medicine. *Science*, 188, 215-220.

Pinkley, H.V. (1973). *The Ethnoecology of the Kofán*. Unpubl. Ph.D. thesis, Harvard University, Cambridge, Mass.

Plowman, T. (1977). *Brunfelsia* in ethnomedicine. *Bot. Mus. Leafl., Harvard Univ.*, 25, 289-320.

Schultes, R.E. (1954). Plantae Austro-Americanae V. De plantis principaliter Colombiae observationes. *Bot. Mus. Leafl., Harvard Univ.*, 16, 241-260.

Schultes, R.E. (1964a). Plantae Colombianae XVIII. De plantis regionis amazonicae notae. *Bot. Mus. Leafl., Harvard Univ.*, 20, 17-324.

Schultes, R.E. (1964b). De plantis toxicariis e Mundo Novo tropicale commentationes I. *Bot. Mus. Leafl., Harvard Univ., 21, 265-287.*

Schultes, R.E. (1969). De plantis toxicariis e Mundo Novo tropicale commentationes IV. *Bot. Mus. Leafl., Harvard Univ.*, 22, 133-164.

Schultes, R.E. (1970). De plantis toxicariis e Mundo Novo tropicale commentationes VII. Several ethnotoxicological notes from the Colombian Amazon. *Bot. Mus. Leafl., Harvard Univ.*, 22, 345-352.

Schultes, R.E. (1975). De plantis toxicariis e Mundo Novo tropicale commentationes XIII. Notes on poisonous or medicinal malpighiaceous species of the Amazon. *Bot. Mus. Leafl., Harvard Univ.*, 24, 128-131.

Schultes, R.E. (1978). De plantis toxicariis e Mundo Novo tropicale commentationes XXIII. Notes on biodynamic plants of aboriginal use in the northwestern Amazon. *Bot. Mus. Leaf., Harvard Univ.*, 26, 177-197.

Schultes, R.E. (1980). De plantis toxicariis e Mundo Novo tropicale commentationes XXVI. Ethnopharmacological notes on the flora of northwestern South America. *Bot. Mus. Leafl., Harvard Univ.*, 28, 1-45.

Schultes, R.E. (1983a). Richard Spruce, an early ethnobotanist and explorer of the northwest Amazon and northern Andes. *Journ. Ethnobiol.*, 3, 139-147.

Schultes, R.E. (1983b). From ancient plants to modern medicine. In *1984 Yearbook of Science and the Future*, pp. 172-187. Chicago, Illinois: Encyclopedia Britannica, Inc.

Schultes, R.E. (1985a). De plantis toxicariis e Mundo Novo tropicale commentationes XXXIV. Biodynamic rubiaceous plants of the northwest Amazon. *Journ. Ethnopharm.*, 14, 105-124.

Schultes, R.E. (1985b). De plantis toxicariis e Mundo Novo tropicale commentationes XXXV. Miscellaneous notes on biodynamic plants of the northwest Amazon. *Journ. Ethnopharm.*, 14, 125-158.

Schultes, R.E. & Cuatrecasas, J. (1972). De plantis toxicariis e Mundo Novo tropicale commentationes IX. A new species of ichthyotoxic plant from the Amazon. *Bot. Mus. Leafl., Harvard Univ.*, 23, 129-136.

Schultes, R.E. & Farnsworth, N.R. (1980). Ethnomedical, botanical and phytochemical aspects of natural hallucinogens. *Bot. Mus. Leafl., Harvard Univ.*, 28, 123-214.

Schultes, R.E. & Hofmann, A. (1979). *Plants of the Gods: Origins of Hallucinogenic Use*. New York: McGraw-Hill Book Co.

Schultes, R.E. & Hofmann, A. (1980). *The Botany and Chemistry of Hallucinogens*, Ed. 2. Springfield, Illinois: Charles C. Thomas.

Schultes, R.E. & Holmstedt, B. (1968). De plantis toxicariis e Mundo Novo tropicale commentationes II. The vegetal ingredients of the myristicaceous snuffs of the northwest Amazon. *Rhodora*, 70, 113-160.

Schultes, R.E. & Holmstedt, B. (1971). De plantis toxicariis e Mundo Novo tropicale commentationes VIII. Miscellaneous notes on myristicaceous plants of South America. *Lloydia*, 34, 61-78.

Schultes, R.E. & Swain, T. (1976). The Plant Kingdom: a virgin field for new biodynamic constituents. In *The Recent Chemistry of Natural Products, including Tobacco*, ed. N. Fina, pp. 133-171. New York: Philip Morris Science Symposium Proceedings.

Spruce, R. (1873). On some remarkable narcotics of the Amazon Valley and Orinoco. *Ocean Highways: the Geographical Review, n.s.*, 55, 1, 184-193.

Science, Industry and Medicinal Plants

Valuing the Biodiversity of Medicinal Plants

Peter P. Principe[1]
U.S. Environmental Protection Agency, Washington, D.C.

Introduction

The preservation of biodiversity has received considerable attention over the last two decades in both the scientific community and in public policy debates. However, much of this attention in the public policy arena has centered on the domestic aspects of the problem. For example, the United States enacted a law (the Endangered Species Act) that established a policy of identifying and preserving endangered species. The international aspect of the loss of biodiversity has elicited less interest from OECD countries, although it was raised as a significant issue at the Stockholm Conference in 1972. While it has remained a widely discussed issue in scientific circles, biodiversity has not been accepted by national policy-makers as a vital global resource warranting preservation at an international level.

A primary reason for this is the considerable uncertainty surrounding the whole issue. The reasons for this uncertainty will be discussed throughout this paper, but they relate to almost every facet of the problem: from determining the size of the plant population and the magnitude of the risk of extinction, through ascertaining of the value of these resources to OECD countries[2] and the likelihood and magnitude of adverse effects resulting from their loss, to identifying potential policies and actions that would resolve the problem. While there appeared to be considerable work on the value of biodiversity to crops, little was available on the pharmaceutical market. It was within this context that the OECD decided to undertake a study of that part of the biodiversity issue concerning medicinal plant species.

The Loss of Species

The world's genetic resources are steadily and rapidly diminishing. During the 300 years between 1600 and 1900, about 75 species of plants and animals became extinct because of human activity. During the first 70 years of this century, about the same number of species became extinct

1. Footnotes, see end of paper – pp 123-124

(Myers, 1979). These numbers, however, are quite small when compared to the number of extinctions projected to occur during the last 20 years of this century.

The earth is on the verge of experiencing an extinction of species unparalleled in human experience. The most widely accepted estimate, by the International Union for Conservation of Nature and Natural Resources (IUCN) and the World Wide Fund for Nature (WWF), is that 60,000 higher plant species could become extinct or near extinct by the middle of the next century if present trends continue. This exceeds earlier estimates of 20,000 or 25,000 threatened species worldwide (as in Lucas & Synge, 1978), which were based mainly on experience in temperate countries. The losses in the tropics are likely to be much greater, both in numbers of species and as a percentage of the flora. The 60,000 figure amounts to about 1 in 4 of all higher plants, since most botanists now believe there are around 250,000 such species on earth. Estimates of the total number of higher plant species (both identified and unidentified) range from 250,000 to 500,000 species (Schultes, 1972) to 750,000 (Balandrin *et al.*, 1985), but since there is no generally agreed upon estimate of unidentified species, no attempt has been made to include unidentified species in the calculations made in this paper.

The primary cause of this loss will be the continuing destruction of the habitats that support these species. Tropical moist forests, which are found almost exclusively in developing countries, are being destroyed at a rate that many scientists believe to be unjustifiable from both an economic and ecological perspective. Yet, for example in Brazil, Balandrin *et al.* (1985) estimate that nothing is known about the chemistry of more than 99% of the flora.

Putting Species Extinction into Perspective

Probably the most daunting aspect of the biodiversity issue is our almost complete ignorance of the magnitude of both the problem itself and the potential benefits that may be lost or retained.

Although this paper will be dealing with the medicinal benefits of higher plants, it must be recognized that there are many other possible benefits, including agricultural products, industrial products (e.g. a plant oil that can substitute for whale oil, thereby lowering prices and saving whales), insight into new processes and mechanisms, a better understanding of nature, and direct environmental/ecosystem benefits (e.g. their role in the carbon cycle, removal of pollution, and prevention of erosion). Clearly, most of these benefits represent knowledge: either insight into known or altogether new products or processes. And every plant species offers different insights and understanding.

This repository of information can be compared to a library. In a library, many of the books are rarely, if ever, used. But the frequency with which they are used is only one measure of their value. Someone (now or in the future) looking for an unusual bit of information may regard the value of a seldom-used book as far higher than a more commonly-used book. The point is that the library, like the body of plant species, is a repository of information whose value can only partially be judged by the current users. It has been said that one of the greatest tragedies in history was the destruction of the library at Alexandria and the effect this had of lengthening the Dark Ages. There is an obvious parallel between this event and the extinction of species: in both cases, the embodied knowledge is lost, with little hope of its reconstruction.

It has been suggested that the loss could be reduced by preserving those books (or species) that are of known value. The problem with this reasoning is that it assumes complete (or something resembling complete) knowledge about the utility of species including those that have yet to be identified. Since this is not the case and since extinction is likely to appear to be a random event, the loss reduction would likely be small, if it exists at all.

There are, then, clear reasons for preserving biodiversity, but the benefits afforded by this action are not cost-free, and any decision to preserve biodiversity must be made fully considering the nature and extent of the costs of such action, who would bear those costs, and the effects of imposing those costs. How this is done is discussed in greater depth later in this paper, but the main purpose of this paper is to establish a framework for estimating the potential value to be realized from preserving biodiversity since unless the benefits are sufficiently large, the process is unlikely to proceed to the point of needing a cost analysis.

The Economics of Extinction

In addition to the ecological, informational, and aesthetic benefits derived from the preservation of plant species, there are economic reasons as well. Biodiversity is of great importance to the world's economy. Plants are used for food, clothing, fuel, building materials, drugs, and many other uses. For example, the world's main agricultural crops are based on fewer than 20 crop species. The availability of diverse germplasm to adapt current crop varieties to changing climatic and disease conditions is of great importance.

The constant need not only to improve the yields of crop species, but to adapt them to different growing conditions can only be met by the availability of alternative sources of germplasm that can be incorporated into current crop genes. The importance of biodiversity to this problem has been generally recognized by both industry and OECD governments (U.S. National Academy of Sciences, 1978) and there is considerable work currently underway at both national and international levels. However, there remains much to accomplish both in terms of identifying new species and preserving germplasm.

One of the considerations that led to this report's focus on medicinal plant species was the considerably greater attention that has been given to the preservation of crop biodiversity. There seemed to be a greater need to examine the pharmaceutical facet of the biodiversity question. This appears to have been the case: the importance of biodiversity to the pharmaceutical industry has not been demonstrated to the same degree that it has been with agricultural crops.

The very different nature of these two aspects of biodiversity suggests that their importance to the public at large must be evaluated from different perspectives. The most obvious difference is that the value of crop biodiversity can be demonstrated within one or two crop cycles. In contrast, the value of a medicinal plant species takes considerably longer to establish, and the conceptual link between the plant and the final drug is far more tenuous than with crops.

Another major problem is that because the number of plants that have been thoroughly evaluated for pharmacological potential is small, the vast majority of plants have no identified market value, either in terms of an existing product or use or in terms of some imputed product or use. However, as a group, these plants will have a positive value because there is a positive probability that their value will be greater than zero.

For the purpose of trying to assess their value, plants can be placed into one of four categories:

1. Those that have not been investigated or that have been investigated but have not been found to be pharmacologically active;
2. Those that have been found to be pharmacologically active and whose active ingredient(s) can be economically synthesized (e.g. ephedrine);
3. Those that have been found to be pharmacologically active and whose active ingredients(s) have not been created synthetically; and
4. Those that have been investigated and found to have useful information (structure or mechanism) that is used in pharmaceuticals.

If, for the moment, only the out-of-pocket costs are considered, then the extinction of Category 1 or Category 4 plants will result in a zero cost.

For the plants in Category 2, a cost will be incurred only if the production cost of the synthetic substitute is greater than that of the natural product. If the synthetic substitute does cost more, then there will be a loss of consumer's and producer's surplus. The loss of Category 3 plants will result in the loss of the total economic value of the plant because society will no longer be able to realize the benefits of the drugs derived from the plant.

However, consideration of only the direct cost of the loss of these plants ignores what is probably their greatest value: the information contained in their genetic structure and the chemicals that this structure produces. Many plants in Category 1 plants are likely to possess some information or produce some chemical that may be needed in the future. The number of plants that have been thoroughly screened is very small, probably fewer than 10,000. Even so, few botanists would agree that no further information can be gleaned from those plants: one can never completely discount any plant, no matter how exhaustively it has been tested for biological effects, as being completely devoid of one or more useful drugs (Farnsworth & Soejarto, 1985).

Public Policy Considerations

While the importance of biodiversity for pharmaceutical purposes is not an especially new topic, there has been relatively little done to consider both the scientific and economic aspects of the problem in a unified fashion. One of the major objectives of this paper is to begin that process and thereby lay the foundation for a proper review of those public policies that affect the biodiversity of medicinal plant species.

In the end, the economic justification for altering these public policies may be inadequate. That must be accepted. However, it will also be a very useful finding since at the moment, the loss of significant economic benefits is often alluded to, without much actual support for this assertion. The fact that the economic support is missing has not gone unnoticed, and this may be why the other arguments are, at times, unfairly discounted. There are powerful reasons why biodiversity should be preserved, and they will appear all the stronger if they are not coupled with a transparently weak attempt at economic rationalization. As described later in this paper, there is also a need to consider the benefits to be derived from policies that do not preserve biodiversity.

However, if there is a sound economic basis for preserving biodiversity (this finding will probably depend upon a proper consideration of all of the benefits of biodiversity), then it is likely that there has been some sort of market failure. This conclusion would be based on the observation that genetic resources are not being properly priced at their long-run marginal cost, a common problem with resources in general and public resources in particular. Consequently, some type of intervention by governments would be logical and desirable, especially if the long-term economic benefits are substantial.

If this proves to be the case, then there will be several delicate policy questions that will require resolution. Most of the genetic resources that require preservation are found in developing countries, but the largest economic effects (at least in the short to medium term) are likely to be seen in the developed countries. There are actions that can be taken by the OECD countries at the national level to aid in the preservation of these resources, but the major actions will be needed in the developing countries.

There are three crucial aspects to this situation: first, it has always been difficult to gain the attention of national governments to address resource problems that are not domestic in nature, even if their impact on domestic policies is great; second, the most efficient actions will be bilateral or multilateral actions with developing countries and will involve considerable cost in the short-term for a long-term goal; and third, the issue of biodiversity is fraught with uncertainty, both as to the potential outcomes of taking action and the magnitude of potential benefits that would accrue to society. Given the short-term orientation of most governments, these three factors combine to make an especially difficult problem for public policy analysis.

This paper will attempt to provide a useful framework for the consideration of these issues. First, the past and current role of plants in medicine is reviewed followed by an analysis of the trends in pharmaceutical research and the possible future role plants in the development of new pharmaceuticals. The economic importance of plant-based pharmaceuticals is then described followed by a review of the economic concepts that apply to the problem and the factors necessary for arriving at a true valuation of biodiversity. Finally, an attempt is made to place these issues within the larger context of public policy development.

The Rôle of Plants in Medicine

Plants have always played a major role in the treatment of human traumas and diseases. India's use of plants for medicinal treatment dates back over 5000 years and has become codified in the Ayurveda, which contains over 8000 herbal remedies. This same system of treatment is still used in over 14,000 dispensaries (Huxley, 1984). At about the same time, a

Chinese emperor was describing 365 herbal remedies, including ginseng, opium (the source of codeine and morphine), and ephedra (the source of ephedrine). During the next millennium, the Assyrians listed 250 medicinal plants, and the Sumerians recorded 1000 plants. The later civilizations of Egypt, Greece, Rome, Arabia, Europe, and America all recorded their use of plants as medicine.

Many of the plants recognized by these societies as having medicinal properties are either still in use or have provided the chemical model for modern derivatives of their natural products. Table 1 lists the 40 higher plant species that were the source of all the plant-based prescription drugs prescribed in the United States in 1980 (Farnsworth & Soejarto, 1985). In addition to these 40 species, there were an additional 58 species or their derivatives that were found in prescription pharmaceuticals that same year in the United States. While the U.S. Food and Drug Administration considers them foods rather than drugs, about 300 species are sold as herbal teas in the United States.

Plants and plant-derived drugs are still widely used in Eastern and less developed countries. India officially recognizes over 2500 plants as having medicinal value, and it has been estimated that over 6000 plants are used in traditional, folk, and herbal medicine, representing about 75% of the medical needs of the Third World, 95% in Africa (Huxley, 1984). This not to say that all of these drugs are efficacious or would be in use if alternative medications were available.

The use of plants in traditional medicine has become the focus of work by the World Health Organisation (WHO). The WHO began in 1978 a study of medicinal plants to determine which ones were effective and to preserve the use of those that were. This work led to the initial identification of 20,000 species of medicinal plants. Of these, 200 have been selected for more detailed study (Levingston & Zamora, 1983).

Until the development of chemical synthesis at the end of the last century, the only way to utilize the active ingredients in medicinal plants was either the direct application of the plant itself or the extraction of the active ingredients. With the development of organic chemistry, many crude vegetal preparations were refined to isolate the active ingredient(s). Many of those isolated were then synthesized. In many instances, this isolation and synthesis process led to a much more effective drug because the purified substances did not have the adverse side-effects of the crude mixtures. In addition, partial synthesis was used to improve the delivery and the activity of the active ingredient(s) or to reduce unwanted side-effects[3]. Today, drugs are not commonly used in their vegetal extract form.

Table 1: Flowering Plants as Sources of Useful Drugs in the U.S. in 1980

Species	Type of Drug Product [a]
Ammi majus	*Xanthotoxin*
Ananas comosus	*Bromelain*
Atropa belladonna	Belladonna extract
Avena sativa	Oatmeal concentrate
Capsicum annuum	Capsicum oleoresin
C. baccatum var. *pendulum*	Capsicum oleoresin
C. chinense	Capsicum oleoresin
C. frutescens	Capsicum oleoresin
Carica papaya	Papain
Cassia senna	Sennosides A + B, senna leaf & pods
Catharanthus roseus	*Leurocristine* (vincristine), *vincaleukoblastine* (vinblastine)
Cinchona calisaya	*Quinine, quinidine*
C. ledgeriana	*Quinine, quinidine*
C. pubescens	*Quinine, quinidine*
Citrus limon	Pectin
Colchicum autumnale	*Colchicine*
Digitalis lanata	*Digitoxin, lanatoside C, acetylgitoxin*
D. purpurea	*Digitoxin, digitalis* whole leaf
Dioscorea composita	Diosgenin
D. floribunda	*Diosgenin*
D. deltoidea	*Diosgenin*
Duboisia myoporoides	*Atropine, hyoscyamine, scopolamine*
Glycine max	Sitosterols
Papaver somniferum	Opium extract (Paregoric), *codeine, morphine, noscapine, papaverine*
Physotigma venenosum	*Physostigmine* (eserine)
Pilocarpus jaborandi	*Pilocarpine*
Plantago indica	Psyllium husks
P. ovata	Psyllium husks
P. psyllium	Psyllium husks
Prunus domestica	Prune concentrate
Rauvolfia serpentina	*Reserpine* alseroxylon fraction, powdered whole root *Rauvolfia*
R. vomitoria	*Deserpidine, reserpine, rescinnamine*
Rhamnus purshiana	Cascara bark, casanthranol. danthron
Rheum emodi	Rhubarb root
R. officinale	Rhubarb root
R. palmatum	Rhubarb root
R. rhaponticum	Rhubarb root
Ricinus communis	Castor oil, *ricinoleic acid*
Veratrum viride	*Veratrum viride* extract, cryptennamine

[a] italics indicate single chemical compounds of known composition
Source: Farnsworth & Soejarto (1985)

The Uses of Plants

Within the realm of pharmacology, there are three major ways in which plants are used (Hänsel, 1972, emphasis added):

"1. *Constituents isolated from plants which are used directly as therapeutic agents.* Examples of such natural substances are digitoxin, strophanthin, morphine, and atropine, which are all still unsurpassed in their respective fields.
2. *Plant constituents which are used as starting materials for the synthesis of useful drugs.* For example, adrenal cortex and other steroid hormones are normally synthesized from plant steroidal sapogenins and the synthetic penicillins from natural penicillin.
3. *Natural products which serve as models for pharmacologically active compounds in the field of drug synthesis.* There are numerous reasons why plant constituents which are potentially useful as drugs cannot be employed directly. The plant material may be either unavailable or available only in limited quantity; furthermore, the plant may not lend itself to cultivation. Frequently, the side effects of a natural product often prevent its use in medicine and can be resolved only by preparation of a synthetic derivative – for example, cocaine, which led to the development of modern local anesthetics, the coumarins as precursors of modern antithrombins, and the modification of colchicine and of podophyllotoxin to obtain anti-tumor preparations."

In addition to these three major uses, a fourth has been added in recent years: the use in research of chemicals whose side-effects are too strong to permit use as an prescription drug (Matthew Suffness, U.S. NIH, pers. comm.). In particular, natural products are used to investigate and characterize biochemical processes and their mechanisms. More than 50 drugs that have anti-cancer properties are not marketed because of their adverse side-effects, but they are widely used in research.

An example of such a drug is taxol, a drug derived from the northwest yew. Taxol was discovered by the U.S. National Institutes of Health (NIH) during their natural products screening program and is currently being produced by NIH for research purposes. It is now undergoing clinical study, and NIH receives 50 requests per month to use it for research. Taxol is an example of a class of plant-derived drugs that have significant research value but little or no commercial value.

While the perceived importance of medicinal plants has waxed and waned through the last century in the Western countries, their importance to both Eastern societies and the Third World countries is widely recognized. There are several factors that contribute greatly to this difference. There is the great difference in individual and societal wealth: richer persons and countries can afford research and products that are simply not within reach for poorer individuals and countries. Another factor is that in the more developed countries, drugs are made available through a system that, for the most part, requires that products be profitable, and most of those responsible for pharmaceutical research believe, correctly or not, that plant-based research is not profitable.

A trend that may be related to the last point has been the replacement of many older products (regardless of source) with newer products, and since most of the research centered on non-plant sources for drugs, the importance of plant-based drugs diminished. In addition, there are many new therapeutic treatments using newer drugs, mostly non-plant-based because of the direction of research. Finally, Eastern medicine (especially Chinese medicine) has emphasized preventative action while Western medicine has until recently concentrated on curative (therapeutic) medicine.

The United Nations attributes some of the decline in doctors prescriptions of plant-based drugs to the marketing efforts of the pharmaceutical companies (International Trade Centre), which are, of course, trying to sell those products on which they have the greatest profit margin or the greatest sunk costs. The growth in consumer demand for plant-based products noted above has been seen in the over-the-counter market, where preparations and herbal remedies can be purchased directly by the consumer and where the costs of bringing a product to market or keeping it on the market are less.

Recent Developments in Plant-based Drugs

Even though plants may not be the favorite area for industry's pharmaceutical research, nine widely-used drugs were developed from plants during the 30-year period from 1950 to 1980. The first was the family of contraceptive steroids derived from diosgenin. Virtually all oral contraceptives used by women are derived from this chemical in a species of Mexican yam. The second group was the antihypertensive and tranquilizing alkaloids reserpine, deserpidine, and rescinnamine that were derived from various species of *Rauwolfia*. These drugs have now been superceded by synthetically-derived drugs. The third group is the two anti-cancer drugs leurocristine and vincaleukoblastine derived from the rosy periwinkle. The fourth group is two laxatives derived from Cassia, sennosides A and B. Finally, there are the cardiac glycosides extracted from *Digitalis* that are used as cardiotonic drugs.

New drugs derived from plants are still being introduced. In 1983, there were two such major new introductions. The first was for Etoposide, an anti-cancer drug derived from the mayapple (*Podophyllum peltatum*) that will be used to treat refractory testicular carcinomas, small cell lung carcinomas, nonlymphocytic leukemias, and non-Hodgkins lymphomas (Balandrin *et al.*, 1985). Etoposide is an example of the second type of use in that it is a slightly modified version of the natural substance. A modification of the natural substance is used because it proved to be most effective in a long series of clinical studies. The second major introduction in 1983 was atracurium besylate, a skeletal muscle relaxant related to the curare alkaloids.

Plants in Pharmaceutical Research

While the pharmaceutical industry in Western countries has been moving toward synthetics and biotechnology in their research, the consumer is expressing a greater and greater interest in natural products. The United Nations concludes that with a few exceptions, the major pharmaceutical companies are no longer investing in the development of new botanical products owing to the time required, and expense incurred, to obtain regulatory approval to allow the product to be marketed with medicinal indications[4] (International Trade Centre). However, the consumers' intentions are evident from the renewed interest in traditional medicine in Asia and in health foods in Europe and North America (International Trade Centre). While the consumption of health food may not provide therapeutic benefits *per se*, it certainly indicates a growing interest in a preventative approach to health care by consumers in some of the OECD countries.

The level of plant-based pharmaceutical research reached its highest level in the United States during the period of 1953 to 1960 (Farnsworth & Soejarto, 1985). By 1974, only one pharmaceutical company was conducting any research directed toward plant-based drugs, and that effort amounted to only $200,000 in that year (Farnsworth & Soejarto, 1985). This represented only about 0.028% of the total U.S. pharmaceutical industry research budget of $723 million in 1974 (Farnsworth & Soejarto, 1985). By 1980, when the industry research budget was about $2 billion (Hall *et al.*, 1985), even that small effort had disappeared. By 1984, the pharmaceutical industry research budget in the United States had reached $4 billion (Hall *et al.*, 1985).

To put this level of research spending into perspective, the annual cost of the U.S. NIH plant screening program (described below) was, at its peak in 1980, about $1.9 million, or about 0.1% of industry research spending that year; by fiscal year 1985, NIH spending was down to $217,000, or 0.005% of the 1984 industry research budget (Spragins & Rhein, 1985).

Fundamental Research Strategy Options

The pharmaceutical industry operates with a development horizon of 15 years or less, meaning that a drug has to be developed, tested, approved by government agencies, and marketed successfully within that period. Given the length of time necessary to complete the testing and approval steps in this process for new drugs, there is considerable pressure on the development stage to produce new drugs as quickly as possible.

Given this intense environment, a strategy of modification – taking an existing drug and modifying it – presents the least risk on a year-to-year basis. This attitude is reflected in the following: (a) research as a percentage of sales has been declining; and (b) very few new classes of drugs have been introduced in the last ten years. In place of more investment in research, many pharmaceutical companies have begun major licensing programs with pharmaceutical companies in other countries. Licensing another companys drug yields healthy profits while significantly reducing both investment and risk.

The pharmaceutical industries of several countries, however, still recognize the importance of screening programs, plant screening programs in particular. Companies in Europe and Japan appear to be doing more plant-based research than their counterparts in the United States. Japan, in particular, has traditionally favored screening research and been a more interested user of plant-based pharmaceuticals because its medical history is rooted in the Chinese medical tradition. This is reflected in the fact that plant-based research in Japan currently represents about 20 to 30% of total pharmaceutical research spending (M. Mutai, Yakult Honsha Co., pers. comm.).

It may be useful to recognize the cyclical nature of pharmaceutical research interests. In the 1950s, plant screening was more important than drug design. Major advances in knowledge and technology led to the current emphasis on drug design. As this research matures, it is likely that the importance of natural sources such as plants will be rediscovered.

Screening Medicinal Plants

Among the traditional factors that have influenced pharmaceutical research in the Western countries away from the direction of medicinal plants include the following:

1. The screening-type of research required to identify a plant of interest and isolate the pharmacologically-active ingredient(s) is tedious,

difficult, very costly, and often unrewarding, for both the researchers and the company;

2. Once the ingredient(s) have been isolated, if it (they) cannot be wholly synthesized, the company then has to be concerned about securing its supply of raw plant material. This has become a considerable concern given that most plants of interest are usually indigenous only to developing countries, and dealing with these countries has become increasingly more complicated (and expensive); and
3. In some cases, especially where the active ingredient cannot be identified or isolated (an increasingly rare occurrence), the patent laws of most OECD countries, which do not permit the patenting of natural products, can present a barrier to marketing the new product. The pharmaceutical firm must either try to create a proprietary drug or modify the active mixture so that it can be patented, otherwise it would not be able to recover the cost of bringing a new drug to market, through all of the laboratory and clinical testing and government regulation, which now costs about $100 million in the United States (Anon., 1985).

The pharmaceutical industry has had a difficult time performing plant-related research in the past. To those who do not like to do this type of research, there are problems with biological variation, inconsistent results utilizing crude extracts, and procurement problems. While there are, in fact, problems performing this type of research, an unfamiliarity with the area will compound the problems at a rapid rate.

With respect to the problems involved with the acquisition of samples, it is important to have a well-conceived sampling methodology, preferably one relying on the folk, or traditional, medicine indigenous to the region where the samples are to be obtained. It is also imperative that the person collecting the samples be well-versed in botanical sampling procedures; several pharmaceutical companies have seen large sampling and analysis efforts go for naught because the raw samples were improperly collected, labeled, preserved, or named.

Considerable specialized knowledge is also necessary to achieve the best results in the laboratory since the analysis of crude plant extracts is a learned art. Technicians and toxicologists familiar only with analyses of synthetic products will quickly reach the wrong conclusions about these extracts. This has, in the past, been a common experience of pharmaceutical companies, whose personnel have been unable to recognize the potential of a crude mixture in which the concentration of the active ingredient might only be 1%. At such a low concentration, the biological activity would likely be quite low, especially when compared to synthetic mixtures in which the concentration of active ingredient is much higher. Faced with such a situation, an uninformed investigator might ignore a very promising lead.

The extent to which past plant screening activities are indicative of what might be expected in the future has been a much-debated topic. As discussed below, those involved in some of these activities point out that there were major flaws in their studies and that the past studies should neither be used as a model nor as a predictor of success for future screening work. Given the number of plants that remain to be screened, it would be well to learn from their experience so that future screening programs can be as effective and efficient as possible.

Probably the largest (and longest) screening effort was that conducted by the U.S. National Cancer Institute (NCI) from 1960 until 1981. During that period, 114,045 individual extracts were tested (Suffness & Douros, 1982) representing between 25,000 and 35,000 species of plants (Suffness & Douros, 1979). Of those 114,045 samples, 4897 (4.3%) representing 3394 species were confirmed as having some biological activity (Suffness & Douros, 1982). About 40 of these biologically-active compounds held some promise of being developed into commercial products.

Two important points must be noted about the NCI program. First, fewer than five screens were used to establish biological activity, all of which were intended to identify anti-cancer activity. Using so few screens (the well-known anti-cancer chemicals vincristine and vinblastine were missed) does not represent an exhaustive screening for anti-cancer activity let alone other useful pharmacological properties (such as analgesics, antiarthritics, and antipsychotics), but no claim was ever made that the screening program was intended to be exhaustive. It should also be noted that modern assay systems (cellular and biochemical) are much more sensitive than older assays (pharmacological). This means that not only would more minor components of mixtures be detected but also that the effects of these lower concentration components could be observed. Second, relatively little attention was paid to folkloric information during the selection process. A retrospective study concluded that the success rate in finding active species could have been doubled by incorporating folklore (Suffness & Douros, 1979).

Unfortunately, the job of screening plants in the future will become more difficult as investigations of the plants that have been widely used by traditional medicine in the Western countries are completed. Consequently, any new successes will have to come from those plants in Western countries that do not have obvious promise or from plants not found in Western countries.

Even if the screening program is successful, there remains considerable work before a commercial product can be produced. Vincristine took about seven years to develop and Etoposide took twelve years (J. Douros, Bristol Myers Co., pers. comm.). Neither of these time periods include the time required to establish the biological activity of the two substances.

The Rôle of Molecular Biology

The monumental advances that have been made in the last decade in cellular and molecular biochemistry, biotechnology, and pharmacokinetics have significantly altered the research plans of most pharmaceutical companies. Some in the industry believe that, based on these developments, one of the more important strategies used in the future will be hypothesizing the structure necessary to achieve the desired effect, creating the proper genetic code structure, inserting that genetic code into the appropriate microorganism, and then placing the microorganism in an environment conducive to their production of the drug (Anon., 1984b; Hall *et al.*, 1985).

Naturally, if this were really going to be the future strategy (or one of the principal ones), pharmaceutical companies have even less incentive to spend research money on screening plants than in the past 20 years, when organic synthesis and screening of micro-organisms were the predominant research strategies. However, there are those who do not believe that this scenario will be commonplace any time in the near or medium-term future. This hesitancy is largely based on the belief that while the recent successes in biotechnology have been directed toward the last two stages of the process – the insertion of genetic code in microorganisms and the inducement of the microorganism to produce the desired drug – the most difficult part of the process is the first two steps – identifying the chemical structure and creating the proper code. "Biotechnology is not concerned with drug discovery, it is concerned with the use of living cells to manufacture complex products."[5]

In spite of the great amount of molecular design that went into the final product, a new antihypertensive drug, Capoten, is based on the venom of a South American snake[6] (Spragins & Rhein, 1985). This example illustrates an important point that many researchers overlook: the chemical models they use in their design work are often natural products.

In addition, many of the most useful pharmacologically-active plant substances are secondary metabolites (Balandrin *et al.*, 1985). While the technology exists for having primary metabolites (proteins, polypeptides, etc.) produced by genetically engineered microorganisms, the creation of the secondary metabolites is far more complex, and all of the processes have not yet been identified. These processes involve many enzymes, some of which may perform functions not directly discernable during the creation of the secondary metabolite (such as detoxifying byproducts that might otherwise create an adverse side-effect), and the synergistic effect of all these enzymes is still not understood. Thus, at least in the immediate future, plants or plant cells will probably continue to serve as the sources for most of these plant-specific materials (Balandrin *et al.*, 1985).

Probably the clearest evidence of this is the inability of the pharmaceutical industry to commercially synthesize the vast majority of the plant-derived drugs currently in use. Of the 76 different chemical compounds used in prescriptions in 1973, only seven can be commercially produced by synthesis (emetine, caffeine, theobromine, theophylline, pseudoephedrine, ephedrine, and papaverine)[7] (Farnsworth & Morris, 1976). Important drugs such as morphine, codeine, atropine, digoxin, and digitoxin cannot be commercially synthesized. Where synthesis is available, it is often far more expensive; so much so that it is not commercially feasible. For example, reserpine can be commercially extracted from natural sources for about $0.75 per gram while the synthesized product costs $1.25 per gram, 67% more. However, the difference is usually much larger, with the synthetic product costing on the order of ten times as much as the natural product.

While the primary metabolites that can be synthesized are certainly useful, the inability to synthesize these secondary metabolites is a serious gap in the industrys ability to utilize its research in molecular biology to the utmost. Even though this research may lead to a faster recognition of useful chemical structures, the source of these chemicals may nonetheless remain natural sources.

The Market Value of Medicinal Plants

In reviewing the available data on market prices for medicinal plants and the drugs derived from them, it becomes clear that the best that can be done is a patchwork of various types of data arranged in what one hopes is a meaningful manner. There is no set of complete data for any one plant or drug, and the pieces that are available are often not exactly comparable. Nevertheless, these data do provide the reader with a feeling for the magnitude of market and of the prices therein.

It should be noted that the numerical data presented in this section represent only market value data and do not reflect the total economic value of either the medicinal plants or the drugs that are derived from them. The distinction between market value and economic value is discussed in the section on the Economics of Genetic Resources. At this point, it need only be said that market value is a subset of economic value, which includes all benefits to society.

In developing estimates of the market value of medicinal plants, it has been argued that the use of market price data inflates the market value thus derived because of the large difference between the raw material cost and the final product cost. This is expected because of the large development and manufacturing costs as well as the incorporation of research costs for failed efforts.

However, while it is clear that only some part of the market price of plant-derived drugs is attributable to the plant raw materials, the existence of consumers surplus indicates that there is a willingness to pay that is greater than the market price. This is due, in part, to the inelasticity of demand for products such as pharmaceuticals. Consequently, while there are flaws in using the final market price, it is probably a better indicator of market value than the market price of the raw material.

The Current Market Value Estimates[8]

In reviewing the available data on the commercial value of medicinal plants, there may be a tendency not to compare like with like because so few data are available. Naturally, this should be avoided. There are three types of data available, none of which is truly complete. First, there are data describing the market for all pharmaceuticals. Second, there are data for raw plant materials; that is, the original plants that are harvested and (often) exported to processors. Third, there are data on the pharmaceuticals derived from medicinal plants.

Every attempt has been made within the text to clearly state which type of data is being used. However, this is simply a caution that discretion and careful reading are necessary before drawing inferences or reaching conclusions.

Natural Products Data

Total world-wide imports of medicinal plants increased from $355 million in 1976 to $551 million in 1980 (International Trade Centre). The Federal Republic of Germany imported 28,326 tonnes of medicinal plants in 1979 worth $56.8 million. While the imports of medicinal plants to the United States declined from $52 million in 1976 to $44.6 million in 1980, the domestic market for the plants was $3.912 billion in 1981. For Japan, imports grew from 21,000 tonnes in 1979 to 22,640 tonnes in 1980, but the value of those imports fell from ¥11 billion ($50 million) to ¥10.8 billion ($48 million).

About 1000 tonnes of periwinkle are used in the United States every year with an unprocessed value of about $1 million. The final products that are derived from the periwinkle had a market value at the wholesale level of about $35 million in 1977 (Farnsworth & Soejarto, 1985). Currently, the ratio (Curtin, 1983) of retail price to wholesale price for the alkaloid derived from the periwinkle is 4:1. If this ratio is used to translate the wholesale value to the consumers level, the commercial value was likely to have been about $140 million in 1977.

Even before the anti-cancer properties of the mayapple were discovered, about 100 tonnes of the plant were being used annually in the United States (J. Duke, U.S. Dept. of Agriculture, pers. comm.). Since the market price for the raw mayapple is about $1.00 per pound, the annual commercial value of this plant was about $200,000. Etoposide, the drug derived from the mayapple, already has annual sales of about $15 million even though it was only introduced in 1984; however, because most current anti-cancer drugs have fairly specific applications, the market for any one such drug is not likely to exceed $100 million (J. Douros, pers. comm.). Any program sponsored by NIH would not be included in this value.

The anti-cancer drug market has been growing at a more rapid pace than the pharmaceutical market as a whole. This sub-market has grown about 25% annually during the last decade, and prices in this sub-market increased 24% in 1983, as compared to 11% for the pharmaceutical market as a whole (Anon., 1984a).

Drug Sales Data

Prescription drugs comprised a total world-wide market in 1984 in excess of $87 billion (in manufacturers prices), an increase of about 1.75% as compared to 1983. Sales in 1985 were projected to increase to over $90 billion (in manufacturers prices). Adjusted to reflect retail prices using the methodology described in Annex 1, the 1985 world sales figure would increase to about $150 billion. OECD countries currently represent about 70% of the world market, or an estimated $105 billion in 1985 sales of prescription drugs.

The break-down of prescription drug sales by world region can be seen in Table 2. As shown in Figure 1, OECD countries represent, by far, the largest portion of the market for pharmaceuticals (about 70% in 1984). Figure 2 shows the relative proportion of the OECD prescription pharmaceutical market represented by the major consuming countries (see also Annex 1).

Table 2 : World Demand for Pharmaceuticals 1980

	(% of total)	
North America	22.1	
United States		18.8
Canada		3.3
South America	6.5	
Western Europe	31.0	
EEC		25.7
Eastern Europe	11.7	
East Asia	19.6	
Central Asia	1.7	
West Asia	2.1	
Southeast Asia	1.6	
Africa	3.0	
Australasia	0.7	

Source: World Drug Market Manual, 1982/1983, as quoted by the International Trade Centre.

One method for estimating the value of plant-based pharmaceuticals in the United States is described by Farnsworth & Soejarto[9] (1985). The data cited below are those used by these authors, except where explicitly noted.

Based on survey data for the years 1959-1973, drugs that contained one or more plant-derived active ingredients represented just over 25% of all prescriptions dispensed from community pharmacies. During that 15 year period, the range of annual percentages varied only between 23% and 28%.

Given the total number of prescriptions that were filled, it is possible to calculate the number of prescriptions filled at community pharmacies that contained plant-derived active ingredients. Since community pharmacies are not the only source of prescription drugs, that value must be adjusted to reflect the additional prescriptions that were filled at these other locations. Farnsworth and Soejarto suggest that the proper adjustment factor is two.

For 1973, there were 1.532 billion total prescriptions dispensed. The average prescription cost to the consumer in 1973 was $4.13. Thus, the total value of prescriptions filled by community pharmacies was about $6.3 billion at the consumer level. Since plant-derived drugs represented 25.2% of the market in 1973, the value of these prescriptions was about $1.6 billion. Applying the factor of two[10] to compensate for other dispensing facilities, the total value of plant-based prescriptions in 1973 in the United States was about $3.2 billion.

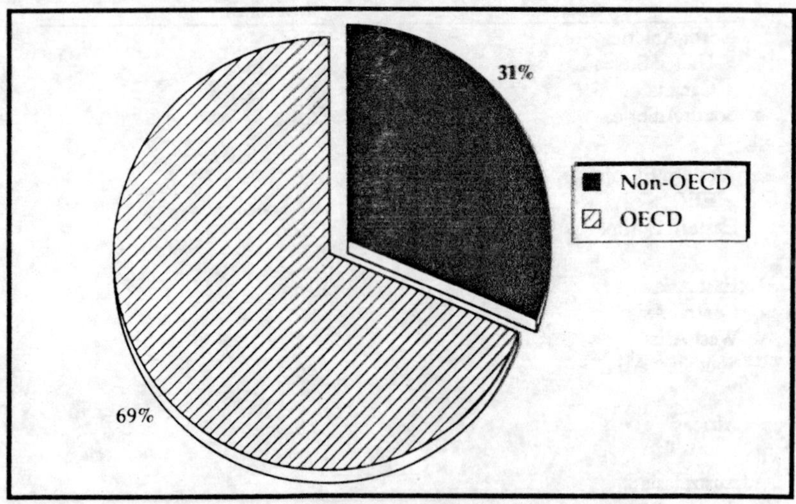

Figure 1. World Prescription Drug Market Shares in 1984

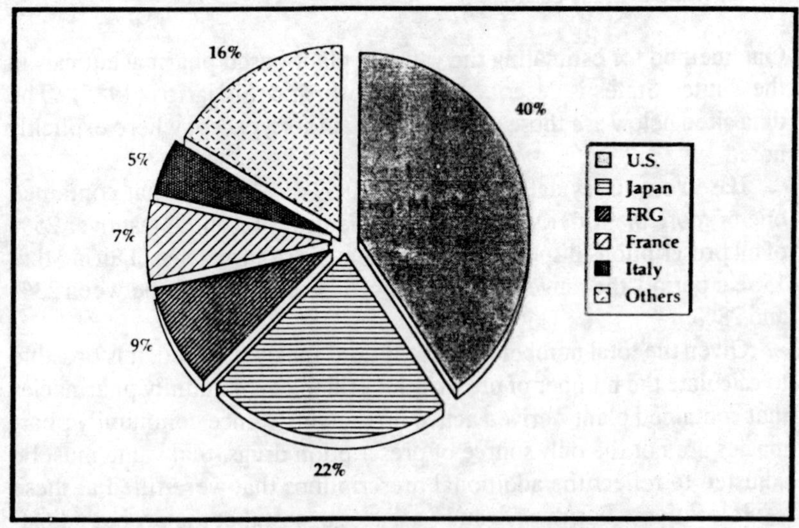

Figure 2. OECD Prescription Drug Market Shares in 1984

For 1980, the total number of prescriptions dispensed was estimated to be about two billion. Using the 15 year average of about 25%, the number of plant-derived prescriptions in 1980 was approximately 500 million. Applying the factor to compensate for other dispensing facilities, the estimated total number of plant-based prescriptions in 1980 was about one billion. Since the average prescription price in 1980 was about $8.00, the total value of plant-based prescriptions in 1980 in the United States can be estimated to have been about $8 billion.

An important limitation of this value, however, is that it does not include three major drug markets: (1) over-the-counter drug sales (i.e., those drugs that can be purchased without a doctors prescription); (2) drugs used exclusively in hospitals; and (3) traditional and herbal medicine sales (including homeopathy, herbalists, etc.). As an illustration of the significance of these markets, the over-the-counter market for laxatives in the United States in 1980 was $331 million, half of which can be attributed to plant-based products (Farnsworth & Soejarto, 1985).

A variation of the Farnsworth and Soejarto algorithm yields a similar value. Applying the 25% factor described above to the sales data in Annex 1, an estimate of about $6.2 billion can be derived for the retail market value of plant-based prescription drugs in 1980 in the United States. Based on estimated sales, this value would increase to about $11 billion by 1985.

Including over-the-counter drugs would significantly increase these values. For 1980, the retail market value in the United States of all plant-based drugs is estimated to be about $9.8 billion. By 1985, this value is estimated to have increased to about $18 billion.

In Japan, 13.3% (208 out of 1555) of the pharmaceuticals found in the 10th edition of the Pharmacopoeia of Japan are derived from plants. In addition, there are 185 Chinese herbal medicines that are compounded as prescription drugs and 210 others that are approved as proprietary or over-the-counter drugs (M. Mutai, pers. comm.).

The demand for these drugs in Japan has been increasing over the last decade, but their production only accounts for 1.5% of the total production of pharmaceuticals in Japan. In 1984, sales of traditional medicines by prescription totaled 69.3 billion ($277 million) (1.7% of total prescription drug sales) and sales of traditional over-the-counter preparations totaled 4.8 billion ($19 million) (2.5% of total over-the-counter sales) (T. Isobe, Suntory Limited, pers. comm.).

However, if all plant-derived drugs are included, the total is much greater. For the three-year period 1981-1983, total sales of prescription plant-derived drugs in Japan were (in current yen)[11] (M. Mutai, pers. comm.):

1981 ¥568 billion ($2.6 billion)
1982 ¥640 billion ($2.6 billion)
1983 ¥655 billion ($2.6 billion)

These values appear to represent between 15% and 20% of the Japanese prescription drug market. This is a smaller percentage than one might expect given the interest in plant-based pharmaceuticals and preparations that is attributed to the Japanese; comparable values are 25% for the United States market and 35-40% for the German market.

While the percentage of plant-based drugs varies from country to country (as discussed below), the 25% factor may be a good estimate of the average. If so, the estimated market value of both prescription and over-the-counter plant-based drugs for all OECD countries is estimated to have been about $43 billion in 1985.

The Present Value of Plant-based Drugs Through 2000

The present value is a parameter that attempts to assign a value in today's market for future events. For example, the value today of a sum of money to be received in five years is less than if the same amount were received today because the same sum received today could be used for the next five years (e.g. it could be invested). This lesser, present value is arrived at by discounting the future payment to reflect the loss of five years of use. In the following example, the present value (i.e. in 1987) of the next 14 years sales of currently used pharmaceutical products is estimated.

As described in Annex 1, the retail pharmaceutical sales data for both prescription and over-the-counter drugs in OECD countries was regressed to estimate sales annual sales for each year through the end of the century. The portion of those sales that would be attributable to plant-based drugs was estimated using the 25% factor described above. Real social discount rates of 2% and 8% were used to develop a range of likely values.

Based on these assumptions, the present value through the year 2000 of current plants-based pharmaceuticals in OECD countries is between $400-600 billion (in 1984 dollars), with the mid-point being about $500 billion, which is equivalent to about three times the annual sales of all pharmaceuticals in OECD countries. In other terms, this is approximately equal to the annual GDP of France or the United Kingdom.

Potential Foregone Market Value[12]

Using the experience of the NIH screening program, between 9% and 13% of the plants had some biological activity, and somewhat more than one in a thousand had biologically-active compounds that had commercial promise. Two opposing observations are relevant: first, the screens that were used to identify biological activity were limited to anti-cancer activity and are now considered inadequate even for that purpose; and second, having commercial promise is not the same as being a commercial product. Based on current experience, it would appear that an reasonable estimate of the probability that any given plant would produce a marketable prescription drug would be in the range[13] of 1:1000 to 1:10,000.

If these ratios are applied to the number of plant species that are estimated to become extinct by the end of the century (60,000), then about 60 species that would produce marketable prescription drugs will become extinct by the middle of the next century.

Recalling that in 1980 the consumer-level market value of plant-based drugs was about $8 billion in the United States and given that this amount was realized from a total of 40 plant species, the average unit commercial value of each of these species was about $200 million (which should be considered a conservative estimate, as discussed immediately below). This value should then be tripled to reflect the value to all OECD countries (see Annex 1), so that each species would have an annual commercial value of about $600 million. Applying this factor to the estimated number of plant species that will become extinct, the market value of the prescription drugs derived from these species that will be foregone in the year 2050 will be about $60 million (in 1980 dollars). To put this into perspective, if these plants become extinct, the market value foregone by OECD countries in this one year alone could be equal to the current OECD market for plant-based pharmaceuticals. It should be noted that this value does not include other benefits, even other pharmaceutical benefits such as insight into biological mechanisms and processes or new over-the-counter drugs.

It must be pointed out that there are two major assumptions in this analysis that may or may not be valid. The first is that the ratio of pharmacologically-active species will be about the same for uninvestigated plants as for previously-investigated plants. Some have argued that those plants that have the highest probability of success have already been investigated and that this ratio can be expected to diminish.

However, with respect to uninvestigated species, botanists believe that the geographic locations of their habitats are the ones that produce the highest percentage of pharmacologically-active plants. Thus, it might be expected that the ratio should increase in favor of discovering more plants with interesting properties. This, in fact, seems to be more likely: investigations tend to concentrate on the easy to obtain data first before moving on to the more difficult. Consequently, the ratio of pharmacologically-active plants that was used is likely to be a conservative value because it includes mostly temperate-climate plants, which would be expected to exhibit a lower ratio of pharmaceutical activity than plants from tropical or arid climates.

The second major assumption is in two parts: first, that an average market value for a medicinal plant species can be estimated by dividing the number of medicinal plant species into the estimated market value of the drugs derived from them; and second, that average market value can be used as an indicator of the likely market value of newly discovered medicinal plant species.

With respect to the first part of the assumption, there really is no other way of estimating the average market value for existing medicinal plant species. For the second part of the assumption, it is clear that no other alternative is really available. However, it might be said that this is likely to be a conservative number as well since the average market value of the existing medicinal plant species includes species that were developed when research costs were much lower than they are now or will be in the future as well as many drugs whose patent protection has lapsed or that have become over-the-counter drugs. Consequently, the average market value of newly discovered plant-based species is likely to be higher because the cost of bringing them to market will be higher.

There are other assumptions embedded in the argument that should be pointed out. Even if these plants did not become extinct, it is not certain that they would ever be discovered or, if discovered, that the necessary research would be done to uncover their pharmacological activity. It may also be the case that substitutes for plant-based drugs could well be developed if there is a demand for a particular type of therapy. The substitute therapy may or may not equal the efficacy of the lost plant-based drug. If it is less effective, then there remains a loss represented by the difference in efficacy. However, with respect to anti-cancer drugs, there is a fairly extensive history showing that substitutes rarely exist. Since there is little certainty as to when these advances will result in new, in-place research methodologies, prudence suggests that we use the past as a conservative basis for estimating the value of these plants.

Finally, it should be re-emphasized that this foregone benefit is not an actual loss but a benefit never realized. But since this is an annual value, OECD countries will be losing this $60 million benefit in every year after 2050 (and the annual foregone benefit will grow as more species become extinct).

Summary

This section has discussed three components of the market value of medicinal plant species: (1) the annual market value of plant species currently used to formulate pharmaceuticals; (2) the discounted cumulative market value (i.e. their present value) through the end of the century of those plants currently used; and (3) the potential foregone market value of plants that are likely to become extinct before the end of the century. The market value of these plant species is based on the retail value of the pharmaceuticals that are derived from them rather than the plants raw, unprocessed value, and the market value does not include non-pharmaceutical uses of the plants nor other aspects of their total economic value.

The estimated value of the world prescription drug market at the retail level in 1985 was about $150 billion. The OECD portion of the world pharmaceutical market is about 70%, representing about $105 billion in retail sales in 1985. Plant-based drugs comprise about 25% of the market in the United States. Extending this percentage to the OECD as a whole, the retail market value of both prescription and over-the-counter plant-based drugs in OECD countries is estimated to have been about $43 billion in 1985.

The discounted cumulative market value of plant-based drugs through the end of the century is estimated to be about $500 billion in OECD countries at the retail level. This is approximately equal to the annual GDP of France or the United Kingdom.

The foregone retail market value for prescription drugs in OECD countries of those plants likely to become extinct by the middle of the next century is estimated to be about $60 million annually. This value, which is for the drugs that might have been derived from those plants, can be compared to the estimated value of plant-based prescription drugs in 1985 in OECD countries of about $43 billion.

There are several points that should be remembered about the foregone benefit value. First, there are tremendous uncertainties involved with the estimation of such a value. Second, this value represents a benefit foregone rather than an actual loss. Third, this is an annual foregone benefit that will recur every year and will increase as additional plants become extinct. Finally, this value only represents market value of the drugs derived from the extinct plants and does not include components of the total economic value of either the drugs or the plants, such as costs to society from not having the drug or plant, the benefit of good health, the possibility of deriving non-pharmaceutical uses from the plants, and the plants existence value.

The Economics of Genetic Resources

There is a considerable body of literature dealing with the economics of natural resources, including the unique problems posed by genetic resources and their preservation. However, as this literature recognizes, there are still considerable difficulties in adequately characterizing the total value of the benefits of preserving genetic resources. This difficulty arises primarily from ignorance: our ignorance of what the alternatives are, not simply the probability of their existence. This ignorance is also a major impediment in the analysis of policy options.

This section will discuss some of the major economic concepts and issues that are relevant to the problem of preserving biodiversity. First, an overview will discuss some concepts relevant to later discussions. The distinction between market value and economic value will then be reviewed and the economic value of plant-based pharmaceuticals will be estimated. Following this is a discussion of the nature and analytical treatment of uncertainty and its relationship to option value. Finally, the factors comprising the total value of the benefits to be realized from the preservation of biodiversity will be discussed, as well as alternative methods for analyzing them.

Overview

There are several points that should be touched upon before proceeding with the more detailed discussions. While all of these points are relevant to the problem of preserving biodiversity, space does not permit a fuller discussion.

The first point is that extinction may be an economically-justifiable alternative. To reach this conclusion, it would be necessary to have a comprehensive understanding of the benefits to be foregone by causing the extinction, and presumably the benefits of proceeding with such an action would result in very large benefits. In such a situation, there could be a positive economic value attributable to extinction. However, as discussed later in this section, achieving a comprehensive understanding of the benefits foregone by extinction would be very difficult. Furthermore, the fact that the extinction decision is irreversible is in itself an important factor. Referring to a dichotomy where extinction is either optimal or not, Clark (1976) describes the situation:

> "Extinction is the optimal harvest policy only because it leads to the largest present value of economic revenues. We are certainly not suggesting that the deliberate extinction of a species is socially or aesthetically desirable just because extinction appears to be the most profitable course of action. Aesthetic or moral questions aside, the decision to exterminate a species is an irreversible decision that can only be justified in economic terms if we are certain that present conditions will persist into the distant future."

One of the major reasons why the comprehension of foregone benefits would be difficult to achieve is that one of the major benefits lost through extinction is the knowledge contained within the lost species. It is impossible to quantify the value of this information, not only because we rarely know what knowledge we are giving up, but because we have no idea how future generations (or even the present generation in a decade) will value that information. It can be helpful to view knowledge, or information, as a public good; that is, a resource whose use does not diminish it, and a resource that is available to anyone. Viewed as such, information has many of the problems common to public goods.

In the previous section, a social discount rate was used to estimate the present value of plant-based drugs through 2000. It should be noted that the social discount rate is almost always lower than the private discount rate. This results from the concept that consumption by future generations is a public good to members of the present generation. (Fisher, 1981) This argument is based on economic efficiency and does not consider intergenerational equity.

This question of efficiency versus equity as applied to natural resource policy has been widely debated. Economic efficiency considerations, by their very nature, are primarily based upon the preferences of present-day consumers. Strong arguments have been put forward favoring the consideration of benefits to future generations in making decisions on the use of natural resources. While this is a valid economic concern, it is also very much a public policy concern.

Related to the equity issue is the proper role of government in the development and conservation of natural resources. The role of government as a steward of a nation's resources is well-established. One of the reasons for this is that government naturally has a longer time horizon than private business, and the time spans involved are too long to rely on private incentives. Consequently, public policy decisions regarding natural resources almost always contain an element of consideration of intergenerational equity.

This is quite relevant to the issue of biodiversity on two levels. First, the governments of OECD countries have an interest, in their role as stewards for the future, in the continued availability of information contained in genetic resources. Second, the governments of those countries that harbour most of the species of concern, likewise in their role as stewards, have an interest in preserving these genetic resources so that future generations in their countries can benefit from their development. Unfortunately, economic and population pressures are diminishing the ability of developing countries to give proper importance to the stewardship of these resources.

Finally, there is the distinction between willingness-to-pay (WTP) and willingness-to-accept compensation (WTA). Both are measures of the consumer's valuation of a commodity. WTP is a measure of how much the consumer would be willing to pay to have a commodity in the future, and WTA is a measure of how much compensation the consumer would require to give up future use of a commodity now available. While the two measures might be expected to be approximately equal, contingent valuation studies have shown that WTA is almost always larger by substantial amounts. With respect to the extinction of species, it would appear more appropriate to use WTA since the species currently exist. In other words, the correct question would be, How much compensation would you require for the loss of all future benefits from the existence of this species? rather than, How much would you pay to ensure that this species would exist in 15 years?

Market Value Versus Economic Value

In attempting to quantify the economic importance of medicinal plants, it is important to distinguish between the market value of a commodity and its economic value. The market value is just that: the value the marketplace attributes to a given commodity or its derivative product(s), as represented in the market price and the quantity of the commodity that is sold. These are the values that were discussed and calculated in the previous section.

Economic value, in contrast, represents all of the societal benefits of a particular type of product, including the market value. Economic value can be viewed as an expression of the total benefits of a product. The absence of observable demand does not imply an absence of value.

The relationship between the economic value of a medicinal plant species and the market price of the drugs derived from it is not a direct one. However, it can be argued that the market prices are minimum valuations assuming that: (1) the demand for the drug is inelastic; and (2) that it is appropriate to value an essential input at its own cost plus the economic rent obtained from it plus the associated consumers surplus.

To illustrate by example, the market value of a stand of forest could be measured by translating the wood volume therein into an equivalent quantity of paper and then taking the market value of the paper[14]. In contrast, the economic value to society includes not only the value of the paper, but also what may be referred to as the benefits of the trees as a forest: that is, the contribution the forest makes to controlling soil erosion, stabilizing the water table, converting carbon dioxide back into oxygen, husbanding all manner of wildlife, and providing recreational opportunities (the last two contributions might be reflected in the market price, depending upon the particular situation). Obviously, the economic value is much larger in magnitude but also much more difficult to quantify.

In the case of biodiversity, market price will be useful as an indicator: large values will tend to lend significance to the arguments in favor of preserving biodiversity. However, the use of market values can be dangerous if it is not recognized that they represent only a small portion of the total value of the plants in question. A common problem in using benefit/cost analysis for public policy decision-making is that those benefits that cannot be quantified are undervalued or ignored. Given the irreversible nature of the decisions involving biodiversity, this problem would be especially acute if market values were used without reference to the potentially large economic values.

The Economic Value of Medicinal Plants

One of the factors that could be used in developing an economic value for medicinal plant species would be examining the current costs to society of a disease whose impact might be diminished in the future by drugs derived from plants. Since several plant-derived drugs have been used in the treatment of cancer, that might be one place to start the process.

Cancer causes about 500,000 deaths per year in the United States and costs about $14 billion annually in treatment and days lost (J. Duke, pers. comm.). If the value of each of these lives is estimated to be between $1.5-8 million (in 1984 dollars), their total value will be between $750 billion and $4 trillion each year. Anti-cancer drugs save about 75,000 lives annually in the United States (an estimated 15% of 500,000 lives) (J. Douros, pers. comm.), and plant-based drugs comprise from 30-50% of the total group of anti-cancer drugs. Combining these estimates, between 22,500 and 37,500 lives are saved annually in the United States as a result of the use of plant-based drugs (i.e. 30-50% of 75,000 lives). Multiplying the range of lives saved by the range of values of life, the annual economic value of plant-based anti-cancer drugs in the United States alone is estimated to be between $34 billion and $300 billion.

Since this estimate reflects only a part of the total economic value of all plant-based pharmaceuticals, two adjustments are necessary. First, this value should be increased to reflect the benefits from these anti-cancer drugs in other OECD countries. Assuming that the market share ratio is a reasonable approximation, this range would be tripled to

$100-900 billion annually to account for anti-cancer applications in all OECD countries. To reflect all of the economic value from both cancer and non-cancer pharmaceutical uses of plant-based drugs, the range could be doubled to $200 billion to $1.8 trillion. Doubling for non-cancer pharmaceutical applications can be viewed as very conservative given the small portion of the market that anti-cancer drugs represent. However, the economic values estimated for anti-cancer drugs may be higher than for much of the rest of the market because of their life-saving attributes. Finally, it should be noted that these values include none of the non-pharmaceutical benefits provided by the plants responsible for these drugs.

Several assumptions made above require discussion. One of the major assumptions is a reasonable estimate of the value of a life saved. For the United States, the range of $1.5 million to $8 million is fairly well-established for on-the-job risks, but since it appears that people place higher values on reducing risks that are involuntary, this range should be viewed as a lower bound for environmental policy assessment. (Violette & Chestnut, 1986). While these estimates were developed for environmental policy analysis, they represent the best available data on the valuation of human life for public policy analysis. Certainly the value of life may be quite different in other countries.

Another major assumption is that the available anti-cancer drugs save lives at roughly the same rate and that they are equally unique (i.e. there are no other anti-cancer drugs that could substitute in their absence). This appears to be a reasonably good assumption for anti-cancer drugs, but may be less valid for non-cancer pharmaceutical applications.

Finally, it should be noted that the development of these estimates is not intended to produce exact numerical values for substitution into benefit-cost equations. Rather this is an attempt to estimate broad indicators of the order of magnitude of these benefits. Clearly, more work is necessary to refine these estimates, but they do not appear to be wholly unreasonable. For example, the current market value of plant-based prescription drugs in the United States is about $11 billion annually. In comparison, the low end of the range of economic values for the United States is $68 billion annually ($34 billion doubled for anti-cancer benefits), which is about six times the market value. The high end of the range is about $600 billion, or about 55 times the market value.

Uncertainty and Option Value[15]

As discussed at several points earlier in the paper, one of the major impediments to estimating either the present or future value of biodiversity is the enormous uncertainty surrounding its potential future value to the pharmaceutical industry. This uncertainty falls into two categories: informational (demand) uncertainty and existence (supply) uncertainty.

Informational uncertainty is of two types: (1) that resulting from the incomplete identification and classification of plant species; and (2) that resulting from the limited extent to which medicinal plants have been investigated for pharmacologically-interesting properties. Both of these subtypes have been discussed earlier, and neither will be easy to overcome: the number of unidentified species of higher plants may be twice to three times as large as the number currently identified, and the number requiring investigation would include all of the unidentified species plus as many as 95% of the identified species. Since the purpose in reducing informational uncertainty is to determine whether these plants would provide potential pharmaceutical (or other) products, informational uncertainty can be thought of as a form of demand uncertainty (i.e. uncertainty as to whether the consumer will be able to demand that particular product).

Existence uncertainty relates to whether the species will continue to survive. The uncertainty arises as a result of both the competitive, ecological pressures of Darwinian selection and the external pressures applied by the demands of man, both in terms of a too-high demand for the plant itself, which results in over-harvesting, and of a competitive demand for the habitat that supports the plant. Since existence uncertainty relates to the future availability of the species, it can also be referred to as supply uncertainty.

The issue of uncertainty is closely related to the concept of option value[16]. Option value is traditionally viewed as a risk aversion premium; that is, the amount a consumer would be willing to pay over and above the expected value of consumer surplus. A concrete example would be the amount of money a consumer would be willing to pay for an option purchased this year to insure that the consumer could visit a certain national park next year. The amount the consumer is willing to pay for the option is the option price, while the difference between the option price and the expected value of consumer surplus is the option value (symbolically: $OV = OP - E(CS)$).

It has been shown (Schmalensee, 1972 & 1975) that in those instances where demand uncertainty exists, option value may be positive, negative, or zero, and its sign cannot be determined *a priori*. However, where supply uncertainty exists, it has been shown that option value will be positive (Bishop, 1982). Biodiversity is an example where supply uncertainty exists, so the option value can be expected to be positive. While the magnitude of the option value is not easily determined, Bishop (1982) argues that:

> "A strong case exists for concluding that, under uncertainty, consumer surplus alone is inadequate. Where demand is certain but future availability of the resource is in doubt, consumer surplus alone would underestimate total consumer benefits from maintaining the asset by an amount equal to option value. Nor is there a strong *a priori* case for arguing that such option values will be relatively small."

As will be discussed below, option value is only one of several values that have to be included among the benefits of preserving biodiversity.

The Net Benefits Equation

To establish the relative utility of the alternatives facing government decision-makers, a relatively simple calculus can be used. As with most such simple devices, it will be seen that having the calculus is one thing and having the data to use in it is quite a different matter. What must be done is to assess the relative value of the various alternatives, taking into account the benefits lost by selecting one alternative over another, a process not dissimilar to benefit/cost analysis.

An example of such a situation might be a tract of tropical moist forest. This forest might be considered for three development alternatives as well as preservation (which could include development carefully managed to avoid any species extinction but this will not be included in the interests of simplicity):

1. Conversion of the area to agricultural uses or to softwood forestry;
2. Sustainable or non-sustainable harvesting of the hardwoods with no forest management;
3. Use of the trees and shrubs for fuel; and
4. Preservation as a reserve for safeguarding the indigenous ecosystems.

Considering the three development alternatives in turn, the assumption underlying the agricultural conversion option is that the development value to be realized by the indigenous population from the agricultural or sylvan commodities harvested will produce a net gain either through their own use or by exporting the commodities. However, three factors must be subtracted from the realized benefits: (1) the costs of clearing the land; (2) the foregone benefits from preserving the resource; and

(3) the foregone benefits from use of the forest for fuelwood.

For the hardwood harvesting option, the benefits to be derived will result primarily from the export of the hardwood to developed countries for furniture. The three cost factors described above would be the same for this option.

The fuelwood option is basically the use of the forest resources by the indigenous population. The cost of development may not have a price *per se* in which case the cost of a substitute fuel could be used. The loss of the preservation benefits would be the same as for the other options.

Leaving aside for the moment the policy issues that must be considered, the simple economic calculus referred to above only requires that the net benefit to be realized from each of these alternatives be compared and that the alternative with the largest positive net benefit should be the favored alternative. If none of the alternatives has a positive net benefit, then the forest should be preserved. In symbolic terms, the general equation is (Krutilla & Fisher, 1975):

$$NB_d = B_d \; C_d \; B_p$$

where:

NB_d = the net benefit of the alternative under consideration;

B_d = the gross willingness to pay of the benefitting population for the development;

C_d = the costs of the developing the resource; and

B_p = the benefits of preserving the resource.

Clearly, the most important term for the purposes of this paper is B_p. Based on earlier discussions, it is also clear that the quantification of B_p will be difficult. For a public policy decision-making, a more useful parameter would be NB_p, the net benefits of preservation, which can be represented as:

$$NB_p = E(CS) + OV + EV + E(R) - E(C_{pd}) - C_p$$

where:

$E(CS)$ = expected value of consumer surplus

OV = option value

EV = existence value

$E(R)$ = expected value of product revenues

$E(C_{pd})$ = expected cost of product development

C_p = costs of implementing preservation program

The existence value is the utility that consumers receive from simply knowing that something exists, independent of any current or anticipated future use. Existence value embodies a sense of altruism, which can be of three kinds (Randall, 1986):

1. Philanthropic – value attributed because others may want to use the resource;
2. Bequest – value attributed because future generations may want to use the resource; and
3. Intrinsic – value attributed because of the existence of non-human components in the resource (e.g. the person values the continued existence of the wildlife in a forest). This definition of existence value probably includes most of the value that is sometimes referred to as aesthetic value.

The cost of preservation refers to the costs of maintaining the preserved area. For example, there would be administrative requirements to monitor and protect the area.

It should be noted that there is not a specific term included to cover other benefits that might be realized from preserving biodiversity. These benefits could include:

1. The value of the biodiversity for applications other than pharmaceuticals, such as other medicinal uses, agricultural crops, and other commercial uses;
2. The commercial value of tourism, scientific investigations, etc.;
3. The value derived from the prevention of soil erosion, the maintenance of water tables, and the maintenance of navigable rivers;
4. The value to the global climate in preserving the carbon balance; and
5. The value to the global environment in removing sulphur oxides and other pollutants.

It could be argued that these benefits are included in the expected value of consumer surplus, but that possibility is remote. In all likelihood, these benefits are not included in the above equation for NB_P. This could lead to a significant underestimation of NB_P. For example, the present value of the Amazon forest for atmospheric sulfur scrubbing is estimated to be $90 billion (using a social discount rate of 10%) (Farnsworth *et al.*, 1983).

Given the difficulty in measuring the component parts of NB_P, natural resource decisions are often made in favor of development because NB_P is ignored or underestimated. The fact that these benefits cannot be adequately quantified calls into question the utility of the procedure. If only the benefits and direct costs of development can be quantified, the net benefits cannot be truthfully said to have been calculated. And the temptation to make the decision in spite of the absence of the preservation values results in a considerable distortion of the true economics.

This situation led to the proposal of a safe minimum standard of conservation (SMS), which would maintain sufficient population and habitat to insure that the species could survive. (Ciriacy-Wantrup, 1952; Bishop, 1978, 1979, 1980) The implementation of this concept is referred to as the SMS approach, which states that the SMS should be adopted unless the social costs of doing so are unacceptably large. Bishop (1980) argues that:

> "... an efficiency oriented approach is incorrect because it assumes away some of the most important parts of the problem. First, to estimate the benefits of maintaining a species, one must assume that the uncertainty does not exist, whereas dealing effectively with that uncertainty lies at the heart of the problem. Secondly, an efficiency-oriented approach would completely ignore all issues of equity or fairness in the distribution of economic gains and losses, yet endangered species decisions must necessarily confront an important issue of intergenerational equity."

The reader will no doubt recognize that these issues have been discussed earlier in this paper and that they are, in fact, among the major impediments to the completion of a routine benefit-cost analysis. Another issue that is specifically addressed by the SMS but not by benefit-cost analysis is that of the irreversibility of extinction. (For a different view, see Smith & Krutilla (1979).) While a decision to preserve can always be changed, a decision to develop cannot.

One of the weaknesses of the SMS approach is that it leaves undefined the question of how to know when the social costs are too high. The SMS embodies the assumption that there will be positive benefits from preservation of biodiversity, but it is not clear that public policy would be established without a requirement to characterize the magnitude of these benefits in some manner. However, recent work (Fisher & Raucher, 1984) tends to support the belief that the benefits of preservation will be much larger than had been expected based on theoretical analyses. Further, the findings in this paper also support this position.

Summary

This section has described some of the important economic concepts relevant to biodiversity and how they might be applied to the development and characterization of potential remedies.

The relationship between market value and economic value is important to keep in mind when reviewing analyses relating to natural resources. While market value is the one most often used in analyses because it is far easier to quantify, it represents only a portion of the economic value of any resource. The economic value of a resource represents all of the societal benefits that are derived from that resource, including the market value of the products derived from that resource.

With respect to medicinal plant species, there are two aspects of economic value that must be considered. First, the economic value of the drugs derived from these plants includes not only the market value but also the societal benefits from increased good health (e.g. wages not lost, health care costs averted, the value individuals place on better health, etc.). Second, the non-pharmaceutical uses and benefits that the plants provide (i.e. the informational and environmental benefits). Only the first of these parts is approximated in this paper, and it is estimated that the pharmaceutical economic value of plant-based drugs ranges from $200 billion to $1.8 trillion annually for all OECD countries.

However, for government officials to have a more complete understanding of the value of these resources, the second part of the economic value should also be quantified. One way of doing this is to describe the net benefits of preservation (NB_P). This value is the sum of the expected consumer surplus, the option value, the existence value, and the expected value of product revenues minus the expected costs of product development and the costs of implementing the preservation program. The major problem is that the first three values are very difficult to quantify. Furthermore, several specific benefits of biodiversity are not included within any of these terms, so even if NB_P were calculated, it would be a considerable underestimation of this portion of the economic value of these plants.

The safe minimum standard of conservation (SMS) was developed to address this very situation. The SMS would maintain sufficient population and habitat to insure that species would survive, unless the social costs of doing so are unacceptably large. This approach begins to address the issue of intergenerational equity, which almost always arises in natural resource questions. However, the SMS leaves open the question of when social costs grow too large.

Biodiversity and Public Policy

While the importance of biodiversity is clear, it is impossible to separate biodiversity from the large issues within which it is subsumed.

Consequently, any attempt to reduce this loss must address larger and even more complex problems than are presented by the loss of biodiversity. Some of these larger issues have been discussed above. These issues and several others will be briefly reviewed below.

Forests, Development and Biodiversity

Biodiversity is clearly only part of a much larger problem: the preservation of endangered ecosystems (such as tropical moist forests) that are the habitats for most of the species in danger of extinction. If it is desirable to preserve this biodiversity, then the habitats must be preserved as well. However, the benefits to be derived from the preservation of biodiversity are only a part of the total benefits that would be realized from the preservation of the whole ecosystem.

Naturally this considerably complicates public policy development. To preserve habitats means that development must be restricted in those countries harbouring these ecosystems. And those countries are almost exclusively developing countries, to whom a suggestion to restrict development is not likely to receive much of an audience unless there is a *quid pro quo*. In most of these countries, government stewardship is not a strongly formed concept or is under considerable economic and population pressures. Any major change in this situation would require OECD countries to place a significant value on these resources in recognition of the large benefits these forests and other ecosystems provide, or may provide, to their citizens. Once this hurdle is overcome, any number of possible strategies could be implemented (e.g. encouraging investment in these countries by the pharmaceutical industry to develop the countries' capability to identify new plant-based drugs, thereby either leading to a recognition by these countries of the value of their own resources or providing them the wherewithal to provide substitutes for the food and fuel their citizens now obtain through damaging these ecosystems).

The key to proper valuation of these ecosystems by OECD countries is a better understanding of the benefits these ecosystems provide and of the value of these benefits to OECD countries. The pharmacological value of biodiversity, which this paper has attempted to quantify, to OECD countries is only a very small portion of the total package of benefits these ecosystems provide. Given the very large benefits that appear to be attributable to just this one portion, the total benefits that are likely to be foregone in the absence of preservation action is certain to be immense. And it appears that to prevent the loss of these benefits, OECD countries will be required to work with developing countries in a manner not commonly found in the past.

Public Policy-making under Uncertainty

The degree of uncertainty surrounding the loss of biodiversity has been discussed in some detail earlier in this paper. Given this large uncertainty, it is difficult to acquire and sustain the attention of government officials who establish policy. Governments, in general, do not deal very effectively with uncertainty, especially when there is a requirement to take positive action. There are so many competing issues that appear to have a more immediate need for attention that those issues without a sense of immediacy are assigned a lower priority. However, unlike most such issues, not acting on biodiversity has irreversible consequences that must be highlighted. Actions to preserve biodiversity will also preserve options for the future. By not taking action to preserve biodiversity, those options will be irretrievably lost.

Like most global problems, biodiversity is not an issue that can be resolved by unilateral action; it will require the concerted action of many countries, both developed and developing. Such issues are always difficult to address. There has not been a strong history of international problem-solving and dispute resolution in these areas, and progress is always slow. Uncertainty as to the nature and magnitude of the problem has contributed to the delay, but the very nature of the problems require taking action under uncertainty. Acting under uncertainty is difficult for governments when they are dealing with unilateral issues, and much more difficult on the international level.

In both of the above issues, what is required is that governments decide upon, in a positive rather than passive manner, the level of risk they are willing to accept. As indicated above, it is not uncommon for governments to either not address or debate overlong difficult issues until events overtake them. This reduces the available options and forces the governments to take hasty, less efficient action. While it is easy to identify this flaw, it will be more difficult to develop a mechanism for overcoming it so that decisions on risk acceptance or risk avoidance can be rationally made. Clearly, issues such as biodiversity require a new method of presentation so that the normal inertia of governments can be overcome.

This paper does not propose what new presentation methods might be successful, but it does provide one of the key elements for any such presentation: better information about the nature and magnitude of both the current benefits from biodiversity and the benefits that are likely to be foregone if action is not taken to preserve biodiversity. There is a clear need for additional data and additional analysis. However, as is evident from this analysis, the marginal utility of this additional data collection and analysis must be carefully weighed against the very real (and large) cost of delaying action.

Annex 1 : Pharmaceutical Sales and Projections

There are pharmaceutical sales data for prescription drugs in manufacturers prices available for many countries of the world. These data are the foundation of this analysis. However, these data do not include over-the-counter drug sales, which are an important part of the market as a whole and for plant-based drugs in particular. In addition, to place the sales figures within a more familiar context, it is necessary to convert the manufacturers prices to retail prices.

The data shown in Figure A-1 are prescription drug sales figures for the ten year period 1976-1985 for the world and OECD countries in constant 1984 dollars and in manufacturers prices. While there is an apparent falling off of constant dollar sales in the 1980s, this may be attributable to both a slowing of growth in the market and the significant increase in the value of the dollar, which would tend to diminish the value of sales in other countries.

To convert these figures to retail values, consumer expenditure data contained in Trapnell *et al.*, was compared to data provided by the Pharmaceutical Manufacturers Association (PMA). The PMA data were the basis for retail estimates in Trapnell *et al.* The resulting manufacturers price/retail price ratios are shown in Figure A-2.

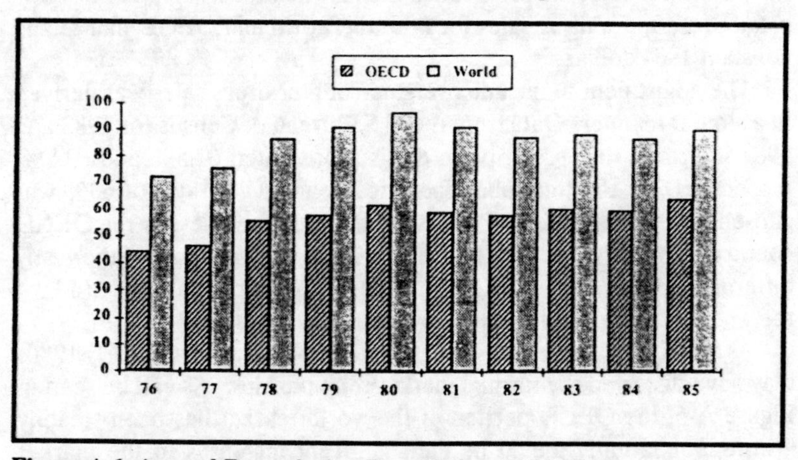

Figure A-1. Annual Prescription Drug Sales, 1976-1985 (in manufacturer's prices, 1984 US$ billions)

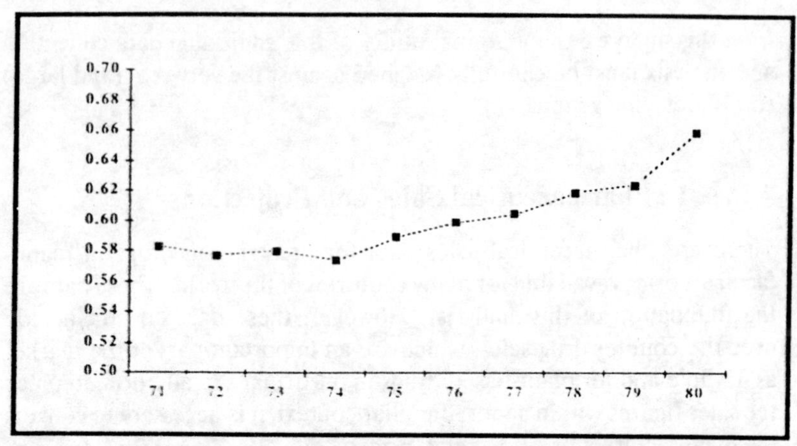

Figure A-2. Manufacture:Retail Price Ratios in the United States for the Period 1971 to 1980

As can be seen in the figure, manufacturers prices were fairly stable during the ten year period at about 60% of retail prices, although the ratio seemed to be rising toward the end of the period. The ratio for 1980 (66%), however, is substantially above what might have been expected, and it is not clear whether this is an anomaly or the beginning of a new trend. Unfortunately, the analysis of Trapnell *et al.* has not be updated, so the ten-year average of 60% was chosen to convert manufacturers prices to retail prices. Since similar data were not available for other countries, the 60% ratio was used for all countries, even though it may not be indicative of the price ratios in those markets. Figure A-3 shows retail prescription drug sales for both the world and OECD markets in constant 1984 dollars.

The adjustment to include over-the-counter drugs sales was derived in a similar manner. Data from the U.S. Bureau of Census for 1983 and 1984 indicated that prescription drugs represented 62.5% and 63.8%, respectively, of the total pharmaceutical market. A ratio of 63% was chosen to adjust prescription drug sales to total drug sales for the OECD countries, and these data can be seen in Figure A-4. Because of the vastly different markets in non-OECD countries, no attempt was made to adjust the world sales figures to include over-the-counter sales.

As mentioned above, the strength of the dollar during the early 1980s may have distorted the normal market relationships. As can be seen in Figure A-5, the OECD portion of the world market has been steadily growing, apparently fueled by the significant increases in the market shares of the United States and Japan. The market shares of the three largest European markets show the effect of the strong dollar. This

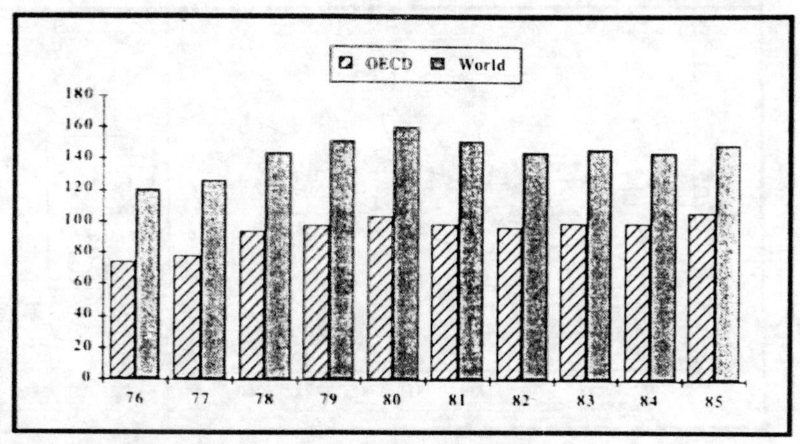

Figure A-3. Prescription Drug Sales, 1976-1985 (in retail prices, 1984 US$ billions)

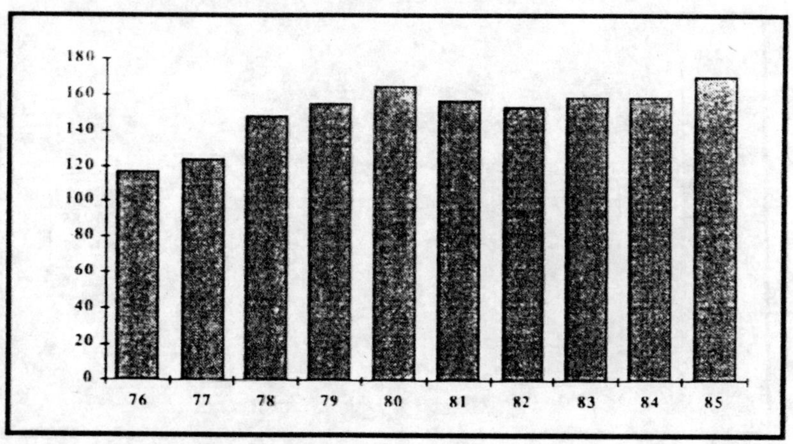

Figure A-4. Total OECD Drug Sales, 1976-1985 (in retail prices, 1984 US$ billions)

makes the increase in the Japanese market share that much more noteworthy. However, these trends may be the result of an unidentified structural demand shift. While that is unclear for the European markets, it seems clear that such a shift is taking place in Japan.

In projecting U.S. data to the OECD, a factor of three has been used. This choice may result in underestimating the OECD totals if historical data (which would suggest a factor of 3.33) are true or overestimating OECD totals if the more recent trends (which would suggest a factor of 2.5) are true. The factor of three would appear to be a reasonable compromise between the two.

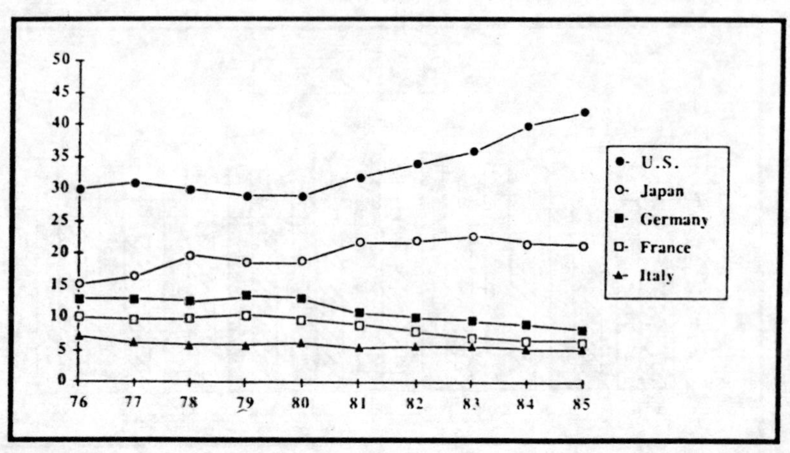

Figure A-5. Domestic Pharmaceutical Sales as Percentage of OECD Total

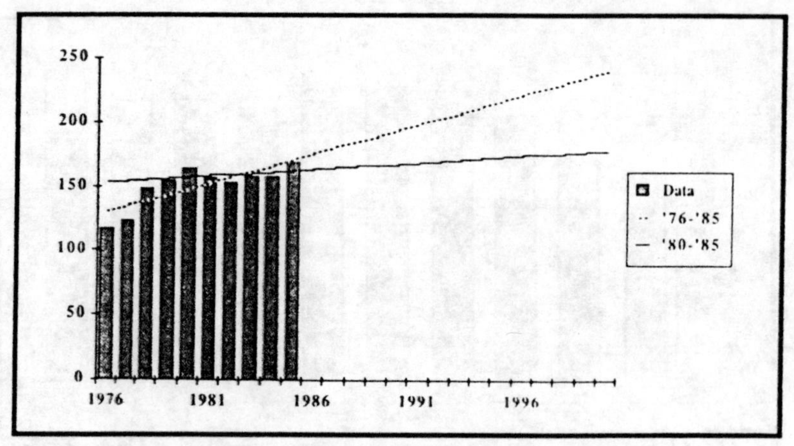

Figure A-6. Projected Total OECD Pharmaceutical Sales (in retail prices, 1984 US$ billions)

For the purpose of developing an estimate of the present value of plant-based pharmaceuticals through the end of the century, it is necessary to regress the available sales data and then, using the equation derived from the regression, calculate projected values for the remaining years. Upon reviewing these calculations and the underlying data, it became clear that there was more than one alternative. Because of the slower growth of pharmaceutical sales in the 1980s, the regression using the data from 1976-1985 has a somewhat steeper slope than the regression using the data from 1980-1985 (see Figure A-6). The choice between the two regressions is largely a function of whether it is believed that the growth in pharmaceutical market will be essentially flat for the remainder of the century or that growth will tend to return to historical levels. Because of the apparent distortions caused by the strength of the dollar in the early 1980s, the regression using the whole range of data was chosen for the purpose of deriving the present value of plant-based drugs.

Bibliography

Anon. (1984a). *Business Week*, 29 October, p. 136.

Anon. (1984b). The drug industry moves from discovery to design. *The Economist*, 7 January, pp. 71-74.

Anon. (1985). *Business Week*, 2 September, p. 77.

Ayensu, E.S. & DeFilipps, R.A. (1978). *Endangered and Threatened Plants of the United States*. Washington, D.C.: Smithsonian Institution and World Wildlife Fund, Inc.

Balandrin, M.F., Klocke, J.A., Wurtele, E.S. & Bollinger, W.H. (1985). Natural plant chemicals: sources of industrial and medicinal materials. *Science*, 228, 1154-1160 [7 June 1985].

Bishop, R.C. (1978). Endangered species and uncertainty: the economics of a safe minimum standard. *Amer. Journ. Agric. Econ.*, 60(1).

Bishop, R.C. (1979). Endangered species, irreversibility, and uncertainty: a reply. *Amer. Journ. Agric. Econ.*, 61(2).

Bishop, R.C. (1980). Endangered species: an economic perspective. *Transactions of the 45th North American Wildlife and Natural Resources Conference*. Washington, D.C.: Wildlife Management Institute.

Bishop, R.C. (1982). Option value: an exposition and extension. *Land Economics*, 58(1).

Ciriacy-Wantrup, S.V. (1968). *Resource Conservation: Economics and Policies*, 3rd edn. Berkeley and Los Angeles: California Division of Agricultural Sciences. (First published 1952).

Clark, C.W. (1976). *Mathematical Bioeconomics: The Optimal Management of Renewable Resources*. New York: Wiley.

Curtin, M.E. (1983). *Bio/Technology*, 1, 649.

Farnsworth, N. & Morris, R. (1976). Higher plants – the sleeping giant of drug development. *Amer. Journ. Pharmacy*, 147(2), 46-52.

Farnsworth, N. & Soejarto, D.D. (1985). Potential consequence of plant extinction in the United States on the current and future availability of prescription drugs. *Econ. Bot.*, 39(3), 231-240.

Farnworth, E.G., Tidrick, T.H., Smathers, W.M. & Jordan, C.F. (1983). A synthesis of ecological and economic theory toward more complete valuation of tropical moist forests. *Int. Journ. Envir. Studies*.

Fisher, A. & Raucher, R. (1984). Intrinsic benefits of improved water quality: conceptual and empirical perspectives. In *Advances in Applied Micro-Economics*, Vol. 3, eds. V.K. Smith & A.D. Witte.

Fisher, A.C. (1981). *Resource and Environmental Economics*. Cambridge: Cambridge University Press.

Hall, A. *et al.* (1985). The Race for Miracle Drugs. *Business Week*, 22 July, pp. 92-97.

Hänsel, R. (1972). Medicinal plants and empirical drug research. In *Plants in the Development of Modern Medicine*, ed. T. Swain.

Huxley, A. (1984). *Green Inheritance: The World Wildlife Fund Book of Plants*. London: Collins/Harvill.

International Trade Centre, UNCTAD/GATT (no date). *Markets for Selected Medicinal Plants and Their Derivatives*.

International Union for Conservation of Nature and Natural Resources (IUCN), United Nations Environment Program (UNEP) and the World Wildlife Fund (WWF) (1980). *World Conservation Strategy: Living Resource Conservation for Sustainable Development*. Gland, Switzerland: IUCN.

Krutilla, J. & Fisher A. (1975). *The Economics of Natural Environments*.

Levingston, R. & Zamora, R. (1983). Medicine trees of the tropics. *Unasylva*, 35(140), 7-10.

Lucas, G. & Synge, H. (1978). *The IUCN Plant Red Data Book*. Morges, Switzerland: IUCN.

Myers, N. (1979). *The Sinking Ark: A New Look at the Problem of Disappearing Species*. Oxford: Pergamon Press.

Pharmaceutical Manufacturers Association (no date). *Annual Survey Report 1983-1985: U.S Pharmaceutical Industry*.

Randall, A. (1986). Human preferences, economics, and the preservation of species. In *The Preservation of Species: The Value of Biological Diversity*, ed. B. Norton. Princeton, New Jersey: Princeton University Press.

Schmalensee, R. (1972). Option demand and consumer's surplus: valuing price changes under uncertainty. *Amer. Econ. Review*, 62, 813-824.

Schmalensee, R. (1975). Option demand and consumer's surplus: a reply. *Amer. Econ. Review*, 65, 737-739.

Schultes, R.E. (1972). The future of plants as sources of new biodynamic compounds. In *Plants in the Development of Modern Medicine*, ed. T. Swain.

Smith, V.K. & Krutilla, J.V. (1979). Endangered species, irreversibilities, and uncertainty: a comment. *Amer. Journ. Agric. Econ.*, 61(2), 371-375.

Spragins, E.E. & Rhein Jr., R. (1985). When nature has the answer: scientists forage for new drugs in the wild. *Business Week*, 22 July, p. 97.

Suffness, M. & Douros, J. (1979). Drugs of plant origin. Chapter III in *Methods of Cancer Research*, Vol. XVI. Academic Press.

Suffness, M. & Douros, J. (1982). Current status of the NCI plant and animal product program. *Journ. Natural Products*, 45(1), 1-14.

Tippo, O. & Stern, W.L. (1977). *Humanistic Botany*. New Jersey: Norton.

Trapnell, G.R., Genuardi, J.S. & O'Brien, J. (no date). *Consumer Expenditures for Ethical Drugs*. Pharmaceutical Manufacturers Association.

United States National Academy of Sciences (1978). *Conservation of Germplasm Resources: An Imperative*. Washington, D.C.: National Academy Press.

Violette, D.M. & Chestnut, L.G. (1986). *Valuing Risks: New Information on the Willingness to Pay for Changes in Fatal Risks*. United States Environmental Protection Agency, EPA-230-06-86-016.

Notes

1. The views expressed herein are those of the author and do not necessarily represent those of the U.S. Environmental Protection Agency. This paper was originally prepared under the direction of Dr Michel Potier, Chief of the Environment and Economics Division, as a part of the work program of the Environment Directorate of the Organization for Economic Cooperation and Development (OECD). The OECD held workshops in 1984 and 1986 to bring together internationally-recognized experts (including academic and government economists, botanists, ecologists, and representatives of the pharmaceutical industry) to review the paper. Finally the OECD's Group of Economic Experts reviewed the paper in 1987. The author would like to thank Dr Potier for the opportunity to develop the paper and the World Wide Fund for Nature (WWF) for the opportunity to participate in the Consultation.
2. The 24 OECD countries are: Australia, Austria, Belgium, Canada, Denmark, Finland, France, W. Germany (FRG), Greece, Iceland, Ireland, Italy, Japan, Luxembourg, the Netherlands, New Zealand, Norway, Portugal, Spain, Sweden, Switzerland, Turkey, the United Kingdom, and the United States. Yugoslavia is an associate member, but has not been included in these analyses.
3. Aspirin, the very first synthesized drug, is a perfect example of the modification of natural chemicals; until it was acylated, the salicylic acid was too strong to be taken orally.
4. It should be noted that the same constraints apply to non-botanicals as well, but this statement does reflect the sentiments of many in the pharmaceutical industry.
5. Professor Thomas F Burks, Head, Department of Pharmacology, University of Arizona College of Medicine, quoted in *Scrip*, no. 1110, 11 June;1986, p. 23.
6. In its haste to attribute to design what properly belonged to discovery, another source neglected to state that Capoten was derived from a natural product.
7. All of the information in this paragraph is from this citation. The situation remains unchanged even though more than a decade has passed since this was written.
8. Unless explicitly noted, the values given are current year not constant values.
9. Farnsworth and Soejarto limited their analysis to the United States. For the purposes of this paper, their methodology has been extended to cover the entire world, and any distortions that result from that extension cannot be attributed to these authors.
10. While this factor corrects for the number of prescriptions filled, it probably overstates the price paid for the prescriptions because the prices charged by other dispensing facilities, mainly hospitals and health maintenance organizations (HMO's), is usually lower than community pharmacy prices. However, the use of this factor does result in an estimate of the value of these prescriptions had they been sold through community pharmacies.
11. It is unknown whether these figures represent manufacturer's prices or retail prices. The US$ equivalents are the same for each year because of changes in the US$/Y exchange rate during this period. Consequently, the US$ values do not reflect the real increase in sales.
12. The methodology used in this section is an extension of the one used by Farnsworth & Soejarto (1985).

13. This range of probabilities was developed by representatives of the pharmaceutical industry at one of the workshops held by the OECD to review this paper. The range is based on the direct experience of the industry representatives in developing new drugs from higher plants. Subsequent reviews of the paper produced general agreement that it represented a reasonable estimate, with advocates for each end but no suggestions to extend the range in either direction.
14. There is, of course, the added value of the processing of the wood into paper that is included in the market price of the paper. However, it will be argued later that the market price of drugs may be a better indicator of their value than the market price of the raw material.
15. This section relies heavily on a paper written by Professor David Pearce, University College London, for the OECD, "The Economic Value of Genetic Materials: Methodology and Case Illustration", June 1983.
16. There is a large body of literature discussing option value. However, the descriptive material in this paragraph was taken largely from articles by Bishop (1978, 1979, 1980, and 1982).

Economic Aspects of Exploitation of Medicinal Plants

Dr Akhtar Husain,
Director, Central Institute of Medicinal and Aromatic Plants,
Lucknow, India

Higher plants have been used as a source of drugs by mankind for several thousand years. In fact, ancient man was totally dependent on green plants for his day-to-day needs of medicaments. With the development of modern medicine, synthetic drugs and antibiotics, the importance of plants as raw material for drugs decreased considerably. However, plants were used as a source of some of the most important drugs, even in the modern system of medicine. With the advancement of synthetic organic chemistry most of the active constituents of plants used in medicine were synthesised. At one time it was thought that ultimately all the plant drugs would be obtained from synthetic sources. However, in spite of phenomenal development in the development of new drugs from synthetic sources and the appearance of antibiotics as major therapeutic agents, plants continue to provide basic raw materials for some of the most important drugs.

Although data are not available for all countries, a study carried out in the United States by Farnsworth and his colleagues between 1958 and 1980 indicated that although the number of prescriptions issued by community pharmacies in the United States increased considerably, the percentage of prescriptions containing one or more plant products remained constant at a figure of 25%. It has been found that in highly developed countries like the United States more than 100 chemical constituents of definite structure derived from 41 species of plants were used in modern medicine. It has also been estimated that in addition to these active constituents, more than 96 crude extracts were also used in the United States. The American public paid more than 8 billion dollars during the year 1980 for the cost of plant drugs alone. This figure must have increased considerably because of inflation. If the amount of drugs used by other countries of the world is comparable to that in the U.S.A., the cost of plant drugs used by entire population of the world would come to a staggering figure of more than 160 billion dollars.

During the last 10 years, the whole world, especially the developed one, has been swept by a green wave. With the realisation of health hazards and toxicity associated with indiscriminate use of synthetic drugs and antibiotics, there has been a general realisation in the West as well as the East that anything in nature is safer than synthetics. This tendency has resulted in the appearance of a large number of health food shops all over the West, which sell numerous crude extracts and herbal teas. According to Tyler (1986), American health food shops sold herbs worth more than 360 million dollars during 1981 alone. This trend is increasing and is bound to increase in future.

In Eastern Europe, where traditional medicine has been amalgamated with modern medicine, the proportion of prescriptions containing plant drugs is more than 60%. The Chinese system of medicine, which is widely practiced in China and is also popular in other South East Asian countries like Japan, Korea, Hong Kong and Malaysia, obtains more than 80% of its medicaments from higher plants. The situation is similar in countries like India, Pakistan, Bangladesh, Sri Lanka and Nepal, where highly developed traditional systems of Ayurveda, Siddha and Unani are practiced, and all of these are mainly based on drugs derived fom plants.

Both the Chinese as well as the Indian traditional systems of medicine use more than 200 species of plants in their pharmacopoeias. It is difficult to estimate the actual value of drugs from traditional systems of medicine. However, it is very well known that the drugs from the Ayurvedic and Unani systems (as well as Chinese drugs) are sold widely in Europe, U.S.A. and Japan, where common people affected by the new green wave buy these drugs, even when there is no scientific proof of their efficacy.

Although more than 100 different plants are today used in modern medicine, an attempt will be made to discuss only those which are used very widely and which have a large market throughout the world. Statistical data are not available for all countries. However, according to the International Trade Centre, the total value of imports of medicinal plants increased from 355 million dollars in 1976 to 551 million dollars in 1980. These import figures cover only OECD countries, Japan and U.S.A. They do not take into consideration the large amount of trade in China and communist countries; if so the present figure of total trade in medicinal plants would be much higher.

Although more than 100 plants are used in modern medicine in various parts of the world, the list of most important ones along with their pharmacological properties is given in Tables 1 and 2. A list of plants which are used widely in crude form, essential oils or oleoresins is given in Tables 3 and 4. Only those plants which have a large world market and considerable potential use in future are discussed in detail.

Sapogenin-bearing Yams

Steroidal drugs are one of the most important therapeutic agents used in modern medicine throughout the world today. They constitute one of the most important group of drugs sold in the world market, which also include sulpha drugs, antibiotics and vitamins. Some of the common drugs belonging to this group are corticosteroids, sex hormones, oral contraceptives and anabolic agents. Although these drugs can be obtained from a number of sources, including total synthesis, animal by-products and plant products, steroidal sapogenins obtained from plant species still constitute one of the major raw materials for steroidal drugs. The most important steroidal sapogenin, which serves as a starting material for the production of various steroidal drugs, is diosgenin from tubers of various species of *Dioscorea*. At one time diosgenin obtained from various species of *Dioscorea* used to provide more than 80% of the raw material for synthesis of these drugs. However, the situation has changed during the last 10 years because of the widescale use of stigmasterol from Soyabean oil and sitosterol from other plant sources, as the starting material for the synthesis of these drugs. At present the use of diosgenin for these drugs has gone down to 50% or lower. However, it is estimated that more than 200 tonnes of diosgenin is still being used for the synthesis of these drugs and the plant will continue to be an important source for steroidal drugs at least in the developing countries for years to come. The species most widely exploited commercially are *Dioscorea composita* Hemsl., *D. floribunda* Mart. & Gal. and *D. belizensis* Lundell in Mexico and other Central American countries, *Dioscorea deltoidea* Wall. in India and *D. nipponica* Makino in China.

Hecogenin, another steroidal sapogenin, is the only other raw material used for synthesis of these drugs and it is used to a very limited extent in Tanzania. Hecogenin is a by-product of the fibre industry from *Agave sisalana* Perrine ex Engelm.

Solasodine – a steroidal alkaloid – is a product from various species of *Solanum*, and can be used as the raw material for synthesis of steroidal drugs. However, the use of solasodine, which is exploited only in the U.S.S.R. from *Solanum laciniatum* Ait., has gone down considerably and the scope of this plant as a raw material for steroidal drugs is very limited for economic reasons. The other species of *Solanum*, especially *Solanum khasianum* Clarke, which was at one time the subject of experiments in India, has not had commercial success because of the high cost of production.

Opium Poppy (*Papaver somniferum* L.)

Because of its alkaloids, Opium poppy is one of the most important plants used in medicine today. The plant contains more than 24 alkaloids, of which morphine, codeine and papaverine are used extensively in medicine. Morphine is used because of its powerful analgesic property. It is employed as a sedative and analgesic to relieve extreme pain in cases of myocardial infraction, cancer and other afflictions where severe pain is involved. It is also employed for treatment of typhoid fever, traumatic shocks and in combination with atropine sulphate for relieving renal and intestinal colic and coronary thrombosis. In addition to analgesic properties, codeine is used as a most important antitussive agent to control persistent coughs.

Papaverine is a smooth muscle relaxant and used to control gastrointestinal spasms and asthmatic attacks. It is also used to relieve pain in coronary and arterial diseases. However, most of the alkaloid is obtained from synthetic sources. Nascopine, another important alkaloid found in opium, is used as an antitussive agent and in modified form also used to control various other diseases. Crude opium in various forms is also used as analgesic and antispasmodic to control various intestinal and respiratory diseases.

The most important opium-producing countries are India, Turkey, Bulgaria, Yugoslavia, U.S.S.R., Australia, France and Spain. The plant is also illegally cultivated in Pakistan, Afghanistan, Laos, Burma and Thailand. In India, opium is only obtained from lancing of the poppy capsule; the other countries obtain alkaloids by processing straw and do not produce opium. The world requirement of opium is estimated at around 200 tonnes of alkaloids. At one time India was one of the major opium-producing countries and was producing more than 2100 tonnes of crude opium. However, because of increase of production in Turkey, Australia, France and the Netherlands, India's production has gone down to less than 1200 tonnes.

Plants Containing Tropane Alkaloids

A number of solanaceous plants have been used in medicine, mainly because of their active constituents, the tropane alkaloids. These include atropine, hyoscyamine and hyoscine. The most important plants used in different parts of the world either as a source of pure alkaloids or in the form of crude extracts are Belladonna (*Atropa* spp.) and various species of *Hyoscyamus, Duboisia* and *Datura*.

Belladonna (*Atropa* spp.)

The drug is obtained from the flowering tops and roots of *Atropa belladonna* L. or *A. acuminata* Royle, referred as Indian belladonna. It is one of the most important drugs used in the form of crude extract or total alkaloids for treatment of a number of diseases. The plant contains a number of alkaloids. However, only three of them are important in medicine. These include hyoscyamine, hyoscine and atropine. The plant is rarely used today as a source of pure tropane alkaloids, except for production of total alkaloids of Belladonna. Most of the drug produced in the country today is employed in the form of crude extract, tincture, plaster and ointments. Because of its anticholinergic properties, belladonna is used in intestinal disorders as an antispasmodic drug and also to check spasms of urinary bladder. It is also used in asthma and peptic ulcers. In the form of plaster and ointments, it is employed as an analgesic and anti-inflammatory. It is also used to control convulsions in cases of parkinsonism and epilepsy.

The plant is indigenous to Western Europe and cultivated in England, East Germany, U.S.S.R., U.S.A. and India.

Hyoscyamus spp.

There are two species of *Hyoscyamus* containing tropane alkaloids, which are used in medicine in different parts of the world. These are Black henbane (*Hyoscyamus niger* L.) and Egyptian henbane (*Hyoscyamus muticus* L.). Black henbane is cultivated to a limited extent in U.S.A., U.K. and India. The drug is used in the form of crude extract or tincture, mainly for treatment of intestinal and respiratory ailments. The flowering tops of the plant contain 0.01-0.1% total alkaloids of which three-quarters is hyoscyamine and the rest hyoscine.

Egyptian henbane has been used as a source of pure alkaloids – hyoscyamine and atropine – for many years. The plant is indigenous to Egypt. Flowering tops of the plant contain 0.5-1% total alkaloids with more than 85% hyoscyamine and the rest hyoscine. The raw material is collected from wild sources in Egypt and about 500-600 tonnes of the drug are exported to West Germany for processing. It is also cultivated to a limited extent in the U.S.A. and India.

Duboisia spp.

Species of *Duboisia* are indigenous to Australia and are the major source of tropane alkaloids today in the world. There are two species of *Duboisia* which are exploited as raw material for the production of hyoscyamine, hyoscine and atropine. These are *D. myoporoides* R.Brown and *D. leichardtii* F.Mueller. Leaves of these two species are used commercially for the production of tropane alkaloids. The leaves of *D. myoporoides* contain 0.5-3.5% total alkaloids with 70% hyoscine and 30% hyoscyamine. The leaves of *D. leichardtii* also contain 0.5-3% total alkaloids with more than 70% hyoscyamine and the rest hyoscine. Some hybrids between *D. myoroides* and *D. leichardtii* containing more than 3% alkaloids have been developed in Australia. The present production in Australia is around 700-800 tonnes of dry leaves.

The plant has recently been introduced to India and is cultivated to a limited extent for internal use.

Datura spp.

Three species of *Datura – Datura stramonium* L., *D. innoxia* Mill. and *D. metel* L. – have been used in medicine to a limited extent in some countries as a source of tropane alkaloids. However, because of their low alkaloid content, the use of *Datura* is very limited for economic reasons. Another species *D. sanquinae* Ruiz & Pav., containing 2% hyoscine, has recently been used as a source of alkaloids in Ecuador.

Psyllium Seed and Husk (*Plantago ovata* Forsk.)

Psyllium husk is obtained from the ripe seeds of *Plantago ovata* Forsk., called Blond psyllium, or *Plantago psyllium* Decne., called Spanish psyllium. India is one of the main producers of this drug, which is cultivated in the State of Gujarat. Approximately 22,500 hectares are under cultivation, producing around 20,000 tonnes of psyllium seed and husk.

Psyllium husk contains mucilage which is colloidal in nature. Mucilage with various flavours is used in the U.S.A. and other developed countries for controlling constipation, diarrhoea and dysentery. The husk is also used as the safest bulk laxative. The mucilage absorbs water in the digestive tract which it also lubricates. It absorbs bacterial toxins and also improves pristalsis causing relief in dysentery and diarrhoea.

In addition to its medicinal use, psyllium husk is also used in various industries. It is also employed in ice-creams, chocolates, candies and for sizing textiles, and is used as a base in cosmetics.

Cinchona (*Cinchona* sp.)

Cinchona is another important plant which is widely used in medicine all over the world. Various species of *Cinchona* exploited commercially are *Cinchona officinalis* L., *C. ledgeriana* Moens, *C. succirubra* Pavon, *C. calisaya* Weddel and hybrids among the various species. The bark of the tree contains 6-7% total alkaloids. Although it contains a large number of alkaloids, the most important ones used in medicine are quinine and quinidine. While quinine is used as an antimalarial, quinidine is used as a cardiac depressant to control auricular fibrillation. Quinine is also used as a bitter in tonic waters and carbonated drinks. The main *Cinchona*-producing countries are Indonesia, Zaïre, Tanzania, Kenya, Rwanda, Sri Lanka, Bolivia, Colombia, Costa Rica and India. These countries produce approximately 400-500 tonnes of alkaloids obtained from about 8000-10000 tonnes of bark produced annually.

At one time the use of quinine as an antimalarial had gone down and production of *Cinchona* bark as well as the alkaloids had decreased considerably. However, because of the development of resistant strains of the malarial parasite, use of quinine as an antimalarial has increased recently and it is expected that with the re-appearance of malaria in Asia and Africa, especially because of resistant strains to synthetic antimalarial compounds, the use of quinine will increase in those countries. Similarly, use of quinidine as antiarrhythmic compound will also increase with the development of better health facilities in the developing countries of the world.

Plants Containing Cardiac Glycosides

A number of plants are used as a source of cardiac glycosides. Out of these, *Digitalis* species are the most important. The two important species used in medicine are *D. purpurea* L. and *D. lanata* Ehrh. The use of *D. purpurea* is very limited and in most of the developed countries glycosides are obtained from *D. lanata*. The latter species contains a number of glycosides of which digoxin and lanatoside-C are important. Both these are used as cardiotonics. It is estimated that approximately 1500 kg of digoxin and 100-200 kg of digitoxin are being used in the world today. The major *Digitalis*-producing countries are U.S.A., United Kingdom, the Netherlands, Switzerland, East Germany and the U.S.S.R.

Licorice (*Glycyrrhiza* spp.)

Licorice is obtained from dried roots and rhizomes of *Glycyrrhiza glabra* L. var. *typica* Reg. & Herd., called Spanish licorice, *G. glabra* L. var. *glanduliflora* Wald. & Kit., called Russian licorice, and *G. glabra* L. var. *violacea* Boiss., referred to in commerce as Persian or Turkish licorice. Species of licorice are herbaceous shrubs indigenous to southern Europe and Asia Minor. Most of the supply comes from the U.S.S.R., Spain, Turkey, Syria, Iraq and Afghanistan. A major portion of the commercial supply comes from wild sources and there is only a limited area under cultivation. The total production is around 40,000-50,000 tonnes per year.

The main active constituents of the drug are saponin-like glycosides of which glycyrrhizin (5-10%) is the most important. In the form of crude extract, it is used as an expectorant and anti-inflammatory as well as a demulscent to improve the flavour of various bitter preparations. Purified extract and carbenoxolone is also used for treatment of gastric and duodonal ulcers. Pure glycyrrhetic acid is used for the treatment of skin diseases as well as for peptic ulcers.

In addition to its medicinal use, licorice has a number of industrial uses. It is used as a flavouring agent in chocolates, carbonated bewerages and tobacco. It is also used to stabilise foam in fire extinguishers. The extract is also used as a flavour in beers, cocoa, pudding, cakes, etc. As a cosmetic, it is used in toothpastes, mouth-washes and breath-purifiers.

Serpent Wood (*Rauvolfia* spp.)

The drug is obtained from the roots and rhizomes of *Rauvolfia serpentina* (L.) Benth. ex Kurz. Although there are more than 24 alkaloids in the drug, the most important ones used in medicine are rescinnamine, reserpine and deserpidine. These three alkaloids are used as a hypotensive and tranquilizer.

Although exact data are not available, approximately 400-500 tonnes of roots are being exploited mainly in India, Thailand, Bangladesh and Sri Lanka. Because of its highly toxic nature, use of *Rauvolfia serpentina* has gone down considerably and a sister species, African serpent wood (*Rauvolfia vomitoria* Afz.), has been exploited much more for the world market. Approximately 800 tonnes of roots are exploited from wild sources in the western coast of Africa, mainly Zaïre, Mozambique and Rwanda, from where it is exported to Italy and West Germany. African serpent wood is much richer in *Rauvolfia* alkaloids. The roots of this plant contain a fairly large quantity of ajmaline, a compound which has become much more popular as a hypotensive agent than reserpine.

Ipecac (*Cephaelis* spp.)

Ipecac consists of the dried roots and rhizomes of *Cephaelis ipecacuanha* (Brotero) A.Richard, known as Rio or Brazilian ipecac, or *C. acuminata* Karsten, referred to as Cartagena or Nicaragua or Panama Ipecac. *C. ipecacuanha* is a perennial herbaceous shrub indigenous to Brazil, and most of the present supply of the drug comes from that country. It is also cultivated to a limited extent in India and Malaysia. *C. acuminata* is indigenous to Colombia, Panama and Nicaragua. The roots contain a number of alkaloids of which emetine and psychotrine are the most important. The roots contain 2-2.5% total alkaloids of which two-thirds are emetine. The drug is used in the form of a crude extract as emetic and expectorant. However, it is mostly used as the source of emetine, which is used to treat amoebic dysentery.

About ten years ago the demand of ipecac went down considerably because of the development of highly effective synthetic antiamoebic agents. However, because of continuous use of synthetic drugs, most of the strains of amoeba responsible for dysentery have developed resistance to these drugs and there has been considerable increase in demand for ipecac during the last 4-5 years. At present there is a worldwide shortage of the drug. All the cultivation of this drug is confined to West Bengal in India producing only 7-10 tonnes of the drug. Total production of ipecac in the world is approximately 100 tonnes per year, most of which comes from Nicaragua, Brazil and India.

Senna (*Cassia* spp.)

The drug senna is obtained from leaves and pods of *Cassia angustifolia* Vahl, called Indian senna, and *C. acutifolia* Delile, referred as Alexandrian senna. While Indian senna is cultivated in the State of Tamil Nadu, Alexandrian senna is exploited from wild sources in Sudan. At present approximately 4,000 hectares are under cultivation of this crop producing about 2,500 tonnes of leaves and 500 tonnes of pods per year. Pods and leaves of both the species contain glycosides, called senna glycosides or sennosides. These glycosides are used mostly as laxatives all over the world. About 700-800 tonnes of leaves and pods of Alexandrian senna are exploited in Sudan.

Periwinkle (*Catharanthus roseus* (L.) G.Don)

The drug is obtained from the leaves and roots of *Catharanthus roseus* (L.) G.Don. The plant is indigenous to Madagascar and is being cultivated in India, Madagascar, Israel and the United States.

Roots and leaves of the plant contain more than 100 alkaloids of which ajmalicine from the roots (raubasine) is important in medicine and is used as a vasodialator for controlling auricular fibrillation. The two leaf alkaloids which are most important in medicine are vinblastine and vincristine, which are used in the treatment of cancer. Vinblastine sulphate is used in the treatment of Hodgkin's disease. It is also employed for treatment of lymphosarcoma, choriocarcinoma, neuroblastoma, carcinoma of the breast, lungs and other organs and in acute and chronic leukaemia. Vincristine sulphate is used particularly to treat acute leukaemia in children and lymphocytic leukaemia. It is also used against Hodgkin's disease, Wilm's tumour, neuroblastoma, rhabdosarcoma and reticulum cell sarcoma.

Ammi majus L.

The drug is obtained from seeds of the plant *Ammi majus* L., which is a native of Egypt. The plant has been recently introduced in India. Most of the drug is produced in Egypt and India and the total world production is approximately 700-1000 kg of xanthotoxine. Xanthotoxine is used for the treatment of leucoderma and as a component of suntan lotion.

Berberis spp.

The drug is obtained from roots and rhizomes of various species of *Berberis* and *Mahonia*. Roots of the plant contain 1-2% berberine. About 600-700 tonnes of roots of the drug are exploited in India.

Berberine is used extensively for the treatment of tropical diarrhoea in Japan and other South East Asian countries. To a certain extent berberine is also used for controlling certain eye diseases. Production of berberine is erratic in India, ranging from 2 to 7 tonnes per year.

Papaya (*Carica papaya* L.)

Papain is a most common peptolytic enzyme obtained from *Carica papaya* L. It is used in medicine to a limited extent. Papaya is cultivated as a fruit tree all over the tropical world. The fruits contain a proteolytic enzyme papain in the latex. The enzyme is produced by a number of countries which include India, Sri Lanka, Zaïre, Uganda, Kenya, Philippines, Cameroun and the Caribbean Islands. The African countries Kenya, Zaïre and Uganda have been the major producers and suppliers of the enzyme to the world market. Papain, the most common proteolytic enzyme used in medicine, is employed as a digestive enzyme for treatment of dyspepsia, intestinal and gastric disorders. This enzyme is also used for the treatment of diptheria for dissolving the diptheric membrane. It is often used in surgery to reduce incidence of blood clots where thrombomiasma is undesirable. It is also used for local treatment of buccal, pharyngeal and laryngeal disorders.

However, only a small portion of the enzyme is used in the pharmaceutical industry and most of it is used in the food industry. One of the most common uses of papain is as a meat and fish tenderizer. For this the most common method used is injecting the beef animals with papain just before slaughtering. This method is used to a large extent in the U.S.A. It is also used in extraction of meat and fish protein. In brewing, the enzyme is used for removing haze from beers. In the leather industry it is used as a bating compound. In the textile industry it is used for degumming silk. It is also used in laundry and dry cleaning as a constituent of washing powders.

Total production of papain is approximately 500 tonnes annually. The present demand is mainly from countries like the U.S.A., Japan, Germany, France, England and the Netherlands.

Some New Potential Plants for Modern Medicine

During the last 5 years researches carried out throughout the world have resulted in some new candidates to be adopted in modern medicine. These include *Valeriana* species as a source of valepotriates, *Silybum marianum* as a source of silymarin, *Podophyllum* species as a source of podophyllotoxin, *Commiphora mukul* as a source of saponins, and *Artemisia annua* as a source of artemisinin.

Valepotriates are obtained from the roots and rhizomes of *Valeriana officinalis* L., found growing wild in Europe, and *V. wallichii* DC., found growing wild in Asia. Major producing countries are Belgium, France, the Netherlands, U.S.S.R., China, East Europe and India. Both the species are used in pharmaceutical preparations as a sedative and flavouring agent and both contain similar constituents, including iridoid valepotriates and a sweet-smelling essential oil. The drug is used as an extract in the form of total valepotriates, which are used as sedatives and tranquilizers. However, the drug is mostly used in West Germany and has not been universally accepted.

Silymarin, a flavanoid obtained from seeds of *Silybum marianum* Gaertn., is used as antihepatotoxic agent in West Germany to a limited extent. Podophyllotoxin is obtained from *Podophyllum hexandrum* Royle, found growing in Europe, and *P. emodi* Wall., round growing in Asia. A derivative of podophyllotoxin has been found as a successful anticancer agent in the U.S.A. in the form of etoposide.

Total steroids, called guggulipid, obtained from the gum of *Commiphora mukul* Engl., a tree growing in India, has been found to be effective as a hypolipidaemic agent for lowering cholesterol in the blood and has been registered in India as a drug. It is also an anti-inflammatory substance used in various types of arthritis.

Artemisinin is obtained from the leaves and flowering tops of the Chinese plant Qing Hao (*Artemisia annua* Linn.). It has been found to be a very effective antimalarial, specially against cerebral malaria in China, and now it is being developed as a drug of choice for the treatment of resistant cases of malaria by the World Health Organization.

Adaptogenic Drugs

There are some drugs which have no specific pharmacological activity. However, they have been used for a long period of time as general well-being drugs to improve resistance and treat various kind of weaknesses. Now they have found a wide market in the world, although they are not adopted in modern medicine. The two most important plants which are in use today are *Panax* sp. (Ginseng) and *Eleutherococcus senticosus* (Siberian ginseng).

Ginseng is the best known of all the Chinese medicinal products and has been used in China, Japan and Korea for thousands of years as a well-being drug. The drug is obtained from two different species – a) *Panax ginseng* C.A.Meyer, which is indigenous to northern China and Korea, and is cultivated in China, Japan, the Republic of Korea and the U.S.S.R., and b) *P. quinquefolium* L., which is indigenous to Canada and the United States and is cultivated in the U.S.A.

Both the species contain a large number of saponins (triterpene glycosides). There are about 12 primary glycosides, known as ginsenosides, found in the two species which are supposed to be the active constituents. Ginseng is prescribed as a tonic, stimulant and aphrodisiac. It is also used in cases of neurastenia, dyspepsia and asthama. It is also incorporated in tonics for treatment of amnesia, headaches, convulsions and dysentery. It has been claimed to enhance the natural resistance and recuperative power of the body, having both stimulant and sedative activity. It is also called adaptogenic because it maintains the body stamina at a regular level without detrimental side effects.

Eleutherococcus senticosus M. is a common shrub growing in Eastern Siberia, certain parts of China and the Republic of Korea. It is supposed to have the same quality as the two species of Ginseng, but does not contain the same constituents. It has been used in the U.S.S.R. for the same purpose for which ginseng is used. The Republic of Korea, China and the United States are the main producers of Ginseng. Korea produces approximately 900-1000 tonnes of roots of Ginseng from an area of 1600 hectares. Today Korea exports Ginseng and Ginseng products worth approximately 300 million dollars. The United States is another major producing country producing approximately 200 tonnes of roots of Ginseng.

Essential Oil Plants Used in Medicine

Some of the most important essential oils used in medicine are Japanese mint oil (*Mentha arvensis* L.) as a source of natural menthol, Peppermint oil (*Mentha piperita* L.) as a source of Peppermint oil, Eucalyptus oil (*Eucalyptus* sp.) as a source of Eucalyptus oil and cineol, Cinnamon leaf oil (*Cinnamomum zeylanicum* Blume) as a source of eugenol and isoeugenol, and oil of *Cinnamomum camphora* Nees & Eberm. as a source of natural camphor. The most important of these is the Japanese mint oil as a source of natural menthol, a chemical which is used extensively in a number of pharmaceutical preparations. Approximately 4000 tonnes of Japanese mint oil and 2000 tonnes of menthol are produced in the world today. The major producing countries are Brazil, Paraguay, Taiwan, Japan, China, India and Thailand. The total production of Peppermint oil is approximately 3000 tonnes, of which a small portion is used in medicine. The major producers of Peppermint oil are the U.S.A., the U.S.S.R., Italy, East and West Germany, the U.K. and India. The major producer of natural camphor is Taiwan, while the major producers of Eucalyptus oil are Australia, India, Spain and Portugal. Cinnamon leaf oil is produced in Sri Lanka and India.

Table 1. Important Active Constituents Used in Medicine

Active constituents	Plant raw material	Pharmacological activity
1. Steroidal hormones (synthesised from diosgenin, hecogenic or solasodine	*Dioscorea* sp., *Agave* sp., *Solanum* sp.	Anti-inflammatory, antiarthritic, hormonal
2. Morphine, codeine, papaverine	*Papaver somniferum*	Sedative, antitussive, smooth muscle relaxant
3. Quinine, Quinidine	*Cinchona* sp.	Antimalarial antiarrhythmic
4. Hyoscyamine Hysoscine Atropine	*Datura* sp. *Hyoscyamus muticus* *Duboisia* sp.	Parasympatholytic
5. Digoxin, lanatosides	*Digitalis lanata*	Cardiotonic
6. Reserpine Rescinamine Deserpidine	*Rauvolfia serpentina* *R. canescens* *R. vomitoria*	Hypotensive Vasodialator
7. Ajmaline	*R. vomitoria*	
8. Ajmalicine	*Catharanthus roseus*	Vasodialator
9. Vincristine, Vinblastine	*C. roseus*	Anticancer
10. Caffeine	*Camellia sinensis*	CNS stimulant
11. Cocaine	*Erythroxylum coca*	Anaesthetic
12. Ephedrine, Pseudoephredine	*Ephedra* sp.	Sympathomimetic
13. Philocarpine	*Philocarpus jaborandi*	Parasympathomimetic
14. Emetine	*Cephaelis ipecacuanha*, *C. acuminata*	Antiamoebic
15. Ergometrine, Ergotamine, Ergotoxine.	*Claviceps purpurea*	Oxytocic, vasoconstrictor, vasodialator
16. Psyllium mucilage	*Plantago ovata*	Laxative
17. Vincamine	*Vinca minor*, *Voacanqa africana* (obtained from tubersonine)	Vasodialator
18. Glycyrrhetic acid	*Glycyrrhiza glabra*	Antiinflammatory
19. Sennoside	*Cassia angustifolia* *C. acutifolia*	Laxative.
20. Berberine	*Berberis* sp.	Antidiarrhoeal

Table 2. List of Active Constituents with Future Potential in Medicine

Active constituents	Plant raw material	Pharmacological activity
1. Valepotriates	*Valeriana* sp.	Sedative, tranquilizer
1. Silymarin	*Silybum marianum*	Antihepatotoxic
3. Podophyllotoxin (Etopsoside)	*Podophyllum* sp.	Anticancer
4. Guggul saponins	*Commiphora mukul*	Hypolipidaemic
5. Artemisinin	*Artemisia annua*	Antimalarial

Table 3. List of Plants used as Crude Extracts in Medicine

1. Belladonna	*Atropa belladonna*
2. Ipecac	*Cephaelis ipecacuanha*
3. Opium	*Papaver somniferum*
4. Henbane	*Hyoscyamus niger*
5. Stramonium	*Datura stramonium*
6. Cascara sagrada	*Rhamnus purshianus*
7. Liquorice	*Glycyrrhiza glabra*
8. Rhubarb	*Rheum officinale* *R. palmatum*
9. Valerian	*Valeriana wallichii*
10. Podophyllum	*Podophyllum peltatum* *P. emodi*
11. Capsicum oleoresin	*Capsicum annuum*
12. Digitalis	*Digitalis purpurea*
13. Aloe	*Aloe* sp.

Table 4. List of Important Essential Oils and Terpenes used in Medicine

1. Japanese mint oil (source of menthol)	*Mentha arvensis*
2. Peppermint oil	*Mentha piperita*
3. Eucalyptus oil	*Eucalyptus globulus*
4. Aniseed oil	*Pimpinella anisum*
5. Clove oil	*Eugenia caryophyllata*
6. Cinnamon leaf oil	*Cinnamomum zeylanicum*
7. Lemongrass oil (As a source of Vitamin-A)	*Cymbopogon flexuosus*
8. Camphor	*Cinnamomum camphora*

References

Anonymous (1982). *Markets for selected medicinal plants and their derivatives*. Geneva: International Trade Centre, UNCTAD/GATT. 206 pp.

Balandrin, M.F., Clocke, James, A., Wurtele, E.S. & Bollinger, W.H. (1985.) Natural plant chemicals: sources of industrial and medicinal materials. *Science*, 228, 1154-1160.

Farnsworth, Norman (1984). How can the well be dry, when it is filled with water? *Econ. Bot.*, 38, 4-13.

Farnsworth, N. & Soejarto, D.P. (1985). Potential consequence of plant extinction in the United States on the current and future availability of prescription drugs. *Econ. Bot.*, 34, 231-240.

Ryler, V.E. (1986). Plant drugs in the twenty-first century. *Econ.Bot.*, 40, 279-288.

Industry and the Conservation of Medicinal Plants

A. Bonati,
Inverni della Beffa, Milan

Medicinal plants and their parts are a primary source of products for the pharmaceutical industry. Large quantities are used in the preparation of infusions and decoctions both in developing countries, where traditional medicine is still of great therapeutical, social and economic importance, and in the industrialised countries, where an ever-growing proportion of the population is using crude drugs for self-medication. However, an even greater amount of medicinal plants are used by the pharmaceutical industry in the preparation of a wide spectrum of derivatives ranging from traditional extracts through extracts with a high and standardized content of active constituents to chemically pure products, the latter being used directly as medicaments or as starting materials for the manufacture of other, pharmacologically active, substances.

The experience gained over the last few decades in the role of a privileged observer, a role granted to a company producing medicinal plant derivatives, enables us to make some observations:

- In the EEC countries all the proprietary medicinal products – prescription drugs and Over-the-counter drugs – must be validated by 1990, whatever the origin of their active constituents, in order to comply with the new EEC rules. Even if some countries, such as France and Germany, will restrict the validation of widely used traditional medicinal plants and extracts to the analytical field, certainly many proprietary medicinal products containing 3, 4 or more drugs or extracts as active constituents will be withdrawn from the market as the scientific and economic efforts to update their chemical and pharmaceutical documentation are not justified by their limited commercial importance and, often, by the lack of a therapeutical rationale. A contraction of the consumption of some medicinal plants is foreseeable, but this is of little importance because the products are already scarcely used.
- On the contrary, these same updating requirements have persuaded the industry to promote a whole series of pharmaco-toxicological,

clinical and analytical studies to be conducted for proprietary medicinal products of great commercial interest, in order to provide scientific proof of their traditional activities and to assure their quality consistency. As a result, many medicinal products containing crude drugs, extracts as well as pure products of plant origin, will remain on the market, viable on a sound basis, in fields of remarkable therapeutical and commercial importance – cardiovascular diseases, dermatology, laxatives, mild sedatives, liver and blood vessel diseases, cough products.

- The launch into the market of "new" natural products has produced significant therapeutical and commercial results. Among the most well known are the antileukemic alkaloids from leaves of *Catharanthus roseus*, etoposide, the anthocyanoside complex of *Vaccinium myrtillus* fresh fruits, the escin from horse-chestnut seeds, silymarine from *Silybum marianum* seeds, the proanthocyanidin oligomers from grape seeds, the triterpene derivatives from *Centella asiatica*, the liposoluble complex of *Prunus africana* bark, the liposoluble complex of *Serenoa repens* seed, the diosmine from Rutaceae, the standardised *Ginseng* or *Ginkgo biloba* extracts, etc. Not to mention the now well-established reserpine, ajmaline, ajmalicine, vincamine, enoxolone and carbenoxolone, thiocolchicoside.
- Researches aimed at isolating new active constituents from medicinal plants used in folk medicine are under way all over the world; it is realistic to believe that some of them will be successful and that new natural substances will be marketed in the near future.
- The use of natural derivatives has increased in industries other than the pharmaceutical one, especially in the cosmetic, dietetic, alcoholic and soft drink industries. New products have appeared here too, for example the steviosides from the leaves of *Stevia rebaudiana*.

These general observations lead to two considerations:

- Today, the use of medicinal plants is enormous even if it is difficult to quantify. It will certainly increase further, extending also to less traditional fields than that of pharmaceuticals.
- Whatever the field of application, all natural derivatives (pure products, extracts and crude drugs) must satisfy one definite requirement: they must be of a standardised quality, because only a standardised quality can ensure the consistency of their pharmacological and/or organoleptic properties. If the concept of standardisation for pure products is equated with that of purity, for crude drugs and extracts, standardisation is the result of a series of parameters, i.e. origin of the

drug, period of harvest, identification and quantitative determination of the active or characteristic constituents, chemico-physical properties (total solids pH, etc.), external contaminants (toxic metals, weed-killers, pesticides), micro-biological content and potential impurities.

These two considerations – increasing demand and quality requirements – lead to one conclusion: the use of medicinal plants from the wild, inevitably heterogeneous and often of limited availability, must be abandoned or limited as far as possible. In their place, cultivated plants must be used or alternative technology must be applied.

There are many advantages to be gained from this logical and inevitable decision:

a) In ecological terms, the continued and indiscriminate use of wild-collected plants may cause incalculable damage to the environment, fauna included. This is always true: the danger is still more serious when parts of medicinal plants such as the roots, seeds and flowers – which are essential to the survival of the plant itself – are used. An example is the root barks of *Rauwolfia vomitoria*, the main source for the manufacture of reserpine and ajmaline.

b) In terms of quality, wild plants are often heterogeneous in terms of age, zone and period of picking, drying methods and, consequently, active constituent content.
   With cultivated plants, on the other hand:
   - Cultivation can be carried out under homogeneous climatic and soil conditions;
   - The age of the plants and exact information on their active constituent content, and therefore on their period of harvesting, are readily available;
   - The plants can be picked "at the right time" and quickly;
   - They can be dried under controlled conditions of time and temperature. We must never forget that a homogeneous and correct drying often represents the most delicate and essential step in the entire manufacturing process.

   Also, cultivated plants are usually no more expensive than wild-collected ones.

c) The continuous growth in the use of many plants can no longer be satisfied in terms of quantity by wild-collected plants. This is a quite separate consideration from the problems of ecology and quality mentioned above. Cultivation, preferably in areas having the same climate and soil as the natural habitat, is the inevitable and realistic alternative. In many cases, this may prove to be a profitable and important agricultural activity not only for developing countries, and could be the starting point for local production of basic industrial products (such as extracts and essential oils) having an added value, to replace the export of plants.

d) Correct selection and fertilisation techniques may improve the cultivation of medicinal plants both in quality and quantity. This is also extremely important from the economic point of view as it allows better control of the cost of natural derivatives than in the case of wild-collected plants whose availability, quality and yield are unpredictable.

e) There is one disadvantage in selective and improved cultivation of medicinal plants. The use of weed-killers and pesticides may represent a risk to the environment and to people if precise regulations are not rigidly followed. Therefore, it is essential that their use in terms of both quantity and frequency is limited to what is strictly necessary and all the controls are put into effect to minimise the risk of pollution, especially to the water supply.

If these are the logical conclusions on the use and production of medicinal plant derivatives, what then is the true situation? Is there a greater use today of wild-collected plants or of cultivated plants? Does the cultivation of medicinal plants represent the only alternative to wild-collected plants, and are medicinal plants, whether cultivated or wild, the only raw material for the preparation of natural derivatives?

I believe that the problem must be looked at from two points of view: traditional medicine and industrial uses.

Particularly in the developing countries, the use of wild-collected plants in traditional medicine is today virtually unavoidable; nevertheless, it is to be hoped that even in such cases everything possible should be done to institute controls to ensure the identity and quality of the drug. Bearing in mind the importance of traditional medicine in these countries, both in a medical and social context, the ideal situation would be to have centres which could undertake the gathering, checking, preparation and wholesale distribution, and which would be able to provide these quality guarantees.

On an industrial basis, today a very large proportion of the medicinal plants used comes from cultivation: the plants are usually grown in their natural habitat where organisations have been set up to cultivate, collect and dry large quantitities of these plant materials. The plants used have not usually been selected for particular characteristics.

There are numerous cases (e.g. *Digitalis* and Solanaceae) of selective cultivation where, alongside the improvement of content and quantity, there have been successful results in modifying the "natural" composition, aiming at the biosynthesis of the active ingredients of greatest industrial importance (digoxine, hyoscyamine and scopolamine) at the expense of other components. It seems obvious from what we have said briefly on the subject of selective cultivation that the process is significant and useful above all for plants to be used in the production of particular active constituent in pure form and for which the plant is selectively cultivated.

Cultivation, on the other hand, is the *only* alternative to collecting plants from the wild for the preparation of total or purified extracts which have a complex composition, probably essential in total to obtain the therapeutical or organoleptic properties characteristic of that drug.

There are new opportunities for the production of natural derivatives by using special techniques providing interesting production methods: fermentation (ergot alkaloids), semi-synthesis or total synthesis (vincamine, ajmalicine, thiocolchicoside), cell or tissue cultures (digoxine, tropane alkaloids, ginsenosides). In the main, these technologies – except for semi-syntheses which use as intermediates plant materials – respect the environment in that they avoid the exploitation of medicinal plants.

However, these technologies are still very expensive and there are only a few areas where they can be applied as a realistic alternative to medicinal plants, which are consequently still the best source of natural derivatives. This is the reason why, in a growing market, with highly specific quality standards and where respect for the environment is a fundamental pre-requisite, the cultivation of medicinal plants must increasingly provide a substitute for the uncontrolled exploitation of wild plants.

Information Systems and Databases for the Conservation of Medicinal Plants

Hugh Synge
Plants Programme Consultant, WWF[1] and
Vernon Heywood
Chief Scientist, Plant Conservation, IUCN

Introduction

As the papers in this book show, there is a great deal of very diverse work on medicinal plants, and it is generating an ever-increasing amount of information. The work is remarkable in that it spans a wide range of disciplines and it is being done by a very widely scattered set of researchers, from the medical, chemical and botanical fraternities. It is a field in which there is little co-ordination of effort and in which its practitioners tend to work in isolation.

It is vital, therefore, to find a way of bringing the resulting information together and fitting it into a coherent whole, so it can be made widely available to those who need it. The pharmacognosists need to know the correct identity of the plants they study; the medical users need to know the correct formulations; the conservationists need to know which plants to conserve.

In today's world, this requirement can best be met by the development of a co-ordinated network of computer databases. Computers have been used in science for at least a generation and for handling bibliographies for almost as long. However, their use in databases on individual plants is much more recent. In botany, most of the earliest databases started in the 1970s, for example the (then) IUCN Threatened Plants Database (started in 1974-5), NAPRALERT (1975), and the Flora de Veracruz (1974-5). Even so, most botanical databases were created only very recently and are at an early stage of their evolution. We are now at a pioneering stage where the norms and standards for the next generation are rapidly being set.

Although much has been learnt, there is still much to discover. One of the clearest lessons so far is that databases are expensive to make and,

1. Since 1989, freelance consultant (49 Kelvedon Close, Kingston-upon-Thames, Surrey KT2 5LF, U.K.)

with rare exceptions, have to be subsidised. It has often been possible to find money for design and initiation, but rarely for maintenance and updating. In both conservation and botany, the users of a mature database usually do not pay enough to cover maintenance of the database, let alone pay back the historic cost of creation. So WHO, IUCN and WWF, and their partner agencies, should be cautious about starting a major initiative on medicinal plant databases unless long-term funding can be assured.

A second lesson emerging more strongly each year is the benefit of networking. Although a central co-ordinating facility is needed, much of the data can be gathered locally, using PCs. This can reduce costs, as many national or specialised research institutes may offer to build sections of the database at their own expense, providing standards are set to ensure compatibility. In return they could receive access to the whole dataset.

In this paper, we outline the various components of a possible database network on the conservation of medicinal plants, concentrating not so much on the details of implementation, but on the principles involved. Our own knowledge is on the botanical and conservation aspects, rather than on the medical side, and so a similar review is needed on the medical and pharmacological aspects.

Guidelines for databases

It comes as a surprise to many that the design and management of databases are not self-evident. Experience shows that databases need constant discussion and rigorous thinking at every stage to be effective. More than most human endeavours, they seem surprisingly difficult to run and can easily disintegrate into chaos.

Frost-Olsen and Holm-Nielsen (1986), creators of the exemplary Flora of Ecuador database, have proposed some guidelines for newcomers to databases. We take them and extend them as follows:

1. **Keep it simple.** This is the most important maxim of all, and is so often forgotten. Over-complexity is the main reason why some databases become white elephants or, at best, reach a stage of arrested evolution. Unless the computer programs are written in a very structured and exemplary way, the more complicated they become, the more difficult it is to change them, even on trivial matters, and the greater the risk of one change upsetting other features. Designers should realise that every new feature added, if it extends the program

code, may make future changes, even unrelated ones, more difficult, not less so.

2. **Design it first.** IBM estimate that it is 1000-times more expensive to correct a fault in database design after construction of the database than before. It is vital to work out the correct hierarchies and relationships of the data elements, and then to make sure that "The World of Records" (the database) is a true reflection of "The World of Plants" (the entities themselves). The more the datafile structures do this, the easier it will be to develop and extend the database in future.
3. **Build up the expert knowledge needed.** Managers and creators of databases do not need to be computer programmers, but they need to understand how computer programs work and so have a feel for what will be easy and what will be difficult to programme. Only then can they evaluate the trade-offs that are inherent in any database design.

   Indeed, the arrival of new software packages in which to make databases, such as Advanced Revelation, means that the computer programmer's work is much reduced and that much of it can now be done by the scientific user or technician. But for a complex application, advice at the beginning from professional system designers can be crucial in choosing the best file structure. This is just as true with the new user-friendly packages as with the earlier computer languages. There is an almost infinite number of ways to programme any application, and it makes a massive difference which one you choose.
4. **Work out the objectives and scope of the database before choosing the hardware.** It is still surprising how often colleagues ask advice on whether or not they should buy some piece of hardware before they have decided what they want to do. Keep an open mind on hardware for as long as possible.
5. **Use standard hardware and software.** It is a great temptation to re-invent the wheel. The commercially available software packages or ones produced by colleagues may not be perfect, but one should remember the massive investment that has gone into them. Even if one could design a better package or program, is one prepared to maintain it and keep it up-to-date in perpetuity?
6. **Allow for future expansion.** History shows that databases often expand and grow faster and further than their users imagine. One of us (HS) started a datafile on threatened plants of Europe in 1975, little imagining that it was the beginning for a world conservation database with hundreds of data files, thousands of text files and hundreds of thousands of records! We had hardware that allowed this natural expansion, but we rightly only added new fields and new features to

the database when we needed them. The step-by-step approach is paramount.

7. **Let the machine do the hard work.** Computing power is cheap, programming power is expensive. Therefore, do not be tempted into very complex programmes and data structures to save the computer work and to give tiny improvements in response time. For example, if you want to store a number of items that are either "yes" or "no", store them as a single byte each, filled with either "Y" or "N", rather than be tempted to store them as single bits, eight to a byte, which would save a little space but require much more programming and in most cases reduce the ease of interrogation of the database.

   Above all, avoid codes that the users have to remember. If a user wants to retrieve data on a plant, let him or her do it by the plant's name, not by some complex code number. It may well be that the system uses codes internally to relate one file to another, but they should be hidden from the user. Conservation computer expert Duncan Mackinder has made this point very clear.
8. **Plan for good quality output.** The almost illegible and repetitively laid-out printouts that one still receives from many databases are simply the result of lack of planning and programming. In the Threatened Plants Unit, investment in a family of flexible, easy-to-read output formats for the most common sets of data extracted (e.g. taxa with country distributions, taxa with botanic garden records) has proved invaluable. With modern computer-based publishing techniques, the output can be greatly enhanced at little extra cost.
9. **Plan for networking, even if you do not use it initially.** Networking is the concept of the future in computing, and is the subject of the next section.
10. **Follow the data standards that have been agreed in your subject.** This is also so important that we have made it the topic of a section in its own right, placed after the one on networking.

The benefits of networking

The best way to set up a global database on a diverse subject such as medicinal plants is to find ways of encouraging individuals and institutions around the world to assemble components of the database, either country by country, or topic by topic. The parts can all be linked together through a co-ordinating centre. This is much better than trying to gather all the data manually at one place because:

- It brings far greater resources and funding to the task of gathering the data;
- The information is more detailed, more up-to-date and more pertinent to local needs;
- The information is far more likely to be used by policy-makers if it can be obtained locally rather than from a single database in a distant country.

Institutes working on medicinal plants probably want to make databases on the results of their work anyway and would welcome the chance to do it as part of a worldwide project. They would be saved the time of designing the database and writing its software, and would also gain by receiving access to the other databases in the project. The main requirement, of course, is that they first agree among themselves the content of the databases.

Our experience in running the database of the (then) IUCN Threatened Plants Unit (HS during 1984-87, VH during 1987-88) showed us the benefit of networking all too clearly. We found that the members of IUCN – the conservation organisations of the world – wanted IUCN to provide them with database standards and expertise, rather than with conservation data. As a result, we tried to shift the role of the Unit towards assisting, promoting and co-ordinating data-gathering by IUCN's constituency rather than gathering all the data ourselves. This meant that the task of identifying which species and areas most need protection worldwide suddenly became achievable in a far shorter time than had ever been imagined before. And our jobs became much more enjoyable and satisfying.

One of the best examples of networking in action comes from The Nature Conservancy (TNC), a conservation group which has set up State Heritage Programs in virtually all the states of the United States. As a basis for planning purchase of key wildlife sites, the staff in each State assembled a database to a strictly controlled set of guidelines. The organisation's headquarters in Arlington, Virginia, maintains an overview of the situation, derived from all the databases that have been made. In recent years, the Conservancy has been sponsoring parallel databases in Latin American countries, a much more difficult task but of vital importance for conserving biological diversity. They have more than proved the validity of the networking concept in nature conservation, and are seen as the leaders in the field. The existing data centers, especially in Latin America, could play a vital role in gathering data on medicinal plants in nature, and have the great advantage that their data are much used by conservation groups.

Another interesting model is that of the Center for Plant Conservation (CPC), another U.S. organisation. The Center's Kerry Walter has prepared a very popular computer program called BG-BASE, which botanic gardens can use to maintain records of the plants they grow. An interesting feature is that when CPC installs the database for a Garden, it can add new features specially designed for that Garden's needs, without changing the core of the database or program, or reducing its compatibility with international standards for botanic garden records. Some computer experts believe that this approach may lead to problems in the longer term, but it has certainly made the program very popular with users.

Another excellent model, this time from the botanical fraternity, is the legume database ILDIS, which is discussed below.

In a networked system the central database does not disappear but becomes an *overview* of the other databases. Its aim is to hold the minimum level of information needed, with pointers to where more detailed information can be found, rather than try to hold every fact relevent to the subject in hand. It will then be used primarily for:

1. Answering enquiries that require a comparison of data between the contents of the local databases, e.g. from one country to another;
2. Acting as a clearing-house for requests, passing them on to the more specialist databases;
3. Providing start-up packs to new databases, to avoid extensive re-keying of information (e.g. a datafile on the countries of the world, or of all the accepted plant genera).
4. Filling gaps in the global and thematic coverage.
5. Providing a test-bed on which to develop the systems and standards needed. One of the clearest lessons of database design is that complex master plans devised without real data usually fail. One has to move ahead slowly, step by step, trying out each component on real data and developing each field one by one.

The staff of the central co-ordinating facility should also be responsible for:

1. Co-ordinating, and if necessary leading, the effort to agree the common data standards that are needed (see below);
2. Encouraging and helping organisations and individuals to set up the feeder databases needed;
3. Providing training facilities for the staff of the feeder databases;

4. Providing simple to use and well-tried software packages to implement the system.

The provision of standards

If databases are to be compatible or if different modules are to fit together, there must be common standards and agreed mechanisms to exchange data. And for the standards and mechanisms to work, they must be rigorously designed and freely approved by the users involved.

The best way to achieve this is by informal working groups or consortia of institutions, meeting under an international aegis. The temptation should be strongly avoided to exclude anyone with whom one does not agree: doing so usually guarantees the creation of a rival set of standards, which invariably prevents any one system, however superior it may be, from becoming widely accepted.

Agreement should be principally on the data rather than on hardware or software. It is certainly neither useful nor necessary to try and standardise hardware – the market is developing far too rapidly for that. Nevertheless, most systems will be designed with the IBM-compatible PC in mind, a machine that most institutes can afford and which can be used for a multitude of other tasks as well.

One can choose to standardise software but it is not necessary to do so. Ideally, a little competition among software producers may be beneficial. But it is worth remembering that our best models of networking in conservation – TNC, CPC and ILDIS – do provide a standard software package.

The absolutely essential points of standardisation are data content and form. By data content, we mean the elements of information to be covered by the database, e.g. that the database will include the name of the plant, its distribution by countries, etc. By data form, we mean the way in which each of these items can be produced by the computer, e.g. that the plant name is separated into Genus, Species and Infraspecies. (In a simple system this is often how it can best be stored in the computer as well.) The rules on data form can be lengthy and complex, requiring judgement on many contentious issues.

The resulting data standard should:

- Explain the meaning of each data item and outline the rules that pertain to it;
- Show how each data item can be categorised and stored, both in manual systems and in a computer;

- Fit each data item into an overall recommendation on database design.

Part of the standard can be expressed in the form of an International Transfer Format. This is a detailed blueprint for how one organisation can transfer data to another on the topic concerned, irrespective of the physical medium used.

The first such International Transfer Format developed was for records of plants grown in botanic gardens and was approved in October 1987 (Botanic Gardens Conservation Secretariat, 1987). It enables botanic gardens to share data on their holdings with each other. It will also enable the Botanic Gardens Conservation Secretariat to monitor the collections around the world and lead towards a position where all the collections are seen as part of a single entity. Such a model may well be appropriate for databases on medicinal plants.

A more sophisticated approach is that taken by Kew's Bob Allkin in the development of XDF, which is to be a general purpose exchange language for any biological database – irrespective of the data held.

The Taxonomic Framework

The heart of any database on medicinal plants must be the basic framework on taxonomy – the names and relationships of the plants concerned.

Fortunately, botanists have been very active in the last few years in setting standards. They know all too well that no one organisation on its own can achieve their prize of a database on all the plants of the world but that if they work together, it could be achieved.

In 1985, a group of botanists formed the Taxonomic Databases Working Group (TDWG) to develop a common approach to data standards. It started as an informal consortium of leading taxonomic institutions, but has since adopted a formal constitution, become the IUBS Commission for Taxonomic Databases and has changed its name to the International Working Group on Taxonomic Databases for Plant Sciences (although its short name is still TDWG). It holds annual meetings, which grow larger and longer each year as more institutions and projects join in the process. One of us (VH) was Chairman of TDWG from its beginning until 1989.

There are two TDWG standards at present which are crucial to the taxonomic framework for a database on medicinal plants:

- The Plant Names module of the International Transfer Format for Botanic Garden Records. Designed for horticultural use, this covers all the different types of scientific names for plants, including cultivars and hybrids, but does not go into the botanical complexities of synonymy. This is probably what is needed for a database on medicinal plants, and so this standard could be a vital component.
- "Names in botanical databases". This is a more generalised version of the above standard, and will form its natural successor. However, it is not yet complete.

These standards have to contend with substantial problems that are not always appreciated outside the world of taxonomic botany:

1. Biological classification is a nested hierarchy. One taxon can only belong to one single higher taxon at any one time, but one higher taxon can have more than one lower taxa under it. In current practice, however, the number of levels used is small:
   - Family
   - Genus and species, used normally in combination
   - Infraspecies.

   (For plants, the International Code for Botanical Nomenclature recognises several infraspecific ranks – subspecies, variety, form – theoretically each at different levels, but increasingly only one is used as is customary in zoological taxonomy.)

   It is strongly recommended to follow these basic levels, which are quite adequate for all practical purposes. Various efforts to create hierarchies on the computer with an unlimited number of levels have all failed, at least in the biological field, and are not necessary anyway.
2. There is no agreement as to the definition of the various levels of taxon and there are basic disagreements as to the rank to be given to particular groups, ranging from Monocotyledons and Dicotyledons to the major subdivisions of the legume family. This leads to problems of synonymy.
3. Synonymy – more than one name for the same plant – is a major problem. There are estimated to be around 1,000,000 names available for an estimated 250,000 species of higher plants! Although many synonyms are not in current use, many others are. In botany, synonyms are of two kinds:
   a) Nomenclatural synonyms, where there is no doubt on the identity of the taxon but it has been placed by taxonomists in different genera or treated at various ranks. The different names it has received because of this are automatically synonyms of each

other. Which rank should be given to a plant is, however, a matter of taxonomic judgement.

b) Taxonomic synonyms, where plants which have been independently described and named are subsequently considered by taxonomists to be so similar that they should be placed together.

In the past some botanists caused havoc by strict interpretation of the rules, which became ever more complex over the years, so as to overturn well-known names for common plants, but this has been widely criticised and botanists are increasingly emphasising stability over an exact following of the rules.

A way round this problem of synonymy is for users to follow the names as used in a standard Flora or Checklist. For example, the Council of Europe has recommended that all work on the conservation of endangered plants in Europe follow the names used in *Flora Europaea* (Tutin *et al.*, 1964-1980).

4. Accurate identification is also a problem. Identifications by non-taxonomists often result in a high proportion of mis-identifications, even in common and easily distinguished plants. In one classic paper on *Digitalis* (foxglove) chromosomes, about half the species were mis-identified, greatly reducing the value of its results. A great problem with medicinal plants is that so much of the information is taken from the literature and there is no way of checking the identity of the plants concerned. Ideally each identification should be vetted by a taxonomist before it is accepted, but this can be difficult or impossible unless voucher specimens are deposited in an established herbarium. At the very minimum, the database should record who identified the plant and when.
5. The sheer biological complexity of many plant groups. Classification, and hence identification, is an inexact science and is especially difficult for many tropical groups which have not been thoroughly studied. On analysis many economic plants turn out to consist of more than one species. In other cases, not enough work has been done to permit identification of the species with any degree of certainty, and here the non-specialist would be rash to identify below the level of genus. But for medicinal plants, it is essential to go to the species and in some cases even the infraspecies as well (subspecies, variety, form, etc.)

These constraints all point to the critical need for more taxonomy and more training of taxonomists. Without adequate taxonomy, scientists in other disciplines of botany can easily waste their time. Yet the taxonomic profession is under greater threat now than it has ever been. Vigorous efforts to reverse its decline have not yet borne fruit.

A particular problem with tropical plants is that the leading experts and institutions are often thousands of miles away from where the plants grow, in Europe or North America, and often little is done to build the local capacity for taxonomy.

This rather bleak analysis of the state of taxonomy is a very serious constraint to the effective use of medicinal plants worldwide. There are, however, some encouraging signs. The Royal Botanic Gardens, Kew, for example have put much of the emphasis of their taxonomic work on generic classifications, working in particular on economic groups like palms, grasses and legumes. This means that non-specialist botanists around the world can at least assign their species to the correct genera. Several large taxonomic institutions, like Kew and Missouri Botanical Garden, are also emphasising the rapid floristic inventory of key sites, rather than the very detailed but slow production of regional Floras.

One of the most exciting initiatives is ILDIS (the International Legume Database and Information Service). Co-ordinated by F R Bisby at Southampton, England, the first stage of this project is to prepare a world checklist of legumes. Numerous individuals and institutions are participating and contributing, mostly using the ALICE software devised by R Allkin (now of Kew) and P J Wingfield. This software package, which runs on an IBM-compatible PC and is designed for those wanting to make species checklists, is now very sophisticated and available for use by biologists.

The ILDIS team have now merged the first of the various "feeder databases" into one, covering 18,000 taxa and 40,000 binomials. The "feeder databases" included African legumes (from Kew), European legumes (from Reading University) and American legumes (from Missouri Botanical Garden's Tropicos system). This is a wonderful example of what can be done.

Perhaps the most novel feature of the ILDIS project is its very democratic system of governance, with an Executive Board and a set of technical subcommittees. It may be time-consuming to run, but it does draw in a very wide representation of botanists from various disciplines. Although still at an early stage, we see it as the best model available so far on how to make a networked database in botany.

At the time of writing, following the success of ILDIS, negotiations are continuing over a World Flora, to be called *Species Plantarum*, an initiative proposed by the Royal Botanic Gardens, Kew. So far seven leading taxonomic institutions have signalled their willingness to join, and more are expected. Kew hopes that *Species Plantarum* will emerge as a co-operative effort to co-ordinate all taxonomic revisions worldwide.

The Plant Conservation Framework

Any database on the conservation of medicinal plants should include not only the names of the plants but also their distribution in nature, how rare or abundant they are, and where cultivated stocks or seed of them might be found. Fortunately good standards and precedents are available on all these matters.

Distribution by major geographical units.

TDWG (see above) has finalised a system of geographical recording units, due for publication in 1990. Developed by R K Brummitt of Kew and S Hollis of Southampton University, the system divides the world up into 600-700 units that cover all terrestrial parts of the world. Most countries are units in their own right, but large countries such as Australia and the United States are divided down to state level. Most islands or island groups are also units in their own right, since their flora often differs from that of their parent country. The units are chosen so that plant records that use it can be merged upwards into either a classification by political units or a classification by agreed botanical regions.

We strongly recommend that all plants in a medicinal plant database record the geographical units where each plant grows using this system. Of course, the database may need to subdivide some of the units further – a database of medicinal plants of Cameroun would, for example, want to record where in Cameroun each plant grew – but the TDWG system should provide the basic framework and should be followed exactly. In fact it follows closely on the geographical system used by the TPU, which has been applied successfully to over 50,000 different plant taxa.

Distribution by sites and protected areas.

Once the medicinal plants are assigned to their broad geographical distributions, it may be possible to record their presence in protected areas (national parks and nature reserves), so as to show where populations of them are conserved.

Under their Plants Programme, IUCN has sponsored the development of an inventory system by A Gómez-Pompa, now Director of the University of California's Mexus Programme. Once developed, this system will enable managers of protected areas to build databases on PCs with an inventory of their flora and fauna. No TDWG or IUCN standard has yet emerged, but there would be few if any intrinsic difficulties in making one.

Any standards in this area should follow the IUCN classification of protected areas into 8 classes, divided by function and degree of human use permitted. This has been used for many years by IUCN's Commission on National Parks and Protected Areas, who are the leading international arbiter in this area. Also any work on this topic should be closely linked to the activities of the Protected Areas Data Unit, part of the World Conservation Monitoring Centre (WCMC), which is a consortium of UNEP, IUCN and WWF. This unit monitors the status of protected areas all over the world, and now has records on around 20,000 of them. The larger and better known sites are further described in text files, which include a short description of the vegetation, and may name a few of the dominant or notable species.

Relationship between plant and place

Recording the distributions of plants – whether they be by countries or by sites – requires another standard that codifies the relationship between plant and place in each case. For example, one needs to specify whether the plant is native or introduced in the area, and to cover situations where the record of its occurrence may be a doubtful one. After much debate TDWG approved a standard to do this early in 1990. It is called POSS, short for Plant Occurrence and Status Scheme, and is due to be published in 1990. It was prepared by the staff of the Threatened Plants Unit, with help from many other experts and following earlier versions by one of us (HS). It should be a small but vital part of any medicinal plant database.

Cultivated status

Data on plants in cultivation, at least in botanic gardens, is more readily available than data on their presence in protected areas. For many years IUCN has monitored the presence of threatened plants in botanic garden collections, work now carried out by the Botanic Gardens Conservation Secretariat (BGCS). Their database now contains over 20,000 records for over 5,000 threatened plant taxa.

The arrival of the International Transfer Format for Botanic Garden Records (see above) and the fact that many gardens have now computerised their plant records, often using CPC's flexible BG-BASE software, means that BGCS will soon be able to update its database electronically rather than by hand. The botanic gardens that are members of BGCS would send a tape or diskette of their database, say once a year, and BGCS would screen it, updating records to their own database for the presence of all threatened species. In return the gardens may receive data on which of their plants are threatened (by comparison with TPU's master file) and in which other gardens each of those plants is grown. This way they can start to see and assess their plants as part of a global collection. Any operation on medicinal plants could work in a similar way.

Degree of Threat

Knowing where the plant can be found, in nature or in cultivation, is only part of the story. To serve conservation needs, a database on medicinal plants needs to show which species are under threat and so need immediate protection.

Botanists and conservationists have been listing threatened plants for nearly 20 years. Many countries have prepared Red Data Books listing their threatened flora, and numerous lists have been published; all those until 1986 are outlined in Davis *et al.* (1987). The Threatened Plants Unit, which is part of WCMC, records all species recorded as threatened on a global scale in its database. This now covers 22,000 threatened species, in a file of 58,000 plant taxa. However, IUCN and WWF predict that as many as 60,000 plants could become extinct or severely depleted genetically by 2050 if present trends continue. The 22,000 threatened species recorded so far seriously underestimate the situation, almost entirely because of lack of knowledge on the tropical floras where two-thirds of the world's plants grow. Nevertheless the TPU database does provide a quick way of seeing whether a plant has ever been recorded as threatened.

The criteria for threatened status are the IUCN Red Data Book categories, which measure the degree of threat. They have been used unchanged since the early 1970s, but are currently under review. The only prominent country which does not use them – or codes that can be converted into them – is the United States, where different categories are defined under the Endangered Species Act. The great advantage of the IUCN categories is their simplicity and the fact that they measure the degree of risk to taxa. Thus they make it easy to assess priorities for action. More complex systems and systems using numerical weightings have proved less useful.

Threatened status means that a species is in some danger of disappearing entirely. It does not include the many widespread species whose abundance is declining due to loss of habitat or over-use. This situation is often prevalent with medicinal plants, and so needs to be addressed. Whether a set of database categories would be useful is debatable. Any classification of the abundance of common species is highly subjective and heavily depends on the amount of habitat remaining for those species.

The International Board for Plant Genetic Resources have carried out very detailed geographical studies on a few widespread crop species, relating the occurrence of the plant in each place to various useful characteristics that may be needed by plant breeders, such as length of stalk, resistance to certain diseases and drought tolerance. These are called ecogeographical surveys and guidelines are available on how to do them (IBPGR, 1985). Although ecogeographical studies on widespread medicinal plants would be useful, especially if they covered the medical effects of the plant or its extract, it is hard to see the resources being available to do more than a handful of them. For this reason, it is not worth building this concept into the design of the medicinal plant database.

The Medicinal Framework

Information on the medicinal properties of plants used in traditional medicine is very complex, and the literature is scattered and diverse. It covers data on the plants concerned, the chemical constituents or compounds they contain, pharmacological and ethnobotanical or ethnopharmacological information.

Data on the chemical constituents of plants are particularly difficult to gather and assess. Experience has shown that the information in different papers on the presence of, say, secondary compounds in particular species is often contradictory. It is not easy to reconcile these differences, which may be due to taxonomic, chemical or environmental variations in the sample used, the analytical techniques involved or even mis-identification of the material.

There a a number of specialised bibliographies available, some national, some regional, some international, outlined by Loub and Farnsworth (1984).

The largest and most comprehensive computerised database is NAPRALERT, short for Natural Products Alert, established by Professor N Farnsworth and colleagues at the University of Illinois at Chicago. It covers "the world literature on the chemical constituents of plant, microbial and animal (primarily marine) extracts. In addition, considerable data are contained ... on the chemistry and pharmacology (including human studies) of secondary metabolites of known structure, derived from natural sources" (NAPRALERT brochure).

The NAPRALERT team screen about 200 journals and record in the database any article containing information on the presence of secondary chemical constituents in living organisms. From this, they can prepare a wide range of printouts providing, for example, ethnomedical profiles, pharmacological profiles and phytochemical profiles for any given plant, animal or microbe. The database is outlined by Loub *et al.* (1985).

By 1988 the NAPRALERT database contained information on 38,726 species of organisms, of which 30,590 were flowering plants, 607 gymnosperms, 887 ferns and fern allies, 337 bryophytes and 65 algae (not marine) (N Farnsworth, pers. comm., 1988).

NAPRALERT provides a vital component towards the design of any global database on medicinal plants. Ideally, some link between the conservation efforts of IUCN, TNC and WCMC with those of NAPRALERT could lead to the development of modules for use in countries on the conservation and sustainable utilisation of medicinal plants. There are also a few smaller regional databases, such as NAPRECA founded in 1984 and covering African plants.

Another interesting model could be the ILDIS Phytochemical Module, in which Chapman Hall Publishers are funding the creation of a database to link verified taxonomy with verified nomenclatural data through to verified chemical names, synonyms and structures, for the Chapman Hall Organic Compound Dictionary (R. Allkin, pers. comm.)

The TDWG Newsletter (No. 5, January 1990) lists Mrs F Cook, Royal Botanic Gardens, Kew, Richmond, Surrey TW9 3AB, U.K., as the contact point for the development of ethnobotany descriptors. Progress will undoubtedly be difficult in this area, since uses of plants, medicinal or otherwise, are notoriously difficult to classify and categorise.

Conclusion

An informal meeting of experts convened by UNIDO suggested that databases on medicinal plants contain the following:

- Botanical Aspects
- Ethnopharmacological Aspects
- Chemical Aspects
- Agrotechnological Aspects
- Technological Aspects
- Chemotaxonomic Aspects
- Market Aspects
- Other relevent Aspects

Such a comprehensive approach is probably a mistake, as few institutes could design let alone complete a database on all these topics. Some, such as markets, may not be suitable for a database anyway. It would be better to start with well-understood components, such as the chemistry, and the botanical and conservation aspects, as outlined above, and proceed from there.

Today is an exciting and stimulating time to develop a medicinal plants database. A wide range of hardware and software is available, with capabilities undreamed of years ago. And even more important, we are able to learn from the experience of those who have pioneered this complex and challenging field.

There are several good models to follow:

- TNC and ILDIS for networking
- TDWG for the setting of data standards
- TPU for global plant conservation data
- NAPRALERT for plant chemistry and pharmacology

We now have to face the challenge of bringing these models together to plan a database and information system that will meet all our needs in the most effective way possible.

After considering an earlier version of this paper, the International Consultation at Chiang Mai agreed that a small interdisciplinary team should be set up to consider the need for databases on conservation and sustainable use of medicinal plants. It was agreed that IUCN should carry this initiative forward once funding had been obtained.

Acknowledgments

We gratefully acknowledge the help and encouragement given to us in preparing this paper by many experts, in particular R Allkin (Kew), R K Brummitt (Kew), C Leon (WCMC) and F Bisby (Southampton).

References

Allkin, R. (1988). Taxonomically intelligent database programs. In *Prospects in Systematics*, ed. D.L. Hawksworth. Oxford: Clarendon Press.

Botanic Gardens Conservation Secretariat (1987). *The International Transfer Format for Botanic Garden Plant Records*. Plant Taxonomic Database Standards No. 1. Version 01.00. Pitsburgh: Hunt Institute for Botanical Documentation, Carnegie-Mellon University. 64 pp.

Davis, S.D. *et al.* (1986). *Plants in Danger: What do we Know?* IUCN, Gland, Switzerland, and Cambridge, U.K.

Frost-Olsen, P. & Holm-Nielsen, L.B. (1986). ***A Brief Introduction to the AAU-Flora of Ecuador Information System.*** **Reports from the Botanical Institute, University of Aarhus, No. 14. Denmark. 39 pp.**

IBPGR (1985). *Ecogeographical Surveying and in situ conservation of crop relatives.* Rome: IBPGR. 27 pp.

Loub, W.D. & Farnsworth, N.R. (1984). Use of computers in the development of natural products. *Impact of Science on Society*, No. 136, 343-351.

Loub, W.D., Farnsworth, N.R., Soejarto, D.D. & Quinn, M.L. (1985). NAPRALERT: Computer handling of natural product research data. *J. Chem. Inf. Comput. Sci.*, 25, 99-103.

Tutin, T.G. *et al.* (Eds) (1964-1980). *Flora Europaea.* 5 vols. Cambridge: Cambridge University Press.

Voss, E.G. *et al.* (1983). *International Code of Botanical Nomenclature.* Regnum Vegetabile, vol. 111. Bohn, Scheltema and Holkema, Utrecht and Antwerp, and W. Junk, The Hague and Boston. 472 pp.

Techniques to Conserve Medicinal Plants

Agronomy Applied to Medicinal Plant Conservation[1]

Dan Palevitch
Department of Medicinal, Spice and Aromatic Plants, Agricultural Research Organization, the Volcani Center, Bet Dagan, Israel

Higher plants are still an indispensable source of drugs and of galenic preparations, despite the progress made in synthetic organic chemistry and biotechnology. The main source of higher plants used as drugs in modern and traditional medicines is the wild flora in developing countries. Only a small portion of the raw material is produced under cultivation. The collection of plants in the wild has many disadvantages (Table 1), so interest has grown in applying modern methods to the cultivation of medicinal plants. The possibility of cultivating medicinal plants that have been genetically improved under suitable environmental conditions enables us to obtain plants which are rich in desirable active compounds. The development of modern methods of processing and preserving the raw material of medicinal plants helps to maintain their quality for a longer time. In order to obtain high quality products attention should be paid to cultivation, post-harvest handling, processing and comprehensive quality control.

Table 1. Collection from the wild *versus* cultivation of medicinal plants

	Collection	*Cultivation*
Availability	Decreasing	Increasing
Fluctuation of supply	Unstable	More controlled and quality
Quality control	Poor	High
Botanical identification	Sometimes not reliable	Not questionable
Genetic improvement	No	Yes
Agronomic manipulation	No	Yes
Post-harvest handling	Poor	Usually good
Adulteration	Likely	Relatively safe

1 Contribution from the Agricultural Research Organization, The Volcani Center, Bet Dagan, Israel. No. 2160-E, 1987 series.

Domestication Strategy

The process of domestication of medicinal plants is divided into two different activities: at the natural or at the cultivation sites.

The investigations and observations at the natural location include: Systematic botany, climatic conditions, soil properties, plant development and plant physiology, reproduction, natural propagation and susceptibility to pests and diseases.

The studies at the cultivation site include the following aspects: Propagation methods, genetic improvement, advanced cultural practices, protection against weeds and pests, optimal timing of harvest, advent of mechanization, post-harvest treatments, handling of the raw material, quality control and phytochemical analysis.

Agronomic Contribution

The breeding of new cultivars with desirable agronomic and chemical traits by classical and modern breeding methods makes it possible to conserve highly valuable germplasm in seed banks or in botanic gardens.

Cultivation of medicinal plants permits production of uniform quality raw material whose properties are standardized and from which the crude drugs can be obtained unadulterated.

Studies on the agronomic aspects of medicinal plants are few compared to those of other categories. A search in *Horticultural Abstracts* (1986 and 6 months of 1987) showed that agronomic studies amounted to only 8-9% of the total publications on medicinal plants. These agronomic studies dealt with 79 different plant species, while the overall list of medicinal plants related to other research disciplines comprises more than 300.

Improving the performance of medicinal plants by agronomic techniques can be summarized under the following headings: Genetic improvement, Optimal environmental conditions, Cultivation under modern cultural practices, and Post-harvest treatments.

Genetic Improvement

The breeding of improved cultivars adapted to different agro-ecological regions will enable us to cultivate medicinal plants under a wide range of conditions outside the present sites of collection. The recent development of highly specific and sensitive analytical techniques such as the radioimmunoassay (RIA) or the enzyme-linked immunosorbentassay

(ELISA) provide new prospects for breeding cultivars with a high content of a specific compound and/or the desired spectrum of secondary metabolites (Arens *et al.*, 1987). With such analytical methods, both conventional and novel breeding methods take less time than before.

Once a suitable cultivar or clone has been developed or identified, the plants can be rapidly multiplied by conventional or by new methods like micropropagation (Holdgate, 1982).

The main agronomic and chemical traits of a suitable cultivar are as follows: *Agronomic traits*: Uniform seed germination (lack of inhibitors); high biomass yield; optimal plant architecture (resistance against lodging, high proportion of the desired organ(s), branching habits, suitability to mechanical harvest); growing pattern (determinate or indeterminate growth); resistance to pests, diseases and weeds; wide adaptation to environmental conditions and allelopathic action. *Phytochemical traits*: Desirable spectrum of and high concentration of the active compound(s). The most common breeding techniques that are used in genetic manipulations are *selection, hybridization, mutation* and *polyploidy*.

Selection

Most of the medicinal plants are still grown as wild plants with a wide range of genetic variability. Thus, the selection of suitable genotypes for desired traits can be achieved relatively easily. Selection for suitable cultivars must take into consideration both the biomass yield and the phytochemical content. The heritability of these two different and complex traits is generally different. The content of the chemicals is generally a dominant factor however; the integrated yield of biomass and the yield of the chemical compounds have a much lower heritability rate. An example to the potential of the selection method in the breeding program of *Papaver bracteatum* is presented in Table 2 (Levy *et al.*, 1979). Moreover, the genetic variability in the wild population enables us to breed cultivars with a high thebaine content and with close capsules of the non-shattering type (Levy, 1985).

Table 2. Selection of genotypes of *Papaver bracteatum* with high thebaine content

	% Thebaine			
	ARYA I (IS)		*ARYA II (IS)*	
	Capsules	Roots	Capsules	Roots
Mean	2.2	0.79	2.0	0.87
Range	0.4 - 4.2	0.39 - 1.23	0.6 - 4.1	0.37 - 1.35
Number of plants observed	925		98	

Crosses and Hybridization

Successful crosses can combine desirable traits in a stable genotype. Use of hybrid cultivars had two advantages: a potential increase in yield due to heterosis, and the possibility of protecting the seeds under a patent.

An example of the economic benefits of hybridization is shown in *Catharanthus roseus* (Table 3). Heterosis effect contributed to both root yield and ajmalicine concentration in the hybrid *White x Red-Eyed* (Levy *et al*., 1983).

Table 3. Heterosis and correlation analysis of the root weight and ajmalicine content in roots of the medicinal plant *Catharanthus roseus*

	Parents			*F-1*		
	Pink (P)	White (W)	Red-eyed (RE)	PxRE	WxRE	PxW
Root weight (g/pl	39	34	40	50	49	38
Ajmalicine (%)	0.30	0.37	0.36	0.34	0.41	0.28
Ajmalicine yield (g/pl)	0.11	0.13	0.14	0.17	0.20	0.11

Mutation breeding

The induction of mutation caused by chemicals or radiation is one of the most promising breeding approaches for the development of better genotypes. However, only a limited application of this method has been achieved in plant breeding largely due to the lack of adequate rapid screening methods to identify the desired genotypes from the large populations used in this technique. Hence, attention has been directed towards easily identifiable qualitative traits (e.g. morphological characters such as early flowering and seed germination). Only very limited work has been done to screen better genotypes of medicinal plants rich in active compounds. The mutagenesis method plays an important role, especially in the case of plants which are propagated vegetatively and have limited natural genetic variability. The benefits of mutation in breeding medicinal plants are presented in Table 4 (Kak *et al*., 1982).

Table 4 Improvement of medicinal plants through induced mutations

Plant species	*Genetic improvement*
Solanum khasianum	High yield, high content of glycoalkaloids, reduction of spine frequency
Solanum laciniatum	Early flowering, high yield of berries, high content of glycoalkaloids
Dioscorea spp.	Increased yield and diosgenin content
Datura spp.	Increased yield
Papaver somniferum	High morphine content

Polyploidy breeding

Induced polyploidy can be used in two basic ways: *Autoploidy*, the induction and the elevation of the chromosome number within the species; and *Alloploidy*, the induction and the elevation of the chromosome number following hybridization between two species.

It is apparent that different types of crop plants respond differently to induced polyploidy. The performance of induced artificial polyploids in crop plants in general and in medicinal plants in particular has consistently fallen short of expectations (Table 5).

Table 5. Genetic improvement of medicinal plants by using polyploidy breeding techniques

Plant species	*Polyploidy*	*Crop improvements in the polyploid*	*Reference*
Solanum khasianum	4n	Increase of solasodine content by 35-50%	Bhatt & Heble, 1978
	4n	Increase of plant height and total fruit yield	Bhatt, 1977
Catharanthus roseus	4n	Increase in ajmalicine content of the roots, root weight, total alkaloids in the leaves and roots	Krishnan *et al.*, 1985
	4n	Larger stomata, broader leaves, slower growth rate, higher pollen sterility	Dnyansager & Sudakaran, 1970
	4n	Increase in total number of branches per plant, average leaf weight, seed weight and seed number per follicle	Kulkarani *et al.* 1984
Papaver bracteatum	4n	Increased thebaine content	Wold *et al.*, 1984
	4n	Increase in thebaine content but not the thebaine yield per plant	Milo *et al.*, 1987
	3n	High thebaine content only in the first season	Ibid.
Trigonella foenum-graecum	4n	Luxurious growth with slightly lower disgenin content	Anis & Aminuddin, 1985
Costus speciosus	3n, 4n	Lower diosgenin content	Jankai-Ammal & Nagendra-Prassad, 1984

The effects of polyploidy in breeding *Papaver bracteatum* are shown in Table 6. The highest yield of both capsules and thebaine was obtained from the plants having the regular chromosome number, in spite the fact that the tetraploid plants have the higher concentrations of thebaine (Milo *et al.*, 1987).

Table 6. Morphological characters and thebaine content in diploid and polyploid plants of *Papaver bracteatum*

Character	*Polyploidy level*		
	2n	3n	4n
Number of plants	5	4	3
Capsules			
Number per plant	10	15	9
Weight (g)	2.5	2.4	2.6
Diameter (mm)	25.9	19.8	21.2
Yield (g/plant)	33.0	21.6	13.0
Thebaine			
Content (%)	2.1	2.2	4.8
Yield (mg/plant)	709	470	621

New breeding methods

Allelopathic cultivars: Available evidence indicates that it will be possible to develop allelopathic crop cultivars through breeding or by biotechnology in order to provide control of pests and weeds. Allelopathic cultivars of sunflower and cucumber resistant against weeds have been developed. (Rice, 1984; Leather, 1983). However, to date no study using allelopathy has been carried out with medicinal plants.

Genetic engineering: Genetic engineering is a technique for isolating individual genes from plants and for multiplying (by cloning) these genes in bacteria and yeast. It is hoped that these procedures can be developed to incorporate genes into the genomes of crop plants in order to aid the adaptation of agricultural crops to human needs. One of the goals of using the recombinant DNA technique is herbicide-resistant cultivars. The new methods offer new avenues and prospects for the breeding and conservation of medicinal plants.

Optimal Environmental Conditions

In spite of the fact that alkaloid biosynthesis is gene-governed, environmental conditions play an important role in controlling plant growth and in the formation of the secondary metabolites. Since most alkaloids are formed in young, actively growing tissues, factors that influence plant growth can influence also the production of the secondary metabolites. Environmental factors such as temperature, light intensity, photoperiodism, mineral and water supply, and the altitude above sea level are involved in the production of plant chemicals (Waller & Nowacki, 1978). It should be noted that not all environmental factors act identically in different species of medicinal plants. The improved cultivars must show maximum suitability to a wide range of environmental conditions, such as extreme temperatures, water stress, changed light interception and a change in soil physical and chemical conditions.

An example of the effect of high temperature on yield components and oil composition of the seeds of *Oenothera lamarckiana* is demonstrated in Table 7 (Yaniv *et al.*, 1989). A decrease of the yield components with rise of temperature was noted. On the highest temperature regime a very low content of gamma-linolenic acid was obtained. It can be speculated that in geographical regions where high temperature prevails, both the yield and seed quality will be lower in comparison to cooler regions.

Table 7. Effect of temperatures on yield component and oil composition of the seeds in *Oenothera lamarckiana* (Evening primrose)

Temp. (°C.) Day/Night	Capsule diameter (mm)	Seed yield (g)	Weight 10^3 seeds (g)	Oil composition (%) 16:0	18:0	18:1	18:2	18:3 (gamma)
17/12	4.3a	12.8a	0.56a	7.0a	1.5b	10.0b	71.65c	9.9a
22/17	4.2a	12.5a	0.52a	6.0b	1.5b	9.2b	73.8b	9.6a
27/22	3.4b	9.1b	0.37b	5.8c	1.5b	9.8b	74.6a	8.1b
32/12	2.0c	6.2c	0.31c	7.2a	3.1a	20.1a	61.9d	5.5c

The effect of altitude on the yields and chemical content in *Dioscorea* and *Solanum* are shown in Table 8. This study revealed an increase in the synthesis of steroids at a higher altitude in all the species examined, but since the tuber yields of the *Dioscorea* species was lower at the higher altitude, the yield of diosgenin was the same in the two environmental conditions as a result of compensation between the two parameters. In *S. laciniatum* both the yield of the fruits and the solasodine concentration were greater at the higher altitude, so the yield of the phytochemical at the high altitude was twice that at the lower one (Chatterjee *et al.*, 1983).

Table 8. The effect of altitude on the yield of some steroid-yielding plants

Altitude (m)	Biomass yield (kg/acre)	*Diosgenin / solasodine* (%)	Yield (kg/acre)
Dioscorea composita			
50	5500	3.25	179
500	4800	3.92	188
Dioscorea floribunda			
50	4550	3.86	176
500	4100	4.12	169
Solanum laciniatum (fruits)			
50	625	1.32	8
500	950	1.76	17

Agrotechnical Improvement

Appropriate cultural practices have a great impact on the final biomass and phytochemical yields. The most important cultural practices are: plant propagation; sowing dates; irrigation and fertilization; herbicides and pesticides; and post-harvest treatments. The effect of chemicals as plant growth regulators on seed germination, plant growth patterns, flowering, phytochemical content and quantity, leaf defoliation and post-harvest incubation can be used for manipulation of the phytochemical and/or the biomass yields.

Plant Establishment

Among the procedures which can be manipulated or adopted to ensure good plant establishment are the rates of seed germination and emergence, vegetative propagation, optimal sowing date and optimal plant spacing. Since many of the cultivated medicinal plants originated from the wild, their germination speed and rate are generally erratic and poor. Several growth regulators, mainly the gibberellins, are well known as promoters of seed germination, especially in seeds which exhibit deep dormancy or have a hard seed coat. Gibberellic acid (GA 3) is the most potent growth regulator for overcoming seed dormancy and for increasing the speed and uniformity of germination and emergence. The effects of GA 3 on the germination rate of *Solanum laciniatum* are shown in Table 9. It was found that GA 3 introduced into the seeds by organic solvents increased markedly the emergence speed and rate (Palevitch *et al.*, 1980).

Table 9. Effect of GA 3 (500 ppm) infused into the seeds by the organic solvents dichloromethane (DCM) and acetone on the seedling emergence of *Solanum laciniatum* at constant temperature of 20°C.

Solvent	*GA 3*	*Emergence (%)* 14 days	18 days
-	-	O c	50 d
DCM	-	4 bc	67 c
	+	12 b	77 b
Acetone	-	7 b	71 bc
	+	43 a	86 a

Cultivation procedures

One of the most important environmental and agrotechnical factors is water availabilty. Water plays an important role in the growth and development of plants. Generally speaking, a limited amount of water has a negative effect on plant development. However, in few cases water stress caused an increase in the content of secondary metabolites (Yaniv & Palevitch, 1982). An example on the effect of water availability on the yield and content of the active compounds is presented in Table 10.

Table 10. Effect of moisture stress on leaf yield, solasodine content and yield in two *Solanum* species (Temp. 15°/25° C.)

	Dry leaf yield		*Content (%)*		*Solasodine (g/plant) Yield (g/plant)*	
Species	*Cont-rol*	*Stres-sed*	*Cont-rol*	*Stres-sed*	*Cont-rol*	*Stres-sed*
S. aviculare	151	10	1.63	2.03	2.3	0.2
S. laciniatum	194	38	1.29	2.25	2.5	0.9

The results showed that water stress strongly inhibited plant growth in two leafy species of *Solanum*. The dry leaf yield was low under the stressed conditions, but the solasodine content was higher in the stressed plants. However, the solasodine yield per plant was much higher in the controlled plants, due to the low dry material yield of the stressed plants (Mann *et al*., 1980). Several groups of plant hormones are involved in controlling growth and plant response to environmental factors. The recognition of such growth control mechanisms has introduced the possibility of modifying growth and development of crop plants by manipulating hormone levels in various organs and various stages in the life cycle (Thomas, 1985). Modifying plant growth and the content of active compounds via plant growth regulators for increased production efficiency has proved successful with several horticultural crops. Examples of the effect of growth regulators on the concentration of the essential oils in

Mentha piperita and *Salvia officinalis* are presented in Table 11. It was demonstrated that foliar application of cytokinins increased the essential oil concentration up to two-fold. The increase in oil yield is a result of increased montoterpene biosynthesis (El-Keltawi & Croteau, 1987).

Table 11. The effect of cytokinins on the essential oils of *Mentha piperita* and *Salvia officinalis*

Treatment (ppm)	Essential oil *M. piperita* (%)	*M. piperita* (mg/pl)	*S. officinalis* (%)	*S. officinalis* (mg/pl)
Control -	0.20	34.5	0.18	33.3
Kinetin (4)	0.34	71.7	0.23	64.2
Diphenylurea (10)	0.41	93.5	0.39	67.5
BA (4)	0.24	44.7	0.28	46.3
Zeatin (4)	0.23	47.2	0.28	46.3
L.S.D. (0.05)	0.03	2.9	0.03	3.3

It is well established that GA_3 can substitute for the requirement of cold induction at the onset of flowering, especially with rosette plants. The effect of GA_3 on flowering in *Papaver bracteatum* is presented in Table 12. The results demonstrated a beneficial effect of GA_3 only on the yield of the late-flowering clones. Higher thebaine yield per plant was also obtained in the treated plants (Levy *et al.*, 1986).

Table 12. The response of two genotypes of *Papaver bracteatum* to applications of GA_3

	Early-flowering clone *GA* -	*Early-flowering clone* *GA* +	*Late-flowering clone* *GA* -	*Late-flowering clone* *GA* +
Days from GA treatment to maximum flowering	69	62	82	69
Capsules per plant	13.7	14.1	3.8	10.1
Wt. of capsules per plant (g)	22.7	22.4	11.2	23.3
Thebaine yield per plant (mg)	522	493	146	373

Post-Harvest Treatments

Treatments using growth regulators and enzymes increased very markedly the content of the active compounds of harvested raw material (Table 13). Post-harvest treatments modified the diosgenin content in *Costus spicosus* rhizomes. Incubation with 2,4 D and Na-acetate increased the content of the diosgenin two-fold and more. Prolonging the incubation period to 48 hours but not up to 60 hours resulted in higher concentrations (Chatterjee *et al.*, 1983).

Table 13. Post-harvest treatments modified the diosgenin content in *Costus spicosus* rhizomes

Treatment	Concentration (ppm)	Duration (h)	Diosgenin (%)
Control	-	-	1.56
2,4 D	10	24	2.19
	10	48	3.36
	10	60	3.19
Na-acetate	50	48	3.79
	50	60	3.23

Summary

It can be concluded that the cultivation of medicinal plants under modern cultural practices is the main tool for preserving and maintaining the germplasm of rare, endangered or over-exploited medicinal plants. Genetic diversity based on selection and breeding of wild flora can serve as indispensable source for successful cultivation of medicinal plants. The cultivated plants can ensure high quality raw material for further processing and/or the preparation of galenic products.

References

Anis, M. & Aminuddin (1985). Estimation of diosgenin in seeds of induced autoploid *Trigonella foenum-graecum*. *Fitoterapia*, 36, 51-52.

Arens, H., Deus Neutnan, B. & Zeng, M.H. (1987). Radioimmunoassay for the quantitative determination of ajmaline. *Planta Med*. 1987, 179-183.

Bhatt, B. (1977). Further studies on colchicine-induced tetraploids on *Solanum khasianum* Clarke. *Environ. Exp. Bot*., 17, 121-124.

Bhatt, B. & Heble, R. (1978). Improvement of solasodine content in fruits of spiny and mutant tetraploid of *Solanum khasianum* Clarke. *Environ. Exp. Bot*., 18, 127-130.

Chatterjee, S.K., Nandi, R.P., Sarma, P. & Panda, P.K. (1983). Studies on developmental physiology of medicinally important steroid-yielding plants growing in India. *Acta Hort*., 132, 85-100.

Dnyansagar, V.R. & Sudakaran, I.V. (1970). Induced tetraploidy in *Vinca rosea* L. *Cytologia*, 35, 227-241.

El-Keltawi, N.E. & Croteau, R. (1987). Influence of foliar applied cytokinins on growth and essential oil content of several members of the Larmiaceae. *Phytochemistry*, 26, 891-895.

Holdgate, D.P. (1982). Tissue culture for commercial plant propagation. *Span*, 25, 24-26.

Jankai-Ammal, E.K. & Nagendra-Prassad, P. (1984). Relationship between polyploidy and diosgenin content in different parts of *Costus speciosus* (Koen) Sm. *Curr. Sci*., 53, 601-602.

Kak, S.N., Singh, C., Ram, G. & Kaul, B.L. (1982). Improvement of medicinal and aromatic plants through induced mutations – An overview. In *Cultivation and Utilization of Medicinal Plants*, ed. C.K. Atal & B.M. Kapur, pp. 771-6. India: Council of Scientific & Industrial Research, Jammu-Twai.

Krishnan, R., Chandravadana, M.V., Mohankumar, G.N. & Ramachander, P.R. (1985). Effect of induced autotetraploidy on alkaloid content and root weight in *Catharanthus roseus* (L.) G. Don. *Herba Hung.*, 24, 43-51.

Kulkarani, R.N., Chandrasshekar, R.S. & Dimri, B.P. (1984). Induced autotetraploidy in *Catharanthus roseus* – A preliminary report. *Curr. Sci.*, 53, 484-485.

Leather, G.R. (1983). Sunflower (*Helianthus annuus*) are allelopathic to weeds. *Weed Sci.*, 31, 37-42.

Levy, A. (1985). A shattering-resistant mutant of *Papaver bracteatum* Lindl.: Characterization and inheritance. *Euphytica*, 34, 811-815.

Levy, A., Milo, J., Ashri, A. & Palevitch, D. (1983). Heterosis and correlation analysis of the vegetative components of ajmalicine content in the roots of the medicinal plant *Catharanthus roseus* (L.) G. Don. *Euphytica*, 32, 557-564.

Levy, A., Palevitch, D. & Lavie, D. (1979). Thebaine yield components in selections of Arya I and Arya II populations of *Papaver bracteatum*. *Planta Med.*, 36, 362-368.

Levy, A., Palevitch, D., Milo, J. & Lavie, D. (1986). Effect of gibberellic acid on flowering and the thebaine yield of different clones of *Papaver bracteatum*. *Plant Growth Regulation*, 4, 153-157.

Mann, J.D., Edge, E.A., Lancastar, J.E. & Blyth, K. (1980). Growth and solasodine production by *Solanum aviculare* and *Solanum laciniatum* under moisture stress by different temperatures, *N.Z.J. Agric. Res.*, 23, 361-366.

Milo, J., Levy, A., Palevitch, D. & Ladizinsky, G. (1987). Thebaine content and yield in induced tetraploid and triploid plants of *Papaver bracteatum* Lindl. *Euphytica* 36, 361-367.

Palevitch, D., Levy, A. & Perl, M. (1980/1). Invigoration of seeds of the medicinal plants *Solanum laciniatum* Ait., *Solanum khasianum* Clarke and *Catharanthus roseus* (L.) G. Don. *Israel J. Bot.*, 29, 74-82.

Rice, F. (1984). *Allelopathy*. Orlando, U.S.A.: Academic Press.

Thomas, T.H. (1985). Chemical manipulation of standing crops. *Ann. Proc. Phytochem. Soc. Euro.*, 26, 73-90.

Waller, G.R. & Nowacki, E.K. (1978). *Alkaloid Biology and Metabolism in Plants*. New York and London: Plenum Press.

Wold, J.K., Paulsen, B.S., Ellingsen, D.F. & Nordal, A. (1983). Increase in thebaine content of *Papaver bracteatum* Lindl. after polyploidation with colchicine. *Nor. Pharm. Acta*, 45, 103-109.

Yaniv, Z. & Palevitch, D. (1982). Effect of drought on the secondary metabolites of medicinal and aromatic plants. In *Cultivation and Utilization of Medicinal Plants*. Vol. 1, ed. C.K. Atal & B.M. Kapur, pp. 1-12. India: Council of Scientific & Industrial Research. Jammu-Tawi

Yaniv, Z., Shomroni, C., Levy, A. & Palevitch, D. (1989). Effect of temperature on the fatty acid composition and yield of evening primrose (*Oenothera lamarckiana*) seeds. *J. Exp. Bot.*, 40, 609-613.

Biotechnology in the Production and Conservation of Medicinal Plants

H.M. Schumacher *
Institut für Pharmazeutische Biologie, Universität München, München

Need for Medicinal Plants

Medicinal plants are the oldest source of pharmacologically active compounds. They remained to be the only source of useful medicinal compounds for centuries. It is estimated that even today 2/3 of the world population relies on plant-derived drugs. It was not until the last century that through increasing scientific knowledge in the western world new sources and methods for the production of medicinal compounds were found, such as synthetic substances, antibiotics produced by fungi or bacteria, hormones extracted from animal tissues, or vaccines produced by animals or animal cell culture systems.

Today in a typical industrialized nation like West Germany 72% of all chemically defined substances, contained in medicinal preparations available on the market, are produced by organic synthesis, 11% are obtained from microbes only 5% are produced with animals or animal cell cultures, while 12% are still plant-derived substances. Presently 75 different plant-derived secondary metabolites are in use and the medicines made from them represent a fraction of slightly more than 5% of the total. Substances most frequently used in medical preparations are cardiac glycosides, theophylline, vincamine, atropine, pilocarpine, codeine and scopolamine. Additionally a lot of important steroid compounds and hormones are derived semisynthetically from plant precursors.

Crude drugs are also still of considerable importance for medical treatment. Medicines made from crude drugs represent about 15% of the total on the German market. This might give an impression of commercial importance of plant-derived pharmaceuticals but to assess properly the medical importance of these drugs one has to focus on special indications. Crude drug preparations represent 25% of all available antihypertension and antitussive medicines. Up to 30% of the available

* Present address: DSM-Deutsche Sammlung von Mikroorganismen und Zellkulturen GmbH, German Collection of Microorganisms and Cell Cultures, Mascheroder Weg 1B, D-3300 Braunschweig, Germany.

hypnotic and tranquilizing medicaments and those for heart and arteriosclerosis therapy are made from crude drugs. In the field of laxatives as much as 50% of all available medicines on the German market are crude drug preparations. Furthermore in the domains of antitussive and heart therapy medicines virtually no preparation can be found which does not contain either crude drugs or chemically defined plant derived substances (calculated from Bundesverband der pharmazeutischen Industrie, 1987).

It has to be noted that some of the most important crude drugs and plant substances are supplied from outside Germany (see Table 1).

Table 1: Origin of plant-derived drugs, substances, extracts and oils imported into West Germany (Data from Statistisches Bundesamt, 1985)

Imports into Germany	Most important suppliers
Crude plant material	
Matricaria chamomilla	Argentina, Egypt
Mentha piperita	Jugoslavia, Egypt
Cinchona succirubra bark	Zaire, Guatemala, Kenya
Glycyrrhiza glabra roots	China
Isolated substances	
Theophylline	China
Theobromine	USSR, Italy
Caffeine	China
Quinine	Indonesia, Zaire
Rutin & deriv.	Argentina
Pilocarpine/Scopolamine	China, Jugoslavia, USSR
Plant extracts	
Opium	India
Aloe extract	Kenya
Glycyrrhiza extract	USA, Iran, China
Plant oils	
Mentha piperita oil	China
Eucalyptus spec. oil	China
Syzygium spec. oil	China, Indonesia
Conifer oils	China, Jugoslavia, USSR

Although they are of great medical and commercial value, most of the medicinal plants are still of wild origin. Two thirds of the different species are still collected in the wild and the cultivation of some important drug-plants like *Gentiana lutea, Valeriana mexicana, Echinacea* and *Arnica* did not start until the last decade (Franz, 1986).

Reflecting the standard of medicinal plant cultivation, plant breeding has only been performed with the commercially most important plants like *Papaver somniferum*, *Papaver bracteatum*, *Cinchona* sp., *Chamomilla recutita* and *Mentha piperita* (Schieder, 1984).

The present situation causes several problems for continuous drug supply:

- Shortage caused by crop failure,
- Shortage caused by increased demand for special drugs,
- Heterogeneity of drug raw material, especially in the case of collected plant material (here mistaken identity of plants is an additional problem) (Franz, 1982),
- Heterogeneity of drug raw material, even in the case of cultured plants (Franz, 1986),
- Danger of extensively collected wild plants becoming extinct (Franz, 1982).

Improvement of Medicinal Plants

The biotechnological methods presently available for plant improvement reflect a set of goals which is well known in conventional plant breeding:

- The selection of desirable traits,
- The increase of variability,
- The achievement of homozygosity,
- The introduction of new genes into cultivars.

Furthermore, biotechnology provides new methods for the mass production of elite plants as well as for the *in vitro* production of plant raw material.

Selection of Desirable Traits

Presently the poorly developed standards of plant breeding with many medicinal plants gives a remarkable opportunity for the selection of individuals with a high secondary metabolite content or other desirable features. Unfortunately very often suitable detection methods for desired products are missing. In our department the radioimmunoassay (RIA) technique has been used in several cases for this purpose.

When in the 1970's diosgenine, obtained from wild–collected *Dioscorea* species had fallen into short supply, there was a need for an alternative. Solasodine from *Solanum laciniatum* proved to be a substitute. Since other analytical techniques gave insufficient sample throughput, a radioimmunoassay was developed. By this method amounts of even 0.7 ng of solasodine or its glycoside can be detected with high specificity. It has been possible to investigate the solasodine distribution in small parts of a single plant as well as in a plant population. While the average content of the plant was 1.6 % of dry weight, individuals with a content of 2 - 3 % (see Fig. 1) of solasodine were found (Weiler *et al.*, 1980).

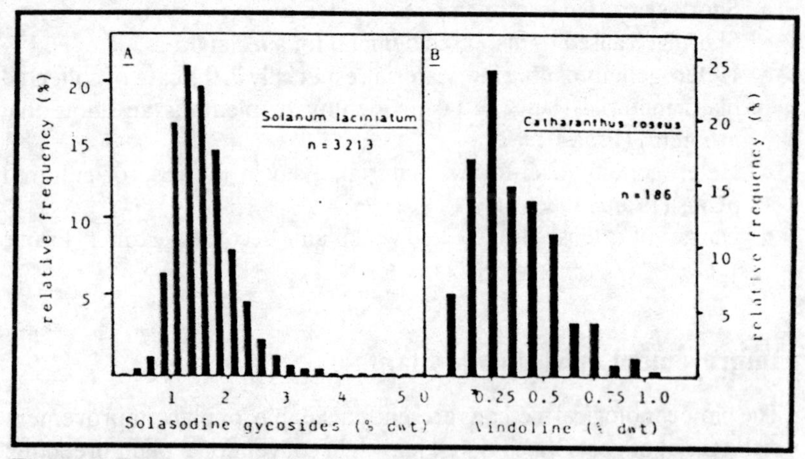

Figure 1. Frequency distribution of A: Solasodine concentration in leaf discs of *Solanum laciniatum*, B: Vindoline within a population of *Catharanthus roseus* plants

In the same way a RIA-method for the detection of vindoline in *Catharanthus roseus* plants was established. Vindoline can serve as one monomer in a semisynthetic vinblastine production. Even 110 fmol of vindoline can be detected by this method. In the population tested the vindoline content varied from 0.1% to 1.0% (dwt). A single plant with a vindoline content of 2%, which means 5.5 fold higher than the average level (0.36%), was selected (Westekemper *et al.*, 1980), (see Fig. 1).

The advantages of the RIA-method lie in its high specificity, the high sensitivity and the immense sample capacity. The high sample troughput is in part due to the high specificity of the methods which makes extract-prepurifications unnecessary. Nevertheless it should also be noticed that high specificity can also be disadvantageous when a whole class of substances has to be detected.

From these examples it can be seen that selection procedures can be extremely useful in finding out the starting material for the further application of techniques for plant improvement.

Increasing Natural Variability by Somaclonal Variation

It has been observed with undifferentiated cell cultures that regenerated plants show an increased variation in a wide range of characters, for example morphology, plant habit, photoperiod response, maturity date and yield. These variants, first considered to be epigenetic artifacts of tissue culture, are caused by chromosome mutations and point mutations, or in the case of maize established by transposon activation. With maize, even changes in the mitochondrial genome can lead to somaclonal variation. By selecting somaclonal variants, disease resistant as well as herbicide resistant plants could be achieved.

This phenomenon has not yet been used to improve medicinal plants but several examples of changes in the colour of flowers of ornamental plants indicate that secondary metabolism can also be affected (see Evans *et al.*, 1987).

Achievement of Homozygosity by Gametophyte Culture

To establish homozygous plants by conventional breeding techniques needs an enormous number of backcrossings. By the *in vitro* culture of gametophytes this goal can now be reached within one generation. When pollen grains (to obtain androgenesis) or ovules (to obtain gynogenesis) are excised from the flower buds and subjected to suitable media, embryos or callus cultures develop, from which whole plants can be regenerated. Normally different stages of ploidy occur among the regenerated plants, but haploids can be selected out. By subsequent treatment of these haploids with colchicine di-haploid homozygous plants are obtained, which then can be used for conventional breeding (Schieder, 1984). In practice it is not as easy to establish a gametophyte culture as might be expected from the description above. No standard procedure for different plant species is available and often an unexpected variation occurs among the di-haploid plants (Bollon & Raquin, 1987). Successful application with increased product yield of di-haploids is described for *Datura innoxia* (Schieder, 1984).

With haploid plants, mutants which would have been recessive with diploids can be sorted out directly. That's why the treatment of haploid lines with mutagenic agents can be another chance to achieve altered characters.

Nevertheless, an important drawback for the application of these techniques in breeding programs is the difficulty of producing enough plants for selection procedures (Schieder, 1984), (see also Table 2).

Somatic Hybridization

The term somatic hybridization means the fusion of isolated protoplasts from different plants, which results in the formation of heterokaryotic cells with mixed cytoplasm, plastids and mitochondria. The fusion of cultured protoplasts is the consequence of a breakdown of the outer membrane's negative charge, mediated by treatment with chemicals and high pH values or electrical pulses (Pelletier, 1987). One critical point of this method is the separation of fused protoplasts from other cells. Different methods have been developed for this purpose, like complementation of deficient lines or mechanical sorting out of fusion products between non-green cell culture derived and green leaf-derived protoplasts (see also Pelletier, 1987)

Table 2 : Medicinal plants investigated for the production of haploids

Plant	Reference
Atropa belladonna	Zenkteler (1971)
Catharanthus roseus	Abou-Mandour *et al.* (1979)
Datura innoxia	Nitsch & Noreel (1973)
Datura metel	Iyer & Raina (1972)
Datura meteloides	Kohlenbach & Geier (1972)
Datura muricata	Nitsch (1972)
Datura stramonium	Guha & Maheshwari (1967)
Datura wrightii	Kohlenbach & Geier (1972)
Digitalis lanata	Data from IBPGR (1985)
Hyoscyamus albus	Raghavan (1975)
Hyoscyamus niger	Corduan (1975)
Hyoscyamus pusillus	Raghavan (1975)
Nicotiana alata	Nitsch (1969)
Nicotiana attenuata	Collins & Sunderland (1974)
Nicotiana clevelandii	Vyskot & Novak (1974)
Nicotiana glutinosa	Nitsch & Nitsch (1970)
Nicotiana knightiana	Collins & Sunderland (1974)
Nicotiana langsdorfii	Durr & Fleck (1980)
Nicotiana otophora	Nitsch (1972)
Nicotiana paniculata	Nakamura *et al.* (1974)
Nicotiana raimondii	Collins & Sunderland (1974)
Nicotiana rustica	Nitsch & Nitsch (1970)
Nicotiana senderae	Vyskot & Novak (1974)
Nicotiana sylvestris	Bourgin & Nitsch (1967)
Nicotiana tabacum	Bourgin & Nitsch (1967)
Papaver bracteatum	see Schieder (1984)
Scopolia carniolica	Wernicke & Kohlenbach (1975)
Scopolia lurida	Wernicke & Kohlenbach (1975)
Scopolia physaloides	Wernicke & Kohlenvach (1975)

The method is still not applicable for general use. Most examples of successful employment deal with plants from the Solanaceae. But even here fusion products or regenerated plants often bear undesired characters, like chromosomal aberrations, abnormal growth or sexual sterility (Schieder & Krumbiegel, 1979). Other limits were set by the availability of suitable regeneration methods for fused protoplasts. Nevertheless, in the genera *Nicotiana* and *Solanum* the transfer of disease-resistance genes from wild to the cultivated species, which is one of the classical goals of conventional plant breeding, has been achieved (see Pelletier, 1987). With interspecific *Datura* hybrids a 25% increase in scopolamine content was achieved (Schieder, 1984). This outstanding result demonstrates the future importance of this technique for medicinal plant improvement.

Genetic Engineering

The most advanced way to substitute and supplement conventional crossing is the introduction of foreign genes by the means of genetic engineering. Several strategies to introduce foreign DNA into the plant genome are presently available, like the infection of plant material with *Agrobacterium tumefasciens* carrying modified Ti-plasmids, microinjection of DNA into protoplasts or transformation of protoplasts with osmotic shock or electroporation. For all methods there are still some limitations in the general application to all plants.

Several plant promoters are already available to assure expression of the transferred foreign DNA. The possibility to construct new genes from natural templates, to introduce them into plants and to achieve their expression allows gene transfer between all living organisms. It is also possible to change the regulation of a specific metabolic trait of a plant. An example is the replacement of the natural promotor of the EPSP-Synthase gene by a more active artificial one. This produces *Petunia* plants with a 40-fold increase of the activity of this specific enzyme, leading to herbicide resistance towards the EPSP-Synthase inhibitor glyphosate (see for review Eckes *et al.*, 1987).

To increase the product yield of a specific secondary metabolite in a medicinal plant by methods of genetic engineering is much more difficult. In this case one has to remember that the production of these compounds requires the coordinated activity of several biosynthetic enzymes. Sometimes even special subcellular structures (Amann *et al.*, 1986) or special transport mechanisms are necessary for product formation or end product storage (Deus-Neumann & Zenk, 1984a). Nevertheless, recently an example for altered secondary metabolism with a *Petunia* mutant, accumulating dihydrokaempferol has been described. Transfer of a gene of *Zea mays* leads to expression of an enzyme, which is capable of dihydrokaempferol reduction. The reduced substance serves as an intermediate for pelargonidin formation, which results in red coloured flowers (Meyer *et al.*, 1988). Another possible approach to change secondary metabolism might be to set a certain biosynthetic block by the expression of anti-sense RNA. This could lead to the accumulation of biosynthetic intermediates with stronger or different pharmacological activity than the normal biosynthetic end product.

One has to see that for this kind of work the complete knowledge of the biosynthetic pathway of a certain secondary metabolite is a necessary prerequisite. For many pharmacologically active compounds this knowledge is still lacking. However, if applied properly, this kind of experiments will revolutionize plant production by creation of plants with completely new characters. This may also for medicinal plants include more general aspects of plant improvement like herbicide-, disease-, drought- or salt-resistance.

Improvement of Medicinal Plant Production

Biotechnology provides not only new methods for the creation of better plant varieties but also for the improved production of plant raw material. The *in vitro* propagation of plants and the mass production of cultured plant cells should be mentioned.

Micropropagation

The *in vitro* propagation of differentiated shoot-tip or meristem cultures, termed micropropagation or clonal propagation, offers certain advantages over the propagation of plants by seeds. Meristem cultures can easily be obtained by the cultivation of dissected meristems or shoot-tips on solidified media. Cutting off apical dominance by exposure to cytokinin or kinetin-rich media normally induces axillary branching. From sep- arated axillary shoots new plants can be obtained by induction of root formation with auxin rich or hormone free media. Changes in the cuticule of the epidermis to reduce water loss, adaptation from auxotrophic to autotrophic growth and the growth of functional roots are necessary and performed in greenhouses before the plantlets could be subjected to normal field growth (for review see Boxus, 1987)

This method offers a way to multiply a special phenotypic character directly and thereby shortens the time for introduction of a new plant variety into markets dramatically (Giles & Morgan, 1987). The method is especially useful when certain problems occur with conventional propagation methods. Two examples of application in the medicinal plant field should be mentioned. In the case of *Cinchona* plants, seed propagation leads to a high variety in agronomic characters even in the bark alkaloid content. Since clonal propagation by conventional means has been of little success, *in vitro* propagation systems proved to be useful (Hunter, 1986). *Digitalis lanata* is conventionally propagated by seed production via high alkaloid producing hybrids obtained by the crossing of inbred lines. Methods for the micropropagation of such high producing lines have been established for plant and seed production (Schöner and Reinhard, 1986, Diettrich *et al*., 1982).

Additionally, micropropagation techniques allow the combination of plant production, disease indexing and disease elimination. Especially important and useful is the production of virus-free plant material by *in vitro* propagation methods.

Although there are certain difficulties during adaptation to industrial scale procedures, like lowered multiplication rates compared with those extrapolated from laboratory experiments, "chronic contamination" with bacteria or yeasts and a certain degree of variation, the method is well established and widely used (Constantine, 1986).

Table 3 : Medicinal plants under investigation for micropropagation

Plant	Reference
Agave spec.	Groenewald *et al*. (1977)
Aloe pretoriensis	Groenewald *et al*. (1975)
Aloe vera	Data from IBPGR (1985)
Artemisia dracunculus	Garland & Stoltz (1980)
Atropa belladonna	Chaturvedi (1979)
Catharanthus roseus	Ramawat *et al.* (1977)
Chrysanthemum cineariaefolium	Roest & Bokelmann (1973)
Cinchona spec.	Hunter (1986)
Digitalis lanata	Erdei *et al*. (1981)
Dioscorea alata	Mantell *et al.* (1978)
Dioscorea bulbifera	Uduebo (1971)
Dioscorea floribunda	Chaturvedi (1975)
Dioscorea rotundata	Mantell *et al.* (1978)
Duboisia myoporoides	Data from IBPGR (1985)
Hamamelis virginiana	Data from IBPGR (1985)
Mentha piperita	Data from IBPGR (1985)
Panax ginseng	Data from IBPGR (1985)
Rauwolfia serpentina	Mitra & Chaturvedi (1970)
Solanum laciniatum	Data from IBPGR (1985)
Solanum mammosum	Data from IBPGR (1985)
Stevia rebaudiana	Yang & Chang (1979)
Strophanthus intermedius	Data from IBPGR (1985)
Tabernaemontana ssp.	Data from IBPGR (1985)
Vanilla planifolia	Data from IBPGR (1985)
Zingiber officinale	Data from IBPGR (1985)

Today more than 130 commercial laboratories for micropropagation exist, some with more than 100 employees (Constantine, 1986). In the Netherlands, a total amount of 300 to 500 million plants are expected to be produced by *in vitro* techniques annually by 1990 (Boxus, 1987). Labour costs are largely responsible for the high price of clonally propagated plants. In the future this problem might be overcome by increased mechanization, especially of the multiplication step or by placing the propagation laboratories in "low wage" countries (Giles & Morgan, 1987).

Production of Valuable Compounds by Cell Cultures

Cultured plant cells are able to produce secondary plant products as alkaloids, flavanoids, cardenolides and lignin. Furthermore substances which have not been found in the differentiated plants have been isolated from cell cultures. Examples are the alkaloids voafrine in *Catharanthus roseus* (Stöckigt *et al.*, 1984) and pericine in *Picralimia nitida* (Arens *et al.*, 1982). Even a new class of substances never observed in nature before, the paniculides, have been discovered in cell cultures (Overtone, 1977). Plant cell cultures can be addressed as a source of "dormant genes", not active in differentiated plants, which might provide a multitude of new pharmacologically active substances.

In the past, different strategies have been developed to enhance product yield or to induce product formation, like the variation of medium constituents (Zenk *et al.*, 1975), a two stage culture strategy of growth and production media (Zenk *et al.*, 1977), the selection of high producing cells and through this the establishment of high producing cell lines (Zenk & Deus, 1982). Induction of organogenesis by plant hormones or the occurence of embryogenesis under suitable culture conditions can restore product biosynthesis (Luckner, 1986). Another approach is the culture of differentiated plant roots. In normal root cultures (Anderson *et al.*, 1982) as well as in those obtained from infection with *Agrobacterium rhizogenes* the formation of medicinally important substances (quinine and scopolamine), not produced by suspension cultures, has been observed (Yamada *et al.*, 1986, Flores *et al.*, 1987).

In many cases it is difficult to induce secondary metabolite formation in large quantities in cell cultures, but in the future they may become a source of useful plant enzymes. Strictosidine the biosynthetic precursor of many indole alkaloids like strychnine, ajmaline and quinine has been successfully synthesized from secologanin and tryptamine in gram amounts by immobilized Strictosidine Synthase isolated from *Catharanthus roseus* cell cultures (Pfitzner & Zenk, 1985).

It is likely that plant enzymes with high specificity cannot be replaced by microbial enzymes for the performance of critical steps in a biomimetic synthesis of a certain compound. But microbes might be a better tool for the production of these enzymes than the slow growing plant cell cultures are. Examples for the successful transfer of genes, coding for enzymes of secondary metabolism from plants to microbes are already known. Nevertheless, there is no example of a large scale technical application up to now.

Conservation of Plant Genetic Resources by Cell Culture Methods

Even though it is possible to create new variability by somaclonal variation, the application of biotechnological methods for plant improvement and plant production is in the same way dependent on natural resources as conventional methods for plant breeding and cultivation. Improved methods for the utilization of this variability makes conservation even more important. The tools for the application of biotechnological methods are:

- Meristem or shoot-tip cultures for micropropagation
- Callus or suspension cultures for somaclonal variation or the production of secondary metabolites and enzymes
- Pollen for gametophyte culture
- Somatic embryos
- Genes for genetic engineering.

The most important tools are differentiated meristem cultures and callus/ suspension cultures, consisting of undifferentiated cells. Differentiated as well as undifferentiated cell cultures are normally maintained by continuous subculturing. In a database search of the International Board for Plant Genetic Resources (IBPGR) database performed by Withers and Wheelans (IBPGR, 1985) for medicinal plants, spices and *Solanum* species, in 90% of about 100 institutes, cultures were maintained in this way. The method is, however, very time consuming and expensive. Additionally, it implies the dangers of genetic aberrations (somaclonal variation) and loss of cultures by contamination. Preservation of both types of cultures can be better achieved in two ways:

1. Maintaining the cultures under slow growth conditions,
2. Cryopreservation of cell cultures.

Slow growth storage seems to be rather unsuccessful in the case of callus cultures (Hiraoka & Kodama, 1984), but an example of successful storage under reduced temperature is also known (IPBGR, 1985).

In contrast slow growth storage is well established and widely used in the case of differentiated cultures. Several approaches have been made to preserve meristem cultures, all exposing the cultures to certain stress conditions (for summary see Kartha, 1985). The simplest way is to store the cultures under reduced temperature. This method is successfully used for genebanking in the field of crop plants. At the Bundesforschungsanstalt für Landwirtschaft in Germany 360 different varieties of *Solanum tuberosum* are maintained at a temperature of 10°C. Cell culture maintenance and propagation is performed by a 4 hours daily work by one technician. Nevertheless, the cultures have to be propagated continuously and therefore are in danger of deterioration, infection and genetic changes.

The best way to solve conservation problems is the total suspension of all viable functions, which could be achieved best by storage in a frozen state under liquid nitrogen. Cell damage during the freezing procedure can occur by ice crystal formation inside the cell or by cell dehydration caused by ice formation in the surrounding medium (for detail see Meryman & Williams, 1985). The high water content therefore makes it much more difficult to freeze a plant cell without damage than microorganisms or for example human sperms.

Cryopreservation is especially important for undifferentiated cultures. During subculturing drastic genetic changes can occur and strains selected for high secondary product content can lose this potential when selection pressure is not continuously maintained (Deus-Neumann & Zenk, 1984b). Fortunately, just suspension cultures showed the best results with cryopreservation. Cells from the logarithmic growth phase of a suspension culture with a low water content generally show best survival rates. The water content can be further decreased by a pregrowth phase in high osmotic media before freezing. Most often mannitol and sorbitol in concentrations between 3 - 6% are used as osmotics. A real need for the survival of the cells is the application of cryoprotectants before freezing. Most widely used are DMSO, glycerol and proline. Their main function is to prevent ice crystal formation in the external medium. Very often a mixture of the different cryoprotectants is used instead of a single substance (see Withers, 1985).

The cooling rate plays another important role in freezing plant cells. Best results were achieved cooling down the cells slowly (1°C/min) to a temperature of about -40°C. Holding this temperature for about 30 - 60 minutes allows careful dehydration of the cells. Cell damage can also occur during the cell thawing. Usually fast thawing achieves better viability rates. Regrowth could be achieved best on soft agar medium. Washing procedures to remove the toxic cryoprotectants unexpectedly turned out to be disadvantageous in most cases (for review see Withers, 1985).

At our department, cryopreservation experiments with 28 different secondary metabolite containing cell cultures have been performed. Nine of these cell cultures could be frozen successfully. From the results it can be clearly deduced, that there is no general procedure applicable routinely to all cell cultures. The necessary arrangement varies from culture to culture. Even different cell culture strains from the same plant react differently to the same cryoprotection procedure. Cryoprotectant mixtures and varying pregrowth methods also trigger different reactions of cell cultures. The most important cryoprotectant seems to be DMSO. Unfortunately it has not been possible to freeze cells with a very high secondary metabolite content (Rittgen & Deus-Neumann, 1985). In the case of high producing cell cultures from *Morinda citrifolia* and from different *Galium* species no long term viability could be achieved (see also Table 4).

Table 4: Medicinal plants which have been successfully stored in liquid nitrogen

Plant	Reference
Asperula glabra	Rittgen & Deus-Neumann (1985)
Atropa belladonna	Nag & Street (1975)
Berberis dictyophylla	Withers (1985)
Catharanthus roseus	Withers (1985)
Corydalis sempervirens	Withers (1985)
Datura innoxia	Hauptmann & Widholm (1982)
Datura stramonium	Bajaj (1976)
Dioscorea composita	Rittgen & Deus-Neumann (1985)
Hyoscyamus muticus	Withers (1985)
Nicotiana plumbaginifolia	Maddox *et al.* (1982)
Nicotiana sylvestris	Maddox *et al.* (1982)
Nicotiana tabacum	Hauptmann & Widholm (1982)
Rauwolfia serpentina	Rittgen & Deus-Neumann (1985)
Rivea corymbosa	Rittgen & Deus-Neumann (1985)

Meristem or shoot-tip cultures are widely used for micropropagation and they assure genetic stability. The general problems involved in freezing a meristem culture are identical as in the case of suspension cultures. But shoot-tips and meristems are organized cell clusters and this makes freeze preservation much more difficult and causes certain problems. After the dissection of the shoot-tip a recovery phase to overcome mechanical injury prior to freezing and the use of cryoprotectants is necessary. The cooling procedures has to face the special problem of homogeneous and rapid heat extraction. Several methods have been developed to solve this problem (see Kartha, 1985). A critical point in the cryopreservation of meristem cultures is the regrowth phase. As meristem cultures are used to assure genetic stability callus formation should be avoided. This problem can be solved by the application of media with a defined hormone composition.

The general use of protoplasts, pollen and embryos for the preservation of natural resources might be considered to be of minor importance. Protoplasts can easily be obtained from suspension cultures but the performance of freeze preservation is more difficult (see Withers, 1985). Pollen can be stored with decreased viability loss under reduced temperature, in liquid nitrogen or even by freeze-drying. This might be important for breeding purposes but plants could better be obtained from preserved meristem cultures (for review see Towill, 1985). More important is the cryopreservation of embryos which can substitute for seed storage in the case of plants with recalcitrant seeds. Embryos can serve as a replacement for seeds in plant multiplication (for review see Bajaj, 1985).

Conclusions

It has already been stressed that biotechnology cannot replace, but on the contrary, is highly dependent on natural resources. To what extent can the techniques described support conventional methods to protect nature?

Endangered plant species can be protected in their natural environment in sanctuaries. This may be considered to be the best way by conservationists. But even when the activities of man are avoided the dangers of natural disasters or diseases remain.

Another approach is conservation in botanic gardens. While this is a good method to protect a species in principle, one can hardly imagine that it is possible to protect all the natural occuring variability within a species in this way.

Perhaps the best approach is the conservation in genebanks. Genebanks can easily and economically store large amounts of seeds of a single species, which also assure the conservation of intraspecific variability. But genebanking by seed storage can only be applied when seeds are available in large amounts and can be dried and stored over a long period of time. Even when these preconditions are fulfilled, decrease of seed viability causes the need of propagation which might be difficult with certain plants, for example trees. Furthermore plants with recalcitrant seeds or plants which have to be propagated vegetatively cannot be conserved in this way.

Biotechnological methods have to be applied mainly in such cases. Presently, micropropagation under slow growth conditions is the easiest method. The safest way is the cryopreservation of meristem cultures but the developement of suitable methods is laborious and time consuming. Even the performance of established methods has equal disadvantages.

In contrast, the preservation of undifferentiated suspension cultures is far simpler. Callus and suspension cultures therefrom are readily obtainable. Standard methods can easily be applied and only a very small amount of plant material is necessary. When a culture is established the production of large amounts of cell material is possible, easily and quickly. Unfortunately, the use of undifferentiated cultures implies the danger of genetic and epigenetic changes. It might be possible to overcome or at least to reduce this danger when a suitable freezing-method as well as a suitable regeneration-method is worked out, immediately after cell culture establishment. Nevertheless, the preservation of undifferentiated suspension cultures is presently the method of choice when not a whole plant but a special synthetic capacity or gene should be conserved. In our culture collection, cultures derived from the genera *Berberis*, *Galium*, *Morinda* and *Rauwolfia* are maintained, which retained synthetic capacity for a special product over more than 15 years.

Probably in future times the storage of desired genes will be the easiest method of conservation. It could be done in the comfortable ways of direct freezing of DNA samples or freeze-drying bacteria carrying "plant gene plasmids".

Need for the Conservation of Medicinal Plants

Since the beginning of this century more than half of the worlds tropical forest area has been destroyed and today more than 11 ha of tropical forest are lost every year (ATSAF, 1987). Experts estimate that only 5 - 10 % of all plants in the world have been systematically investigated for their pharmacological activity (Baerheim-Svendsen, 1984) and many of them by old fashioned pharmacological tests in the 1950's. It is likely that most of the uninvestigated plants are to be found in the threatened tropical forests. The antileukemia alkaloids vincristine and vinblastine are the most elaborate examples of the fact that even today important new substances for medical treatment could be discovered in plants. Forskolin, lately discovered in *Coleus barbatus*, will probably become a useful drug for the treatment of asthma and congestive heart disease (Valdes *et al.*, 1987). Recently Phillipson and co-workers (1987) detected several substances with an antiplasmodial activity even higher than that of the synthetic chloroquine. The application of new test systems achieved even the detection of new pharmacological activities in well known medicinal plants as the isolation of immunostimulants in *Arnica montana* (Wagner, 1984). The field of plants as a source of enzymes, which can serve as catalysts in organic synthesis has not even been touched by pharmacy. Considering the magnificient application of microbial and fungal genes and taking into account that only 3,000 bacterial and 60,000 fungal species are known but about 370,000 plant species one can imagine the immense benefits plant genes might provide in the future. Nobody knows how many unknown secondary metabolites, new pharmacological activities or useful enzymes in plants are still undiscovered. When a plant species contains only 10 different secondary metabolites each one synthesized in 10 biosynthetic steps 100 hundred different enzymes are working. For all plants at least 37,000,000 highly specific enzymes can be estimated. All these metabolites and enzymes provide a specific knowledge for mankind created by nature in millions of years of evolution. Therefore we have to strive for the protection of every single plant species to conserve its biotechnological potential for future generations.

Acknowledgement: The work performed in Munich was supported by Bundesministerium für Forschung und Technologie, Bonn, FRG.

References:

Abou-Mandour, A., Fischer, S. & Czygan, F.-G. (1979). Regeneration von intakten Pflanzen aus diploiden Kalluszellen von *Catharanthus roseus. Z. Pflanzenphysiologie*, 91, 83-88.

Amann, M., Wanner, G., & Zenk, M.H. (1982). Intracellular compartmentation of two enzymes of berberine biosynthesis in plant cell cultures. *Planta*, 167, 319-320.

Anderson, L.A., Phillipson, J.D. & Keene, A.T. (1982). Alkaloid production by leaf, root organ and suspension cultures of *Cinchona ledgeriana. Planta medica*, 46, 25-27.

Arens, H., Borbe, H.O., Ulbrich, B. & Stöckigt, J. (1982).Detection of pericine, a new CNS-active indole alkaloid from *Picralimia nitida* cell suspension cultures. *Planta medica*, 46, 210-214.

ATSAF - Arbeitsgruppe für tropische und subtropische Agrarforschung (1987). *ATSAF-Circular*, 15, 27.

Baerheim-Svendsen, A. (1984). Biogene Arzneistoffe - Heute noch oder heute wieder. In *Biogene Arzneistoffe*, ed. F.C. Czygan, pp. 27-42. Braunschweig: Vieweg.

Bajaj, Y.P.S. (1976). Regeneration of plants from cell suspensions frozen at -20, -70 and -196C. *Physiologia Pl.*, 37, 263-268.

Bajaj, Y.P.S. (1985). Cryopreservation of embryos. In *Cryopreservation of plant cells and organs*, ed. K.K. Kartha, pp. 227-243. Boca Raton: CRC Press.

Bollon, H. & Raquin, C. (1987). Haplomethods: A tool for crop improvement. In *Nestle Research News*, ed. Nestec Ltd., pp. 81-93. Vevey: Nestec Ltd..

Bourgin, J.P. & Nitsch (1967). Obtention de *Nicotiana* haploides *in vitro*. *Ann. Physiol. Veg.*, 9, 377-382.

Boxus, Ph. (1987). *In vitro* vegetative propagation of plants. In *Nestle Research News*, ed. Nestec Ltd., pp. 73-81. Vevey: Nestec Ltd.

Bundesverband der pharmazeutischen Industrie (1987). *Rote Liste 1987*. Aulendorf/ Württ.: Editio Cantor.

Chaturvedi, H.C. (1975). Propagation of *Dioscorea floribunda* from *in vitro* culture of single-node stem segments. *Curr. Sci.*, 44, 839-841.

Chaturvedi, H.C. (1979). Tissue culture of economic plants. In *Progress in Plant Research*, vol. 1 M/s, ed. T.N. Koshoo & P.K.K. Nair, pp. 265-288. New Delhi: Today and Tomorrows Printers and Publishers.

Collins, G.B. & Sunderland, N. (1974). Pollen derived haploids of *Nicotiana knightiana*, *N. raimondii* and *N. attenuata*. *J. Exp. Bot.*, 25, 1030-1039.

Constantine, D.R. (1986). Micropropagation in the commercial environment. In *Plant Tissue Culture and its Agricultural Application*, ed. L.A. Withers & P.G. Alderson, pp. 175-187. London: Butterworth.

Corduan, G. (1975). Regeneration of anther derived *Hyoscyamus niger* L.. *Planta*, 127, 27-36.

Deus-Neumann, B. & Zenk, M.H. (1984a). A highly selective alkaloid uptake system in vacuoles of higher plants. *Planta*, 50, 427-431.

Deus-Neumann, B. & Zenk, M.H. (1984b). Instability of indole alkaloid production in *Catharanthus roseus* cell suspension cultures. *Planta medica*, 50, 427-431.

Diettrich, B., Neumann, D. & Luckner, M. (1982). Clonation of protoplast derived cells of *Digitalis lanata* suspension cultures. *Biochem. Physiol. Pflanzen*, 177, 176-183.

Durr, A. & Fleck, J (1980). Production of haploid plants of *Nicotiana langsdorffii*. *Plant Sci. Lett.*, 18, 75-79.

Eckes, P., Donn, G. & Wengemeyer, F. (1987). Gentechnik mit Pflanzen. *Angewandte Chemie*, 99, 392-412.

Erdei, I., Kiss, Z. & Maliga, P. (1981). Rapid clonal propagation of *Digitalis lanata* in tissue culture. *Plant Cell Rep.*, 1, 34-35.

Evans, D.A., Sharp, W.R. & Bravo, J.E. (1987). Plant somaclonal variation and mutagenesis. In *Nestle Research News 1986/87*, ed. Nestec Ltd., pp. 63-73. Vevey: Nestec Ltd..

Flores, H.E., Hoy, M.W. & Pickard, J.J. (1987). Secondary metabolites from root cultures. *Trends in Biotechnology*, 5, 64-69.

Franz, C. (1982). Arzneidrogen - Qualitätssicherung durch Anbau und Züchtung. *Deutsche Apotheker Zeitung*, 122, 1413-1416.

Franz, C. (1986). Züchtung und Anbau - Chance für die Qualität pflanzlicher Arzneimittel. *Pharm. Zeitung*, 131, 611-617.

Garland, P. & Stoltz, L.P. (1980). In vitro propagation of tarragon. *Hort. Science*, 15, 739.

Giles, K.L. & Morgan, W.M. (1987). Industrial-scale micropropagation. *Trends in Biotechnology*, 5, 35-39.

Groenewald, E.G., Koeleman, A. & Weesels, D.J.C. (1975). Callus formation and subsequent plant regeneration from seed tissue of *Aloe pretoriensis* Pole Evans. *Z. Pflanzenphysiologie*, 75, 270-272.

Groenewald, E.G., Weesels, D.J.C. & Koeleman, A. (1977). Callus formation and subsequent plant regeneration from seed tissue of an *Agave* species (*Agavaceae*). *Z. Pflanzenphysiologie*, 81, 369-373.

Grout, B.W.W. & Henshaw, G.G. (1977). Freeze preservation of potato shoot-tip cultures. *Ann. Bot.*, 46, 381-384.

Guha, S. & Maheshwari, S.C. (1967). Developement of embryoids from pollen grains of *Datura in vitro. Phytomorphology*, 17, 454-461.

Hauptmann, R.M. & Widholm, J.M. (1982). Cryopreservation of cloned amino acid analogue-resistant carrot and tobacco suspension cultures. *Plant Physiol.*, 70, 30-33.

Hiraoka, N. & Kodama, T. (1984). Effects of non-frozen cold storage on the growth, organogenesis and secondary metabolite production of callus cultures. *Plant, Cell, Tissue and Organ Cult.*, 3, 349-357.Hunter, C.S. (1986). *In vitro* propagation and germplasm storage of *Cinchona*. In *Plant Tissue Culture and its Agricultural Applications*, ed. L.A. Withers & P.G. Alderson, pp. 291-303. London: Butterworth.

IBPGR (1985). *International Board for Plant Genetic Resources, Database*, University of Nottingham, School of Agriculture, carried out by L.A. Withers & S.K. Wheelans. Date of Survey: 1985.

Iyer, R.D. & Raina, S.K. (1972). The early ontogenie of embryoids and callus from pollen and subsequent organogenesis in anther cultures of *Datura metel* and rice. *Planta*, 104, 146-151.

Kartha, K.K. (1985). Meristem culture and germplasm preservation. In *Cryopreservation of Plant Cells and Organs*, ed. K.K. Kartha, pp. 115-135. Boca Raton: CRC Press.

Kohlenbach, H.W. & Geier, T. (1972). Embryonen aus in vitro kultivierten Antheren von *Datura wrightii* Regel, *Datura meteloides* Dun. und *Solanum tuberosum* L.. *Z. Pflanzenphysiologie*, 67, 161-165.

Luckner, M. & Diettrich, B. (1986). Integration of secondary metabolism in the programms of cell differentiation and developement - Cardenolide biosynthesis in embryogenic strains of *Digitalis lanata. In VI. Int. Congr. of Plant, Tissue and Cell Cult., Abstracts*, ed. A.D. Somers, B.G. Gengenbach, D.D. Biesboer, W.P. Hackett, C.E. Green, pp. 114. Minneapolis: University of Minnesota

Maddox, A.D., Gonsalves, F. & Shields, R. (1985). Successful preservation of suspension cultures of three *Nicotiana* species at the temperature of liquid nitrogen. *Plant Sci. Lett.*, 28, 157-162.Mantell, S.H., Haque, S.Q. & Whitehall, A.P. (1978). Clonal multiplication of *Dioscorea rotundata* Poir. and *Dioscorea alata* L. yams by tissue culture. *J. Hort. Sci.*, 53, 95-98.

Meryman, H.T. & Williams, R.J. (1985). Basic Principles of freezing injury to plant cells: Natural tolerance and approaches to cryopreservation. In *Cryopreservation of Plant Cells and Organs*, ed. K.K. Kartha, pp. 13-49. Boca Raton: CRC Press.

Meyer, P., Heidmann, I., Forkmann, G. & Saedler, H. (1987). A new Petunia flower colour generated by transformation of a mutant with a maize gene. *Nature*, 330, 677-678.

Mitra, G.C. & Chaturvedi, H.C. (1970). Fruiting plants from in vitro leaf tissue of *Rauvolfia serpentina* Benth.. *Curr. Sci.*, 39, 128-129.

Nag, K.K. & Street, H.E. (1975). Freeze-preservation of cultured plant cells. II. The freezing and thawing of plant cells. *Physiol. Plantarum*, 34, 261-265.

Nakamura, A. Yamada, T., Kodatani, N., Itagaki, Y. & Oka, M. (1974). Studies on the haploid method of breeding in tobacco. *SABRO J.*, 6, 107-131.

Nitsch, C. (1969). Experimental androgenesis in *Nicotiana. Phytomorphology*, 19, 389-404.

Nitsch, C. (1972). Haploids from plant pollen. *Z. Pflanzenz.*, 67, 3-18.

Nitsch, J.P. & Nitsch C. (1970). Obtention de plantes haploides a partir de pollen. *Bull. Soc. Fr.*, 117, 339-360.Nitsch, J.P. & Noreel, B. (1973). Effect d'un choc thermique sur le pouvoir embriogene du pollen de *Datura innoxia* cultive dans l'anther. *C.R. Acad. Sci., Paris*, 276, 303-306.

Overtone, K.H. (1977). Biosynthesis of mevalonoid derived compounds in cell cultures. In *Plant Tissue Culture and its Biotechnological Application*, ed. W. Barz, E. Reinhard & M.H. Zenk, pp. 66-76. Berlin: Springer.

Pelletier, M.H. (1987). Somatic hybridisation of higher plants. In *Nestle Research News 1986/87*, ed. Nestec Ltd., pp. 55-63. Vevey: Nestec Ltd..

Pfitzner, U. & Zenk. M.H. (1985). Isolation and immobilisation of strictosidine synthase. In Methods in *Enzymology, 136, Part 3*, ed. K. Mosbach, pp. 342-350. New York: Academic Press.

Phillipson, J.D., O'Neill, M.J., Wright, C.W., Bray, D.H. & Warhurst, D.H. (1987). Plants as a source of antimalarial and amoebicidal compounds. In *XIV.International Botanical Congress, Abstracts*, ed. W. Greuter, B. Zimmer & H.D. Behnke, pp. 291. Berlin: 291.

Ramawat, K.G., Raj Bhansali, R. & Arya, H.C. (1977). Shoot formation in *Catharanthus roseus* (L.) G.Don callus cultures. *Curr. Sci.*, 47, 93-94.

Raghavan, V. (1975). Induction of haploid plants from anther cultures of henbane. *Z. Pflanzenphysiologie*, 76, 89-92.

Rittgen, B. & Deus-Neumann, B. (1985). Stammkonservierungsmethoden für pflanzliche Zellkulturen. In *Pflanzliche Zellkulturen, BMFT Statusseminar*, ed. Bundesministerium für Forschung und Technologie der BRD, pp. 131-149. Jülich: KFA Jülich.Roest, S. & Bokelmann, G.S. (1973). Vegetative propagation of *Chrysanthemum cinearifolium in vitro. Sci. Hortic.*, 1, 120-122.

Schieder, O. (1984). Aktuelle Züchtungsforschung mit Arzneipflanzen: Ergebnisse und Perspektiven. In *Biogene Arzneistoffe*, ed. F.-C. Czygan, pp. 177-201. Braunschweig: Vieweg.

Schieder, O. & Krumbiegel, G. (1979). Höhere Erträge bei Arzneipflanzen durch künstliche Zellfusion. *Umschau*, 79, 545-546.

Schöner, S. & Reinhard, E. (1986). Long-term cultivation of Digitalis lanata clones propagated *in vitro*: Cardenolide content of regenerated plants. *Planta medica*, 6, 478-481.

Statistisches Bundesamt (1985). *Außenhandel, Fachserie 7, Reihe 2, Außenhandel nach Waren und Ländern (Spezialhandel)*, ed. Statistisches Bundesamt. Wiesbaden: Kohlhammer Verlag.

Stöckigt, J., Pawelka, K.H., Tanahashi, T., Danieli, B. & Hull, W. (1984). Voafrin A and B - New dimeric indole alkaloids from suspension cultures of Voacanga africana Stapf.. *Helvetica Chim. Acta*, 66 Frasc. 8, 2525-2533.

Towill, L.E. (1985). Low temperature and freeze-/vacuum drying preservation of pollen. In *Croypreservation of Plant Cells and Organs*, ed. K.K. Kartha, pp. 171-199. Boca Raton: CRC Press.

Uduebo, A.E. (1971). Effect of external supply of growth substances on axillary proliferation and developement in *Dioscorea bulbifera*. *Ann. Bot.*, 35, 159-163.

Valdes, L.J., Mislankar, S.G. & Paul, A.G. (1987). *Coleus barbatus* (*C. forskohlii*) (*Laminaceae*) and the potential new drug forskolin (coleonol). *Economic Botany*, 41, 474-483.

Vyskot, B. & Novak, F.J. (1974). Experimental androgenesis *in vitro* in *Nicotiana clevelandii* Gray and *N. sanderae* Hort.. *Theor. Appl. Genet.*, 44, 138-144.

Wagner, H. (1984). Immunostimulants of fungi and higher plants. In *Natural Products and Drug Developement*, ed. P. Krogsgaar-Larsen, S. Brogger Christensen & H. Kofod, pp. 391-403. Copenhagen: Munksgaard.

Weiler, E.W., Krüger, H. & Zenk, M.H. (1980). Radioimmunoassay for the determination of the steroidal alkaloid solasodine and related compounds in living plants and herbarium specimen. *Planta medica*, 39, 112-124.

Wernicke, W. & Kohlenbach, H.W. (1975). Antherenkultur bei *Scopolia*. *Z. Pflanzen physiologie*, 77, 89-93.

Westekemper, P., Wieczorek, U., Gueritte, F., Langlois, N., Langlois, Y., Poitier, P. & Zenk, M.H. (1980). Radioiummunoassay for the determination of the indole alkaloid vindoline in *Catharanthus*. *Planta medica*, 39, 24-27.

Withers, L.A. (1985). Cryopreservation of cultured plant cells and protoplasts. In *Cryopreservation of Plant Cells and Organs*, ed. K.K. Kartha, pp.243-269. Boca Raton: CRC Press.

Yamada, Y., Hashimoto, T. & Yukimune, Y. (1986). Tropane alkaloid biosynthesis in root cultures: Purification and characterisation of hyoscyaminee-6-hydroxylase. In *VI Int. Congr. of Plant, Tissue and Cell Cult. Abstracts*, ed: D.A. Somers, B.G. Gengenbach, D.D. Biesboer, W.P. Hackett, C.E. Green, pp. 117. Minneapolis: University of Minnesota.

Yang, Y.W. & Chang, W.C. (1979). *In vitro* plant regenaration from leaf explants of *Stevia rebaudiana* Bertoni.. *Z. Pflanzenphysiologie*, 93, 337-343.

Zenk, M.H., El-Shagi & H., Schulte, U. (1975). Anthraquinone production by cell suspension cultures of *Morinda citrifolia*. *Planta medica*, Suppl., 79-101.

Zenk, M.H., El-Shagi, H., Arens, H., Stöckigt, J., Weiler, E.W. & Deus, B. (1977). Formation of the indole alkaloids serpentine and ajmalicine in cell suspension cultures of *Catharanthus roseus*. In *Plant Tissue Culture and its Bio-Technological Application*, ed. W. Barz, E. Reinhard & M.H. Zenk, pp. 27-44. Berlin: Springer.

Zenk, M.H. & Deus, B. (1982). Natural product synthesis by plant cell culture. In *Plant Tissue and Cell Culture, Proc. 5th Int. Congr. Plant Tissue and Cell Culture*, ed. A. Fujiwara, pp. 391-395. Tokyo: Maruzen.

Zenkteler, M. (1971). *In vitro* production of haploid plants from pollen grains of *Atropa belladonna*. *Experientia*, 27, 1087.

Enhancing the Role of Protected Areas in Conserving Medicinal Plants

Jeffrey A. McNeely and James W. Thorsell
IUCN, The World Conservation Union, Gland, Switzerland

Introduction

As we flew over Thailand's denuded hills, we could see visual evidence of what humanity can do to vegetation when unrestricted use is permitted. Much of this destruction of vegetation was carried out by shifting cultivators who were producing a medicinal plant, Opium Poppy (*Papaver somniferum*), for the international market.

But if we had flown slightly further to the west, we would have passed over the forests of the Huay Kha Khaeng Wildlife Sanctuary; and if we had flown slightly further to the east, we would have been able to look down on the lush tropical forests of Khao Yai National Park and the five protected areas of the Petchabun Range. Protected areas are more than just scenic places where tourists go to play. Because they are protected from the worst excesses of mankind, these areas often conserve significant biological resources which are of considerable use to humanity. Many such areas contain medicinal plants which are essential to medical research and practice.

It is apparent even from an aeroplane seat that the diversity of plants and animals contained in these protected areas is far greater than in the radically altered habitats that surround them. Many of the plants and animals in these forests have real or suspected medicinal value, and people frequently enter these areas to harvest species of use. Under present laws, such use is usually considered "poaching" and is punishable by fines and/or imprisonment.

Surely a better approach is required, one which will enable people to benefit in a sustainable way from the natural resources that are being conserved. IUCN's approach to the problem of linking conservation with sustainable development is contained within the *World Conservation Strategy*, which defines conservation as: "The management of human use of the biosphere so that it may yield the greatest sustainable benefit to present generations while maintaining its potential to meet the needs and aspirations of future generations" (IUCN, 1980).

This paper suggests several innovative ways and means for the plant genetic resources contained in protected areas to be conserved for human use. The guidelines it includes shows how protected areas can be selected and managed to ensure the continued maintenance of the medicinal plants they contain.

It should be noted, however, that there is virtually no long-term conservation experience dealing specifically with medicinal plants in protected areas. This contribution to the workshop is, therefore, not based on experience, but rather builds on established modern principles of protected area management aimed at bringing the benefits of conservation to human society.

Protected Areas and Medicinal Plants: Guidelines for Selection and Planning

Managing large areas of natural vegetation is a complex process, involving the establishment of objectives, definition of management steps required to attain the objectives, implementing action, monitoring the results of action, and feeding adjustments back into the management system (McNeely, Miller & Thorsell, 1987). In order to determine whether a given protected area should include the conservation of medicinal plants as a management objective, the following guidelines should be followed:

Review what is known about the distribution of original and existing vegetation in the country

Each country should have an up-to-date inventory of the known vegetation resources found there. Maps of original and existing vegetation cover should be available based on the latest satellite imagery techniques and standardized classification systems for vegetation and climate. National herbaria, botanic gardens, and botany departments at universities hold much of this data; it should be brought together in one place to be entered into one centralized database to support planning efforts at the national level. Rare and endangered species and threats should be emphasised.

Identify medicinal uses of plants

For many countries, especially in Asia, vast numbers of plants are used for local medicinal purposes. Additional knowledge about the use of medicinal plants is certainly held by rural traditional practitioners or by tribal peoples living in the forests; this knowledge needs to be collected in a systematic way. Existing and new knowledge needs to be compiled and systematised.

At the international level, the World Health Organisation should consider supporting the medicinal aspects of a worldwide database on medicinal plants. This would be an interdisciplinary database to which IUCN would contribute the conservation component.

Identify the distribution and abundance of the medicinal plants

Once a nation has reviewed its general vegetation types and specific species of value, detailed information is needed on where these plants can be found. This information can be based on botanical collections; but once the habitat of a species has been identified, predictions can be made about where the plant might occur and might have a population that could support sustainable use. On the basis of this information, specific maps can be prepared showing where key species arc likely to be found.

As far as possible, data on phenology (e.g. time of flowering, fruiting and seed maturation) should also be kept. These are valuable aids to germplasm collection and exchange. Ideally, each medicinal plant found within a nation should have a herbarium sheet, backed up with ecological information and details of use.

Assess the extent to which the existing protected area system incorporates the plant diversity of the nation

Once general vegetation inventories and specific species distribution maps are available, these can be superimposed on maps which show the protected area coverage of the country. Where medicinal plants are known or thought to occur in a protected area, detailed information on the distribution and abundance of the species should be sought. Species which occur only in areas of natural vegetation should receive particular attention, as protected areas are likely to be especially important for them.

Identify the major economic incentives for maintaining natural habitats and wild species within the country

Modern conservation planners try to reduce loss of living resources by showing how conserving biodiversity helps safeguard the sustainable benefits people derive from wild species. Medicinal plants are only one component of this economic valuation process and cannot be disassociated from other related wild species. Further, the pharmacologically active ingredients (alkaloids, volatile oils, etc.) in medicinal plants will vary somewhat within the same species, depending on such factors as soil chemistry, associated vegetation, and presence of insect predators. For this reason, some wild populations of all medicinal plants should be maintained, even when the major source of production is under cultivation. Such wild populations should be considered to have significant economic values, even when they are harvested primarily for reproductive materials rather than medicinal products.

Among the great diversity of species found in tropical forests, some are of greater value to humans than others. Particular conservation attention should be devoted to those of known value or those which have been identified as having future potential. Such species would include:

- Wild plant species related to domesticated food crops
- Species providing utilized or potentially useful drugs
- Wild relatives or forms of domestic animals
- Species with recognized potential for domestication
- Species harvested from nature for food, wild fruits, bush meat, or important for recreational hunting
- Species harvested by man for other forms of utilization, e.g. for dyes, medicine
- Species whose value for food or other useful products is increasing due to increasing rarity
- Fodder plants for domestic animals
- Species vital for fulfilling functions on which other harvests depend, e.g. pollination by bees and bats, natural control of pests
- Species with capacity to improve soil structure, stability or fertility
- Animal species which are useful research models for studies of human behaviour and physiology such as our nearest relatives – apes and monkeys
- Species with a high capacity to modify their environment
- Species with specialized tolerance to extreme living conditions – to salinity, temperature extremes, deep shade, drought, fire and wind.

Prepare a system-level policy on the utilization of medicinal plants in the nation's protected area system

If medicinal plants are of national interest, and the various protected areas in the nation are of particular importance in conserving the medicinal plants, then policies need to be established at the national level for controlling the utilization of these resources. Such policies might cover establishing appropriate objectives for protected areas which are important for medicinal plants; appropriate techniques for recording, harvesting and producing materials from within protected areas; and appropriate training programmes for protected area managers (which should include information on the utility of medicinal plants).

Protected Areas and Medicinal Plants: Guidelines for Management

Once it has been agreed that a given protected area is sufficiently important for medicinal plants and that their conservation should be included as an objective of management, a series of steps should be followed to ensure that conservation of medicinal plants is carried out in an efficient and effective manner, as part of the normal management of the area. These steps should include the following:

Prepare a management plan for utilizing the medicinal plants on a sustainable basis

In close collaboration with the protected area manager, a team of botanists, medicinal plant specialists and park planners should prepare a plan for each area providing details on how the medicinal plants can be utilized on a sustainable basis. Harvest techniques and limits should be established and enforced; mechanisms should be designed for ensuring that a percentage of the profits earned from the medicinal plants are returned to improving management of the protected areas; certain areas should be established which are off limits to harvesting, because they are important for tourism, seed reservoirs, etc.

Many plant species of value for their medicinal properties are confined to early successional stages, and may be replaced by other species as the vegetation proceeds to more mature stages. Conservation of such species will require intervention to maintain optimum populations of the species concerned. The necessary measures to be outlined in the species management plan could include:

- Controlled grazing or mowing to maintain forage populations adapted to the grazing subclimax;
- Controlled burning to maintain fire-climax vegetation;
- Logging to open up the canopy;
- Thinning and removal of competing vegetation (undergrowth or overgrowth) to promote survival and regeneration of particular individuals or populations;

- Control or removal of predators, parasites or competitors to favour a particular species; and
- Enrichment planting to assist regeneration of particular populations (the seed for such planting should always be from the population concerned).

Publicize the role of the area for medicinal plants

Human health is of great concern to society, so it will often be in the interest of the protected area to publicize the fact that it is contributing to human welfare through providing a sustainable supply of material for medicinal uses. The information materials should emphasize that the harvest of medicinal plants is not damaging the objectives for which the protected area was established. As their interests are very much at stake, private industry can often be called on to provide support for the preparation and dissemination of public information materials.

At the international level, WHO should use its contacts and prestige to make Ministries of Health more aware of the contributions of protected areas in conserving medicinal plants.

Review how benefits of conserving wild medicinal plants could accrue to people living around protected areas

While many protected areas have tended to close their doors to local people, it has become increasingly apparent that local people will co-operate best in conservation if their own self-interest is enhanced. A wide range of techniques to maximize local benefits should be pursued (see MacKinnon *et al.*, 1986, Chapter 6). Institutions like WWF should ensure that their field projects include elements dealing with the benefits which can be provided for the local people living around the area.

Design a system for management of harvesting of products from protected areas

Certain categories of protected areas that contain reservoirs of medicinal plants should allow harvesting on a sustainable basis of medicinal plant products and reproductive materials (seed, herbs, budwood, etc.) by *bona fide* researchers, and by commercial and traditional users on a sustainable basis. Where such harvesting is permitted, it should be viewed by those who benefit as a privilege rather than a right, and in this regard some form of payment in cash or kind may have advantages. In any case, when medicinal products from protected areas can be harvested under appropriate control without affecting natural values to a significant extent, public relations are improved and potential illegal and more damaging activities will be discouraged.

Any management strategy for sustainable harvesting needs to incorporate ways and means of controlling over-exploitation, while permitting levels of exploitation which are sufficient to meet legitimate needs. The formula for controlling communal harvesting of medicinal plant products in protected areas should consider the following points:

- Try to introduce a territorial element into harvesting so that a given area is designated as being for the sole use of people from one community, who will defend their privilege against outsiders.
- Try to ensure that every resident household is able to benefit fairly; otherwise those left out will not respect harvesting limits. However, to avoid creating an attractive zone which draws more people, harvesting rights should be granted only to original households and their descendants.
- Encourage those who have harvesting rights to appoint a management committee to decide on cropping and planting schemes, quotas and labour requirements.
- The committee must arrange with harvesters a fair system of labour inputs. Villagers may be requested to provide an agreed quota of workdays themselves, provide substitute labour, or make a financial payment to the management committee to hire labour.
- Harvestable quotas should be decided by the committee with approval by the protected area management authority.
- The management committee should keep records of all materials harvested, and be responsible for marketing any cash products and arranging fair disposal of products.

Of course, many traditional societies already have local rules controlling the use of communal resources and these can sometimes be adapted to control use of resources in reserves. The main advantage of such local systems and tribal laws is that they are familiar to the local people rather than seen as edicts imposed by an outsider, in this case the management authority. The authority should seek to incorporate such local traditions into its own control system as this makes regulation easier. Although protected area managers should consult with local community councils about levels of harvesting, and how they are controlled, the ultimate responsibility for regulation should remain with the management authority.

It should also be noted that many protected areas already are managed to provide benefits to local people, including everything from thatch grass in Chitwan National Park in Nepal, to water in Venezuela's Canaima National Park, to animals in several of Zimbabwe's protected areas (see McNeely & Miller, 1984, and McNeely & Thorsell, 1985, for case studies).

Make special provision in protected area management plans for medicinal plant considerations

In principle, the objective of conserving medicinal plants is compatible with other objectives of most categories of protected areas (e.g. watershed protection, species reservoirs, etc.). However, in some cases some protected area management objectives – such as promoting tourism, conserving climax vegetation, or conservation of maximum populations of elephants or tigers – might conflict with medicinal plant conservation. In such a case, certain areas may need to be zoned and conflicting uses may have to be controlled if medicinal plant conservation objectives are not to be compromised.

Establish research and monitoring programmes on medicinal plants in the most important areas

Protected areas which are particularly important for medicinal plants should be the sites of continuing research programmes on medicinal plants, in collaboration with relevant Ministries of Public Health, Primary Production, and Research. Since many medicinal plants are of particular concern to the pharmaceutical industry, protected area managers might also ensure that their interests are specifically addressed in the management plan, and that private industry is given an opportunity to contribute to the welfare of the protected area and the people living around it. Linkages with local universities and botanical gardens can often be extremely useful. The protected area manager should take special steps to ensure that these linkages are established and promoted.

Management-oriented research should be undertaken on such topics as minimum effective population size, relationship of soil chemistry to the concentration of pharmacologically active compounds in medicinal plants, traditional knowledge, pharmaceutical qualities, boundary maintenance, effects of exotics, etc.

Medicinal Animals and Protected Areas

Most medical practitioners think more in terms of medicinal plants than medicinal animals, but animals play an important role in medicine in many parts of Asia. McNeely & Wachtel (1988) present a useful overview of these uses. For example:

- In India, research has shown that musk from the musk deer stimulates the heart and central nervous system, and acts as an anti-inflammatory agent for the treatment of snake bites.
- A recent study at the Worker's Hospital in China's Hunan Province determined that powdered elephant skin is effective against inflammation, gastric ulcers, chilblains, external hemorrhoids and chronic rheumatoid arthritis.
- Research in the Soviet Union has shown that the velvet of deer antlers contains a substance called pantocrin which has been proven in clinical trials to be useful as a tonic and for accelerating the healing of wounds and ulcers.
- Snake venoms have many medicinal applications (in addition to antivenins), including for hemophilia (Russell's viper), phlebitis (Malaysian pit viper) and myasthenia gravis (many-banded krait).

Other medicinal uses of animal products are not nearly so scientific. In various parts of Asia, various parts of rhinoceroses are used to provide relief against tuberculosis, fever, dropsy, skin lesions, snake bit, ear ache and stomach upset. The official Chinese *Barefoot Doctor's Manual*, which is the basis for the primary medical care received by the majority of that vast country's population, prescribes rhino horn for a wide variety of illnesses including encephalitis B and typhoid.

Various parts of tigers are used for skin disease, to provide courage, to immunize against bullets, to drive away centipedes, to cure laziness and acne, to cure abcesses on hands and feet, to stop persistent watering of the eyes and to cure convulsions. The gall bladder of Asian bears are so prized in Korea as a medicine for chronic fatigue, digestive problems, inflammation, blood impurities, hemorrhoids, and various liver, stomach, and intestinal ailments, that the price of a 3-ounce gall bladder is US$200.

None of the tiger, bear and rhino applications have been shown to have a pharmacological basis, but they are nonetheless real in the minds of the people who use them. McNeely & Wachtel (1988) quote a Chinese traditional apothecary in Bangkok: "Modern drugs are certainly much stronger and can cure some things my medicines can't touch. But, for many illnesses, western medicines are too strong and do as much harm as good. My treatments are aimed at controlling symptoms, keeping the patient alive, if not comfortable, until the disease runs its natural course."

Irrespective of their pharmacological value, the medicinal animals have value to people, and this has contributed to the great over-exploitation of species such as rhinos, tigers, musk deer, bears, monkeys and pangolins. Protected areas may well provide the last sanctuary for these animals which are of great importance to humans, and modern management plans need to incorporate ways and means to ensure that the benefits of these species are made available to society.

Finally, field biologists are finding that many animals also seem to be aware of the value of medicinal plants. Elephants, for example, selectively seek the bark of a tree in Thailand which is known as a powerful anthelmintic agent. While little systematic research has been done in this field, it would seem very worthwhile to encourage field biologists to observe the possible use by animals of medicinal plants. The catalogue of medicinal plants which is recommended above could provide a very useful tool for such fieldworkers. It is also likely that some animals may have found plants which are of medical or veterinary use, but have not yet been discovered by people.

Private Industry and Protected Areas

The guidelines included above are aimed primarily at protected area managers, field researchers, and the public health community. But the pharmaceutical industry also has a role to play, because it has a vital interest in ensuring that the wild reservoir of medicinal plants is managed in such a way that benefits continue to flow to human health. The pharmacutical industry can contribute in at least the following ways:

- Ensure that the technical-level officials in the industry are well aware of the role that protected areas play in conserving medicinal plants.
- Support field research on medicinal plants in protected areas, and ensure that the results of such research are fed back into the protected area management system.
- Design public information materials that include appropriate acknowledgement of the ultimate source of such materials, and encourage continued protection for this source.
- Give sympathetic consideration to requests from protected area managers to provide financial support to public information materials, for distribution both within the protected area and through extension programmes in the surrounding countryside.

Conclusions

As the rising tide of humanity reduces the extent of natural vegetation everywhere in Asia, those who are interested in medicinal plants will need to look increasingly to legally protected areas as the last reservoir of plants and animals with enormous genetic significance. Equally, those who are charged with managing these areas will need to accomodate their management approaches so that sustainable supplies of medicinal plants and animals can be provided. Thus a new partnership needs to be formed between private industry, those institutions involved in medicinal plants (including ministries of health, indigenous peoples, and rural development as well as universities, botanic gardens, and research institutions), and those institutions involved in conservation of biological diversity (including protected area management authorities, donor agencies, and non-governmental organisations).

The major international development institutions, such as FAO, WHO, UNDP and the World Bank, also need to support this linkage between effective management of protected areas and the provision of sustainable benefits to human societies. At the national level, ministries of planning, finance and rural development need to be made aware of this new partnership, and take steps to foster more productive collaboration.

References

IUCN (1980). *World Conservation Strategy.* Gland, Switzerland: IUCN/WWF/UNEP.

MacKinnon, J. & K., Child, G., Thorsell, J. (1986). *Managing Protected Areas in the Tropics*. Gland, Switzerland: IUCN.

McNeely, J.A. & Miller, K.R. (eds) (1984). *National Parks, Conservation, and Development: The role of protected areas in sustaining society.* Washington, D.C.: Smithsonian Institution Press.

McNeely, J.A. & Thorsell, J.W. (ed) (1985). *People and Protected Areas in the Hindukush-Himalaya*. Kathmandu: ICIMOD.

McNeely, J.A., Miller, K.R. & Thorsell, J.W. (1987). Objectives, Selection, and Management of Protected Areas in Tropical Forest Habitats. In *Primate Conservation in the Tropical Rain Forest*, pp. 181-204. New York: Alan R. Liss.

McNeely, J.A. & Wachtel, P.S. (1988). *Soul of the Tiger: The Relationship Between People and Animals in Southeast Asia*. New York: Doubleday.

Udvardy, M.D.F. (1975). *A Classification of the Biogeographical Provinces of the World.* Morges, Switzerland: IUCN Occasional Paper N° 18.

Acknowledgements

The authors are grateful to Vernon Heywood, Hugh Synge and Paul Wachtel for their comments on an earlier draft of this paper.

Annex
The Role of IUCN and the CNPPA

The International Union for Conservation of Nature and Natural Resources (IUCN) is the largest and most representative partnership of conservation, environment and wildlife interest groups in the world. Founded in 1948, the Union includes 60 State members, 125 ministries or other government agencies and 412 major national and international non-governmental organisations and citizens' groups – a total of 597 members in 119 countries. This unique mix of policy makers, administrators and activists helps solve common conservation problems. It also provides an independent forum for conservation debate.

The Union works in close partnership with the United Nations Environment Programme (UNEP) and Unesco and advises the World Wide Fund for Nature (WWF) on conservation priorities, prepares programmes and manages over 300 conservation field projects in 70 countries.

IUCN, primarily through its Commission on National Parks and Protected Areas (CNPPA), has been deeply involved with national parks from its very beginnings. This involvement has included:

- Establishing the Commission on National Parks in 1960. Now enlarged to become the Commission on National Parks and Protected Areas (CNPPA), this body has a current membership of 355 senior park professionals from 126 countries.
- Establishing a system of biogeographic provinces of the world (Udvardy, 1975), now widely used for assessing protected area coverage and suggesting regions for priority attention.
- Publishing lists and directories of protected areas. The Protected Area Data Unit (PADU) was established in 1981 to computerise the data held by IUCN and to promote greater applications of the data.
- Publishing basic conceptual papers dealing with protected area matters such as regional system reviews, legislation guidelines and threats to the world's protected areas.
- Publishing the quarterly journal Parks which provides informative articles on protected area management problems and solutions.
- Cooperating closely with United Nations agencies involved in protected area matters (FAO, UNEP, Unesco), at both planning and field levels. This includes providing technical evaluations of natural sites nominated for the World Heritage List to Unesco's World Heritage Committee and acting as the Secretariat for the Convention on Conservation of Wetlands of International Importance especially as Waterfowl Habitat.

- Holding meetings in various parts of the world to promote protected areas. CNPPA holds two working sessions per year, rotating among the biogeographic realms. IUCN has organised major international meetings on protected areas, including the First World Conference on National Parks in Seattle, Washington, 1962, the Second World Conference on National Parks in Grand Teton, Wyoming, 1972, the International Conference on Marine Parks and Reserves, Tokyo, 1975, and the World Congress on National Parks, Bali, Indonesia, October, 1982.
- Supporting field projects, especially in developing countries, aimed at establishing and managing national parks and protected areas. Funded primarily by WWF, some 1600 projects involving the expenditure of over $US 45 million had been implemented in support of protected areas by 1987.

Botanic Gardens and the Conservation of Medicinal Plants

Vernon Heywood
IUCN, The World Conservation Union, Kew, Richmond, UK

Introduction

Botanic gardens in the western tradition have been intimately involved in the cultivation of medicinal herbs since their earliest beginnings. In fact, the first botanic gardens were medicinal or herbal gardens, attached to faculties or schools of medicine and created for the provision of medicinal plants for study by students and for the actual production of drugs. Such were the 16th century Italian botanic gardens of Pisa, Padova and Florence, the latter still called today the Giardino dei Semplice (Garden of Simples), and in other countries of western Europe (Garbari & Raimondo, 1986; Heywood, 1987; Stafleu, 1969).

These early physic or medicinal gardens are a reminder of the very close relationships that existed between botany, pharmacy and medicine in earlier times. In fact they could not be easily separated. This link between botany and medicine persisted until the 20th century with medical students in British universities attending lectures in botany in their first year of studies; and even today there are Departments of Botany in Faculties of Pharmacy as well as Science in Spanish, French and Portuguese universities.

Many of these 16th and 17th century botanic gardens were very small and in urban settings. Today, they are often at risk of closure and regarded by many as little more than living museums although some of them have made great efforts to modernize and widen their facilities.

The medicinal role of some botanic gardens was sometimes directed to the exterior as in the case of Nantes where the botanic garden was originally in the 18th century the Jardin des Apothicaires, which grew medicinal plants for the medicine chests of the doctors of ships sailing from the port. In turn the garden was at the receiving end of new plants, medicinal and others, from abroad following a Royal Edict signed at Versailles, requiring all captains and merchant mariners of Nantes to bring back all new plants found on their trips to foreign countries and the French colonies of America.

Botanic Gardens and Medicinal Plants Today

In the face of the current renewal of interest in the role of plants in traditional medicine it is pertinent to ask what the reaction of botanic gardens has been. This question cannot be divorced from the movement of renewal and reappraisal through which the botanic garden movement is going at the present time.

Traditional medicines are mainly used as a principal health delivery system in developing countries, especially in the tropics and subtropics. As is well known, the number of botanic gardens in these regions is quite limited, some 230 in all out of a world total of over 1400. This is especially true in Latin America and tropical Africa. In Asia, there are proportionately more botanic gardens with some 36 in India alone. A curious feature, however, is that a medicinal plant tradition did not develop in these tropical botanic gardens until this century: their original role was economic but with an emphasis on potential spices, food crops, timber and plantation species. The exploitation by the west of tropical plant species as a source of drugs is in most cases relatively recent and botanic gardens have so far not played a major part in this. The exceptions are mainly spice species which have subsequenly been found to contain chemical compounds of medicinal value and alkaloid-containing species such as Cola, Cacao, Coffee and Guarana and of course quinine.

Today, we are conscious of the fact that many thousands of species of tropical plants are used in traditional medicine yet not only is our scientific knowledge of them limited but our abilities to cultivate and propagate them equally defective. This is where the cooperation of botanic gardens must be enlisted.

After recent decades of quiescence or neglect, botanic gardens are beginning to appreciate that they have to consider themselves as basic instruments or resources in conservation and development. As Ashton (1984) has noted, "botanic gardens have an opportunity, indeed an obligation which is open to them alone, to bridge between the traditional concerns of systematic biology and the returning needs of agriculture, forestry, and medicine for the exploration and conservation of biological diversity". There needs to be an appreciation, likewise, on the part of the scientific community to recognize that botanic gardens are not just academic luxuries or quaint parks reminiscent of bygone days, havens of nostalgia. It is up to the gardens to reassert themselves and for the conservation movement to recognize the essential part they play in the overall strategy.

For Ashton is right in implying that botanic gardens are uniquely qualified to undertake certain tasks that no other institutions are suitable for. Botanic gardens are centres of scientific study of plants. The plants are labelled and the collections are usually documented. They contain highly qualified and skilled staff who are experts in raising and growing plants, in germinating seeds, propagating plants by diverse techniques, and relating to the outside world. These are all skills that are needed in conservation and particularly in the case of medicinal plants.

Medicinal Plant Gardens

There are today a number of botanic gardens which specialize mainly in the cultivation and study of medicinal plants. In addition, many other botanic gardens contain either a special medicinal plant garden or maintain special collections of medicinal plants (see the Annex).

Some of the major specialized medicinal botanical gardens will now be considered:

The Tokyo Metropolitan Medicinal Plants Garden is an important Centre founded in 1945, as a link in medicinal and drug administration. It collects and cultivates medicinal plants and produces some biopharmaceutical drugs for teaching purposes. It also monitors such drugs that are available on the market and undertakes basic research on medicinal plants and conducts quality control of biopharmaceutical drugs. Another important function of the Garden is to disseminate reserach knowledge about drugs and medicines to Tokyo citizens. It is open to the public and produces instructional material and holds classes on medicinal plants. About 1600 medicinal plant species are cultivated in the Garden.

India has a rich medicinal plant flora of some 2000 species and most of the drugs prescribed in the native systems of medicine, notably Ayurveda, have their origin in these plants. While the demand was originally met by local supplies of plant material, often in home gardens or backyards, the commercialization introduced at the beginning of this century in the face of the increasing demand from a growing population, led to factory scale production of herbal drugs. The build-up of demand led to over-exploitation of several species and this coupled with the large scale deforestation experienced in India led to the introduction of substitute drugs or, worse, adulterated materials.

It was against this background that the Tropical Botanic Garden and Research Institute, Trivandrum, decided to establish a 'gene-pool' of medicinal plants, together with a project on the micropropagation of rare medicinal plants aimed at distributing material to interested institutions and growers. The medicinal plant division of the Garden has so far raised 650 species.

In Mexico, an Ethnobotanical Botanic Garden (Jardín Etnobotánico) was started in 1979 at the Centro Regional Morelos. It occupies 4 hectares, only one of which is currently in use and mostly occupied by plants used in medicine in the State – wild and cultivated, native and exotic. Other botanic gardens involved in ethnobotanical studies include the Waimea Arboretum and Botanic Gardens, and the Amy Greeenwell Ethnobotanical Garden of the Bishop Museum, Honolulu, Hawaii, which focus on Hawaian plants, and the Devonian Botanic Garden, Vancouver, Canada, which grows and undertakes research on plants used by native peoples.

The National Institute of Hygienic Sciences in Japan is a major centre for drug research and related fields. It maintains five Experimental Stations of medicinal plants in different climatic locations in the country. This climatic diversity is used for the acclimatization of medicinal plants. The Institute places emphasis on the experimental cultivation and breeding of medicinal plants of domestic or foreign origin to ensure the supply of raw materials to drug manufacturers and thereby contributes to the rural agrarian economy. The Experimental Stations provide technical information for growers and makes seeds and seedlings available to them. They exchange seeds with botanic gardens throughout the world.

Medicinal Plant Botanic Gardens are particularly common in Eastern Europe – in Bulgaria, Czechoslovakia, Hungary, Jugoslavia, Poland, and the USSR where there is a long-standing tradition of use of native and exotic plants in herbal medicine.

The Polish herb industry, for example, which is the main manufacturer of native medicinal plant drugs, uses about 100 species from native resources plus about 60 of foreign origin. The Institute of Medicinal Plants founded in Poznan in 1947 is a major scientific centre for herb cultivation. It collects and cultivates medicinal plants of native origin and from other areas and grows some two thousand species altogether. Species which are successfully cultivated are then analyzed for their chemical principles and where appropriate introduced into cultivation on a large scale for the industrial production of crude drugs.

Not surprisingly, in view of the very widespread and important part played by traditional medicine in China, there are several botanic gardens in that country which specialize in medicinal plants, such as the Botanic Garden for Medicinal Plants of the Institute for Medicinal Plant Development (IMPLAD), Beijing.

A Conservation Strategy for Botanic Gardens

Medicinal plant conservation is just a special case of the wider role that botanic gardens should and can play in conservation of plant resources. The development of a conservation strategy for botanic gardens has evolved over the past few years following a series of key events such as the 1975 and 1978 Kew conferences (Simmons *et al.*, 1976; Synge & Townsend, 1979) and the Las Palmas conference on Botanic Gardens and the World Conservation Strategy held in 1985 (Bramwell *et al.*, 1987). At the Las Palmas meeting a draft of the IUCN-WWF Botanic Garden Conservation Strategy was discussed and approved and a revised version is nearing completion. One of the main follow-ups to the Las Palmas meeting was the decision by IUCN to establish as from January 1987, a Botanic Gardens Conservation Secretariat based at its offices at Kew, UK.

Of course, many individual botanic gardens have engaged in conservation activities for some considerable time but most gardens and arboreta, especially in North America, have devoted most of their efforts to their traditional roles such as display, education and horticultural research. There is nothing remarkable about botanic gardens growing plants well! What is a new development is the recognition of conservation as an additional priority activity for many such gardens. Indeed a number of North American gardens have incorporated conservation of native species into their charters and accessions policies and there have been similar developments in other countries to varying degrees. In France, for example, Conservatoires for plant species have been set up at Brest, Porquerolles and Nancy in association with the Muséum National d'Histoire Naturelle charged with the mission of cultivating and preserving endangered species of particular regions. In 1987, a fourth was added – the Conservatoire du Mascarin – which is being developed on the island of Réunion in the Indian Ocean. Similarly the Jardín Botánico Viera y Clavijo on the Canary Island of Las Palmas specializes in the cultivation and propagation of endangered species of the Macaronesian region.

The rationale for considering botanic gardens as natural partners of the conservation movement, the sleeping constituency as it has been termed, can be summarized as follows.

In the face of the continuing and accelerating destruction and alteration of the world's plant life, it is recognized that cultivated populations of plants and their propagation must become integral parts of our global strategy for the preservation of the genetic diversity of plant species. This is affirmed in the World Conservation Strategy.

Botanic gardens and arboreta should be major instruments in this strategy

- By serving as rescue centres for the cultivation of samples of taxa which are threatened with extinction in the wild.
- By maintaining gene banks and other collections of germ plasm of endangered species as an adjunct to living wild populations.
- By the provision of seed and other propagules for reintroduction into natural habitats.
- By undertaking research on the reproductive biology of endangered species, notably their germination and establishment.
- By undertaking research into the management of populations in reserves, and minimum size to maintain an adequate representation of genetic variations.
- By contributing to conservation of plant species in natural wild habitats.
- By acting as information and education centres, providing the scientific community and general public with arguments in support of plant conservation.

Botanic gardens and arboreta in many countries are well placed to act as focal centres for plant conservation activities, both *in situ* and *ex situ*. Preservation in the form of seed or other propagules in genebanks may be the only option available in very many instances. A major difficulty, however, is that it has not yet proved possible to identify many of the thousands of species that may be at risk, due to the shortcomings in our floristic and taxonomic knowledge.

Ex Situ Conservation in Botanic Gardens

There is considerable controversy as to the effectiveness of the role that botanic gardens can play in conservation of plant resources. Some go so far as to declare botanic gardens (and plantations) as quite unsuited for the purposes. Others consider *ex situ* conservation as, at best, a poor substitute for *in situ* conservation. The fact is that both approaches are necessary and complement each other.

Conservation of endangered plant species in an obvious and appropriate role for botanic gardens today. The idea is not a new one, having been explicitly enunciated at the first and second International Congresses for Nature Protection held in 1923 and 1931 in Paris.

While most attention has been directed at attempts to conserve rare or endangered species in botanic gardens as living collections, it has to be remembered that a policy of "pre-emptive" action may be equally important in building up information on the cultural requirement of plants that are not at present endangered or suffering from genetic erosion but which since they are known to have medicinal potential are susceptible to population depletion and subsequent endangerment through overcollection. Bringing species into cultivation can not only help provide a ready and continual supply of material for drug manufacturers but may take the pressure off field populations.

While cultivation is an obvious solution, this presupposes that we have knowledge about how to germinate the seeds, how to grow the plants on, under what conditions of shade, temperature, soil, etc, how to propagate the plants. Then there is a need to ensure that a range of useful germplasm is brought into cultivation for agricultural/sylvcultural purposes and as wide a sampling as possible made of the genetic variations involved in the species.

Even then, as has often been pointed out in the case of rare and endangered species, bringing plants into cultivation is only the first step towards saving them from extinction. A second step is to ensure their survival in suitable protected sites in the wild and this may involve reintroductions which will usually require years of monitoring and follow-up to ensure that the reintroduced population builds up and survives.

Until recently most *ex situ* conservation in botanic gardens has been undertaken on a somewhat random or unscientific basis and has been justifiably criticised. The cultivation of small numbers of individuals or even single specimens (in the case of trees) of endangered species in botanic gardens has little more than curiosity value. It is unfortunately true, however, that there are today many instances when such tiny samples are the only representatives left of species which are no longer known in nature.

There is a need to introduce a set of appropriate guidelines and procedures designed to achieve the most effective and long-lasting conservation possible.

Seed Banks of Medicinal Plant Species in Botanic Gardens

The most cost-effective method of *ex situ* conservation is the storage of seed samples under controlled conditions in seedbanks following well-established techniques which have been developed principally for crop plant species and trees by agencies such as FAO and IBPGR.

Very few plant species used in traditional medicine are conserved in seedbanks and there is a clear and urgent need to address this issue. One genebank that does hold material of medicinal plant species is that at Gatersleben in Germany. The genebank as a whole contains more than 60,000 accessions of cultivated plants and their wild relatives and included in the special collection of medicinal plants are more than 2,000 accessions. In addition, there are several hundred accessions of medicinal plants which are traditionally considered primarily as forage and grain crops (e.g. Leguminosae).

Gatersleben is developing a programme of research into the problems of maintaining large collections of medicinal plants in the genebank such as seed storage conditions, cultivation requirements and isolation needs.

The IUCN Plants Office, in association with the Botanic Gardens Conservation Secretariat, is drawing up a set of guidelines for the establishment of a network of seedbanks for wild species to be held in botanic gardens and plans to do this in cooperation with the International Board for Plant Genetic Resources (IBPGR). The numbers of rare and endangered species or other selected groups of wild species that might be considered for inclusion in such seedbanks is potentially very large and greatly in excess of any facilities that are likely to become available in the next few years. There is a need, therefore, to establish a set of priorities and this could include species used in traditional medicine.

Ex situ conservation will always have limitations. It is in most cases, as Ashton (1988) has emphasised, "a refuge of last resort; a high risk refuge, perhaps of no escape". No matter which part of the plant is used for conservation – the whole plant, seed or *in vitro* culture of tissue, meristem, callous, cells, it is simply not possible to conserve species *ex situ* unless the species is already reduced to small number of individuals in which case it can be argued that the original species no longer exists except as relictual fragments. Even if adequate samples of a range of populations is undertaken to ensure that as much genetic variability as possible is represented (and one has to admit that this is normal practice), this will not guarantee the preservation of all the particular character combinations that can occur at a very local level. Selection *ex situ* will be unnatural. Hybridisation is a risk. And of course, *ex situ* conservation of small samples will inevitably lead to a genetic change that cannot be predicted. Ashton is of the opinion that for all practical intent, *ex situ* species conservation leads inexorably and irreversibly to domestication.

We have to accept, however, that such a last-resort policy may be all that is available to us in some cases. In other instances, the introduction of material into cultivation will allow us to study the biological characteristics of growth and reproduction that will facilitate the large-scale cultivation of medicinal plants on a commercial basis, or in some instances complement *in situ* conservation by helping us to reintroduce endangered species back into the wild or by giving us insights into how to manage reserve areas more effectively.

Conclusion

With national and public health authorities turning their attention increasingly to medicinal plants as the source of complex drugs, and with growing concern at the actual and potential loss of genetic resources of medicinal plant species, it is appropriate for botanic gardens to reassess their position and re-establish in modern times their close involvement in the cultivation and study of these plants.

Botanic gardens are by definition centres of plant cultivation, whatever else, and their skills should be fully deployed in the wide range of activities involved in the selection, analysis, assessment, commercial cultivation, conservation and preservation of medicinal plant species.

Botanic gardens should be regarded as a key constituency in this increasingly important field and steps should be taken to involve them as closely and as fully as possible in the traditional medicine movement.

References

Ashton, P.S. (1984). Botanic gardens and experimenstal grounds. In *Current Concepts in Plant Taxonomy*, eds. V.H. Heywood & D.M. Moore. London & New York: Academic Press, pp. 39-48.

Ashton, P.S. (1987). Biological considerations in *in situ* vs *ex situ* plant conservation. In *Botanic Gardens and the World Conservation Strategy*, eds. D. Bramwell, O. Hamann, V. Heywood & H. Synge London & New York: Academic Press. Pp. 117-130.

Ashton, P.S. (1988). Conservation of biological diversity in botanical gardens. In *Biodiversity*, ed. E.O. Wilson. Washington, D.C.: National Academy Press, pp. 269-278.

Bramwell, D., Hamann, O., Heywood, V. & Synge, H. (eds) (1987). *Botanic Gardens and the World Conservation Strategy*. London & New York: Academic Press. Pp. xxxix + 367.

Garbari, F. & Raimondo. (1986). Botanical gardens in Italy: their history, scientific role and future. *Museol. Sci.*, 3 (1-2), 57-81.

Heywood, V.H. (1987). The changing role of the botanic garden. In *Botanic Gardens and the World Conservation Strategy*, eds. D. Bramwell, O. Hamann, V. Heywood, H. Synge. London & New York: Academic Press. Pp. 3-18.

Simmons, J.B., Beyer, R.I., Brandhan, P.E., Lucas, G.L. and Parry, V.T.M. (eds). (1976) *Conservation of Threatened Plants*. New York: Plenum Press. Pp. xvi + 336.

Stafleu, F.A. (1969). Botanical gardens before 1818. *Boissiera*, 14, 31-46.

Synge, H. & Townsend, M. (eds) (1979). *Survival or Extinction*. Kew, U.K.: Bentham Moxon Trust. Pp. 250.

Annex
List of Botanic Gardens Specialising in Medicinal Plants or with specialised collections of Medicinal Plants

Australia

Anderson Park BG, Townsville
Museum Applied Arts and Science Plantation, Ultimo

Belgium

Jardin Botanique Jean Lebeau, Jamioulx
Jardin Botanique National, Meise
Centre Pilote des Plantes Medicinales, Hainaut

Bulgaria

Hortus Plantarum Medicarum, Sofia

Brazil

Hortus Plantarum Medicarum, Fortaleza
Jardim Botânico Rio de Janero

Canada

Devonian Botanic Garden, Vancouver
Royal Botanic Garden, Hamilton
Botanic Garden, University of British Columbia, Vancouver

China

Botanical Garden for Medicinal Plants, IMPLAD, Beijing
Guangxi Botanic Garden of Medicinal Plants
Kunming Botanic Garden
Nanjing Botanic Garden
Xishuangbanna Botanic Garden

Colombia

Jardín Botánico Leandro Agreda, Putumayo

Czechoslovakia

Hortus Plantarum Mdicarum, Fac. Pharm, Univ. Comenianae, Bratislava
Botanicka Zahrada Veterinarin Fakulty, Brno
Hortus Centralis Cultura Herbarium, Medic. Fak. Univ. Purkyniana, Brno
Hortus Botanicus, Fak. Univ. Carolinae, Hradec Kralove
Botanic Garden, Palaky University, Olomouc
Výzkumný Ustav pro Farmacii a Biochemii pracoviště Hloubetín, Praha

Finland

University Kuopio Botanic Garden

France

Jardin Botanique de la Ville & Univ. Besançon
Jardin Botanique de la Ville & Univ. Caen
Jardin Botanique Yves Rocher, La Gacilly
Jardin Botanique de la Ville de Limoges
Jardin Botanique de la Faculté de Pharmacie, Lyon
Jardins Botaniques Municipaux, Marseille
Jardin Botanique de la Faculté de Pharmacie, Paris
Jardin Botanique de Rouen
Jardin Botanique, Universite Medic. de Grenoble, Station Alpine du Lautaret
Jardin Botanique de a Faculté de Pharmacie, Tours

Federal Republic of Germany

Botanischer Garten der Stadt Bielefeld
Botanischer Haupschulgarten, Frankfurt am Main
Botanischer Garten der Stadt Krefeld
Botanischer Garten der Wurzburg

Honduras

Lancetilla Botanic Garden

Hong Kong

Kadoorie Botanic Garden

Hungary

Hortus Botanicus Instituti Plantarum Medicinalium, Budakalász

India

Allahabad Experimental Garden
Indian Botanic Garden, Howrah
Experimental Botanical Garden, Poona
University Botanical Garden, Surat
Tropical Botanic Garden and Research Institute, Trivandrum
National Orchidarium & Experimental Garden, Yercaud

Italy

Orto Botanico dell'Università, Camerino
Orto Botanico Comunale, Lucca
Orto Botanico Università degli Studi, Padova
Mimosa Stazione Sperimentale Agricola Spermentale, Toscolano
Orto Botanico, Urbino

Japan

Tsukuba Medicinal Plants Research Station, National Institute of Hygienic Sciences
  Branches: Hokkaido Experimental Station
  Izu Experimental Station
  Tanegashima Experimental Station
  Wakayama Experimental Station

Kyoto Takeda Herbal Garden
M. Yamanaka Medicinal Plants Garden, Osaka
Tokyo Metropolitan Medicinal Plants Garden
Botanical Garden for Medicinal Plants, Tohoku University
Medicinal Plant Garden, Fac. Pharmaceutical Sciences, Josai University
Experimental Station for Medicinal Plant Studies, Univ. Tokyo
Drug Plant Garden, Nihon University
Medicinal Plant Garden, Hoshi College of Pharmacy
Medicinal Plant Garden, Tokyo College of Pharmacy
Medicinal Plant Garden, Kitasato University
Medicinal Plant Garden, Showa University
Herbal Garden, Toyama Medicinal & Pharmaceutical Univ.
Medicinal Plants Research Centre, Toyama Prefecture
Botanical Garden of Medicinal Plants, Hokuriku Univ.
The Naito Museum of Pharmaceutical Science & Industry Medicinal Plants Garden
Garden of Medicinal Plants, Kyoto Pharmaceutical Univ.
Experimental Institute for Medicinal Plants & Herbal Garden, Osaka University
Setsunan University, Medicinal Plants Garden
Kobe Womens College of Pharmaceutical Sciences, Medicinal Plants Garden
Experimental Station of Medicinal Plants, Hiroshima University
Wakunagaseiyaku Co. Medicinal Plant Garden
Medicinal Plant Garden, Tokushima University
Experimental Station for Medicinal Plants, Kyushu Univ.
Nagasaki University Medicinal Plant Garden
Medicinal Plant Garden, Kumamoto University

Malta

Argotti Botanic Garden

Malaysia

Rimba Ilmu Kuala Lumpur

Mexico

Museo de Herbolaria Jardín Botánico de Plantas, Col Roma
Jardín Botánico de Plantas Medicinales de Ticomán

Nepal

Royal Botanical Garden, Lalitpur

Netherlands

Botanische Tuin, Elsloo
Farmaceutisch Laboratorium, Utrecht

New Zealand

University of Auckland Botanic Garden

Philippines

Pharmaceutical Garden, University Santo Tomás, Manila

Poland

Instytutu Hadowli i Aklimatyzacji Ogrod Botaniczny, Bydgoszcz
Ogrod Roslin Leczinczych, Gdansk-Wrzeszcz
Division of Medicinal Plants, Inst. Pharmacology, Krákow
Hortus Plantarum Medicarum, Lódz
Ogrod Botaniczny, Lódz
Hortus Pharmacognosticus, Lublin
Zaktad Farmakognozji, Poznán
Karedra Zaklad Roslin Leczniczych, Poznán
Hortus Botanicus Instituti Plantarum Medicinalium, Poznán
Hortus Plantarum Medicarum, Wroclaw

Romania

Grădina Botanică, Institutul de Medicină si Farmacie, Tîrgu-Mureş

Sri Lanka

Peradeniya Botanic Garden
Hakgala Botanic Garden
Gampaha Botanic Garden

Switzerland

Jardin Botanique Lausanne

Taiwan

Heng-chun Tropical Botanic Garden

UK

Bradford Botanic Garden
Middleton House, Enfield
Chelsea Physic Garden, London

USA

Mattheai Botanic Garden, Ann Arbor
Riverside Botanic Garden, California
Berkeley University Botanic Garden, California
Dept. Pharmacognosy, Univ. Illinois, Chicago
Des Moines Botanic Garden, Iowa
National Tropical Botanic Garden, Lawai, Hawaii
Waimea Arboretum & Botanic Garden, Hawaii
Amy Greenwell Ethnobotanical Garden, Honolulu
The Cornell Plantations, Ithaca
Greenhouse, College of Pharmacy, Univ. Minneapolis
Marsh Botanic Garden, New Haven
Drug Plant Garden, Rhode Island
Medicinal Herb Garden, Univ. Washington, Seattle

USSR

Main Botanic Garden, Alma Ata
Cheboksary Botanic Garden
Chernovtsy Botanic Garden
Frunze Botanic Garden
Kaunas Botanic Garden
Kishinev Botanic Garden
Krivoy Rog Botanic Garden
Kobuleti Medicinal Plant Garden
Kursk State Medicinal Institute
Komarov Botanic Institute, Leningrad
Leningrad State University Botanic Garden
Altai Botanic Garden, Leninogorsk
Minsk Academy of Sciences Botanic Garden
Moscow Main Botanic Garden
Botanic Garden 1st Moscow Medicinal Institute
Botanic Garden, Kabardino-Balkars University
Botanic Garden, Pedagogue Institute Nezhin
Botanic Garden Agricultural Institute, Omsk
Arboretum Perkal, Pyatigorsk
Botanic Garden Pharmacuetical Institute, Pyatigorsk
State University Botanic Garden, Riga
Botanic Garden of the Tuberculosis Clinic, Rodnyky
Botanic Garden Academy of Sciences of Latvia, Salaspils
Botanic Garden State University Tartu
State University Botanic Garden Tashkent
Academy of Sciences Botanic Garden, Ufa

Botanic Garden of Medicinal Plants, Vilar
Botanic Garden Podolensis, Vinnitsa
Hortus Botanicus Institute Medicinalis, Vitebsk
State University Botanic Garden, Voronezh
Botanic Garden Agricultural Institute, Zhitomar

Yugoslavia

Botanicki VRT, Fakulteta Farmaceutsko-Biokemijskog, Zagreb

The Role of Chinese Botanical Gardens in Conservation of Medicinal Plants

He Shan-an and Cheng Zhong-ming
Nanjing Botanical Garden Mem. Sun Yat-Sen, Nanjing, Jiangsu, People's Republic of China

The Significance of Medicinal Plants in Botanical Gardens

What is the significance of medicinal plants in botanical gardens? "Historically, medicinal plant collections are the cradle of botanical gardens". The plant kingdom is one of the main living requirements for human beings. Because plants are the source of food and medicine, mankind has focussed their attention on plants and their uses. Considering that the main food crops consist of only about 20 species, the utilization of medicinal plants is certainly very important to mankind both as a resource and in recognizing the significance of the diversity of plants. There are several thousand species of medicinal plants that have been used by mankind and each of them has their own function. Furthermore during the early history of mankind medicinal plants were mainly wild ones so we find a very close relationship between the development of botany and the science of medicinal plants. The celebrated botanist Linneaus was a medical doctor who was knowledgeable on medicinal plants and it led to his life-long study of plants.

During 40-90 A.D. the Greek Dioscorides wrote a volume named "De Materia Medica", which is the primary botanical work of ancient times and an authoritative reference work which can be regarded as the prototype herbal and pharmacopoeia. A similar situation exists in Chinese history. In ancient times pharmacology was the same as herbalism and the development of herbalism actually led to the development of botany in China. Many large volumes in herbalism represent the development of botany.

Many of the oldest botanical gardens originated from medicinal plant gardens. The initial development of botanical gardens in many parts of the world originated from collections of medicinal plants. According to the book written by Sima Qian in the Song Dynasty (420-479 A.D.) of China, there was a garden of medicinal plants named Du-Le. In Japan, the Botanical Garden of Tokyo University, or Koishikawa Botanical Garden, was first established in 1684 as a medicinal plant garden. In Europe the important old botanical gardens have also mostly originated from medicinal plant gardens (collections or nurseries). For example, the oldest botanical garden in Italy, Pisa Botanical Garden, was established in 1543 and the Padua Botanical Gardens as well as the Florence Botanical Garden were set up two years later. These all originated from medicinal plants. In the United Kingdom many old botanical gardens, such as Chelsea Physic Garden and Oxford Garden both established in the 17th century, were medicinal plant gardens. The Glasgow Botanic Garden was built up from a medicinal plant garden established in 1705 and the Royal Botanic Garden in Edinburgh was also originally founded to grow medicinal plants for teaching medical students. In the USSR, the botanical garden system is based on the foundation of medicinal plant gardens. Many of the oldest botanical gardens, e.g. Komarov Botanical Garden established in 1705, were previously medicinal plant gardens.

Medicinal plants are important in modern botanical gardens. At present medicinal plants are still one of the main parts or an important section in many botanical gardens, especially in China. Although the history of modern Chinese botanical gardens is relatively short, a little more than half a century, the collection and research of traditional medicinal plants has had a history of many thousands of years, with distant sources and a rich and marvellous cultural legacy. The utilization of traditional Chinese medicines has caused Chinese nations to flourish and prosper. Today still 80% of the total population uses traditional Chinese medicines to cure their illnesses and even in cities the figure is about 30 or 40%. People in towns and the countryside are mostly or entirely dependent on traditional Chinese medicines. It is well known that more than 90% of the traditional Chinese medicines are plants which are always an important component of most Chinese botanical gardens. The total number of taxa used exceeds 10,000 in number. The traditional Chinese medicinal system is vital to the country and is presently followed with interest by worldwide health research organizations. Reviewing its history it can be concluded that the collections and gardens of medicinal plants are the cradle of modern botanical gardens. Even in the face of well developed modern sciences medicinal plants are still an active field of research and an important component of botanical gardens.

The basic task of botanical gardens is to focus on the great diversity of plants and consequently the conservation and display of genetic resources are important functions of a botanical garden. To collect and preserve germplasm of medicinal plants from various regions within a country and to establish medicinal plant gardens (or areas) for the display and popularization of knowledge about medicinal plants are common themes of most Chinese botanical gardens.

Collections of Medicinal Plants in Major Botanical Gardens in China

China contains a wide diversity of ecological environments and is very rich in medicinal plant species. In most major botanical gardens there are special medicinal plant gardens (or areas) where numerous medicinal plants are arranged and planted according to their ecological characteristics, their biological behavior, or their functions relative to the specific organs for medicinal uses. In Nanjing Botanical Garden (NJBG) there are about 800 taxa of medicinal plants; in Hangzhou Botanical Garden (HZBG), 1,200 taxa; in South China Botanical Garden, Chinese Academy of Sciences (SCBG), 700 taxa; and in the Botanical Garden, Institute of Medicinal Plant Development, Chinese Academy of Medical Sciences (MPBG), 1,300 taxa.

		Acreage (ha.)		Taxa in Collections	
Name*	Location	Total	MP**	Total	MP
YTBG	21°10' N, 90°55' E	266	3	2000	800
GXBG	22°51' N, 108°19' E	160	160	2100	2100
SCBG	23°08' N, 113°19' E	100	ca.2	4500	700
LSBG	29°35' N, 115°59' E	293	ca.4	3400	500
HZBG	30°19' N, 120°12' E	226	ca.2	3400	1200
WHBG	30°35' N, 114°17' E	66	ca.2	3000	600
SHBG	31°10' N, 121°26' E	80	1.5	5000	600
NJBG	32°00' N, 118°48' E	186	2	3000	800
XABG	34°15' N, 108°55' E	20	1	2000	1150
MPBG	38°51' N, 116°19' E	11	11	1300	1300

* YTBG: Yunnan Tropical BG; GXBG: Guangxi Mecicinal Plant BG; SCBG: South China BG; LSBG: Lushan BG; HZBG: Mangzhou BG; WHBG: Wuhan BG; SHBG: Shanghai BG; NJBG: Nanjing SYS BG; XABG: Xian BG; MPBG: Institute of Medicinal Plant Development; BJBG: Beijing BG.

** Area devoted to medicinal plants

Conservation is an Urgent Task for Medicinal Plants

Among the 389 rare and endangered species listed in the Chinese Red Data Book, Volume I, there are 77 typical traditional Chinese medicinal plants, of which more than 50 species have been introduced to botanical gardens, though there is still not enough research on their protection.

Undoubtedly, the extensive exploitation and overutilization of plant resources, as well as the construction of cities, highways, factories, plants and mines, are all factors which have damaged medicinal plant resources seriously and even have pushed some of them into endangerment.

Examples are: *Fritillaria cirrhosa* is a famous traditional medicine for coughs, and is distributed mainly in the northwestern part of Sichuan province but now only one or two individuals can be occasionally found over large regions of steppe. Plants of the genus *Dioscorea* are widely distributed all over the country. The total amount of its rhizomes in the wild was estimated in the 1950's at about 0.3 million tons, but the resources have nearly been exhausted during the past 30 years. Nowadays, not only the supply of material for industrial usage becomes difficult to obtain but also the high-content individuals appear much less than before. It demonstrates that germplasm is apparently being lost. Even some cultivated medicinal plants, such as "Huazhou Juhong", which has been identified as a rare and special variation of pomelo (*Citrus grandis* Osbeck var. *tomentosa* Hort.), faces the risk of extinction. Furthermore, *Iphigenia indica*, which possesses a valuable potential as an anticancer agent, is only distributed in Lijiang and Dali counties as well as northwestern part of Yunnan province. Because of the excessive consumption of this plant and its very low natural reproductive ability the species is almost extinct.

Owing to this critical situation Chinese botanical gardens have carried out many projects on introduction and protection. Tissue culture of *Dioscorea zingiberensis, Colchicum autumnale* and *Catharanthus sinensis* has been conducted in NJBG and plantlets initiated from callus have been obtained. In WHBG the micropropagation of *Iphigenia indica* has been carried out and plantlets produced from callus have been grown on into adult plants (Zhang, 1986).

In YTBG the cultivation of *Rauvolfia verticillata, R. serpentina, Vanilla planifolia, Homulomena occulta, Maytenus molina, Gloriosa superba, Piper lalax* and *Amomum xanthioides* have been researched. It has been found that the species which grew poorly and appeared endangered in natural forest conditions may grow well under cultivated conditions (Xu, 1986).

Compilation of Medicinal Floras and Studies on Ancient Chinese Herbalism

Several thousand species have been used as traditional medicinal plants in China. It makes the identification of every species and the determination of its function in curing illness extremely important. Well equipped botanical gardens offer the best opportunity to scientists and also to the public for identification of plant species. They not only contain literature, books and pictures, but also herbarium specimens and, especially valuable, living plant collections. Botanical gardens are often the best place to compile reference books, checklists and Floras. Science has always been an important heritage of mankind and since China has its own ancient civilization, the study of Chinese herbalism has become a unique research field in Chinese botanical gardens.

Professor Pei, Chien (Pei, Jian), the former director of NJBG and Professor Cheo, Taiyien (Zhou, Taiyan) started their own research on Chinese medicinal plants in the 1920's. Based upon investigations and literature studies they have compiled nine volumes of "Icones of Chinese Medicinal Plants" which are well known throughout China and have been published separately since 1951 (Pei & Cheo, 1951-1985). Pei combined the basic research of taxonomy with the applied research of medicinal plants and took advantage of Chinese proverb "one arrow, two hawks" to modify his thought and as a guideline in his research. This means to obtain two purposes with one project. Afterwards, he and his colleagues compiled a series of extensive volumes, such as "Traditional Herbal Medicines in Nanjing" (Cheo *et al.*, 1956), "A Flora of Chinese Medicinal Plants in Jiangsu" and "A Flora of Chinese Drugs in Jiangsu Province" (1965). Recently, NJBG cooperated with KMBG and IMPLAT is compiling another extensive medicinal herbal named "XinHua (New China) Herbal Essentials" including more than 6,000 species. The contents are very concentrated and specialized. It is considered the current authoritative work on taxa of medicinal plants. HZBG has taken part in the compilation of "A Flora of Chinese Drugs in Zhejiang Province" and "A Flora of Chinese Medicinal Plants of the Tianmo Mts"; GXBG, "A List of Medicinal Plants in Guangxi Province"; LSBG, "Chinese Medicinal Plants on Lushan Mts"; and YTBG, "A List of Flora in Xishuanbanna".

Based upon the research works mentioned above a great number of herbarium specimens and information have been accumulated and most species of medicinal plants have been identified. In addition, in NJBG a unique textual research "Studies on ancient bencaology and the history of Chinese medicinal plants" has been established and conducted for more than 20 years. It is the only formal long-term project in this field in China (Huang & Chen, 1988).

All the approaches mentioned above reveal that research work in botanical gardens plays an important role in the utilizaton of medicinal plants.

Introduction and Acclimatization

Plant introduction and acclimatization are basic tasks in botanical gardens. Because 80% of traditional Chinese plant drugs come from natural resources, and because of the present shortage of natural resources, it is necessary to develop the cultivation of medicinal plants At the present time there are altogether about 300,000 ha in China devoted to medicinal plant cultivation. Research on propagation of medicinal plants has become one of the main fields in botanical gardens. Some concerns have been raised by many scientists and medical doctors about the content and quality of effective compounds in plants. But experiments of medicinal plant cultivation have revealed that this concern can be resolved successfully by monitoring the effective components of plants. For example, NJBG has successfully introduced *Fritillaria thunbergii*, a traditional medicine for coughs, from Zhejiang province to Jiangsu province and developed it in 22 ha. The studies on *Dioscorea*, including introduction and breeding for high content of diosgenin and their ecophysiological characteristics, have been going on in NJBG for over 25 years. It has been found that the highest content of diosgenin of some promising clones was about 16.5%, the highest record reported in the world.

Coptis chinensis, native to Yunnan and Guizhou plateau and Sichuan province, has been introduced by Dou (1981) of NJBG to Jiangsu province in the most eastern part of the country, facing the East Sea, and by Beijing Botanical Garden (BJBG) to Hebei province on the loess plateau in the northern part of the country. In BJBG (1976), the failure to fruit of *Andrographis paniculata* was studied and the technique of shortening the photoperiod during the young seedling stage was recommended. An experiment on the introduction of *Mentha* was completed in BJBG and 12 of them were selected as suitable for Beijing and its vicinity. With experimental cultivation of *Silybum marianum* a yield of 1200 kg/ha was produced in BJBG. The domestication of *Polygonum multiforum* and *Platycodon gradiflorus* has been accomplished successfully in HZBG (Shen, 1980).

The introduction of medicinal plants to low elevations from high mountains has been a scientific project in WHBG. Liu (1983) introduced several hundred species of medicinal plants from Shennongjia, western Hubei province. With experiments on *Codonopsis pilosula* and *Rumex officinalis* improved cultivation practices have been proposed.

The frost hardiness research an *Tamarix chinensis, Elaeagnus angustifolia, Hippophae rhamnoides, Lycium chinense, Vitex incisa, Periploca sepium, Alangium plantanifolium, Ligustrum lucidum* etc. was emphasized in the Heilongjian Forest BG and the survival of broad-leaved evergreen woody medicinal plants, including *Pittosporum tobira, Photinia serrulata, Berberis chinensis, Mahonia fortunei, Ilex chinensis* and *Gardenia jasminoides* was studied in XABG. A great interest has been created in growing *Panax quinquefolium* and cooperative groups of NJBG, XABG, LSBG and Heinongjiang Forest BG have focused their attention on it. As a result, the production of *Panax quinquefolium* has been rapidly increasing during the past several years.

In the south, a series of introductional experiments of medicinal plants such as *Santalum album, Rauvolfia vomitoria, Styrax tonkinensis, Melaleuca leucadendra, Myristica fragrans, Hydnocarpus anthelminthicus, Dracaena angustifolia, Strychnos nuxvomica, Sterculia scaphigera, Terminalia chebula* and 40 species of *Dendrobium* etc. have taken place in CSBG during the past 30 years. Many of them were successful and *Santalum album* trees have been extended to Guangxi, Yunnan, Guizhou, Shichuan, Zhejiang and Fujian provinces. With research on local Chinese medicinal plants an improved technique of artificial pollination for high yields of *Amomum villosum* was recommended by CSBG. Since 1956, 16 species of 9 genera of Dipterocarpaceae have been introduced to the Introduction Laboratory, YTBG (1975) and some of them appeared quite cold-resistant. In addition, *Styrax hypoglauca* was introduced at GXBG and the mechanism of the generation of aromatic material has been studied. Domestication (acclimatization) of *Nervilia fordii* was one of the main projects in GXBG also (Hu, 1988). Since the plant can grow only one leaf every year the collectors often dig the entire bulb and hence diminish it in the wild. It has led to the plant now being endangered. Because of its effective functions on releasing fever and stopping coughs it was popularly used as a highly valuable herb in Hongkong, and Macao and an important export commodity. The annual production is only 7 or 8 tons, so that the cultivation of this plant is of significance both in conservation and utilization.

Search for and Utilization of New Plants

Since the 1950's a comprehensive project on *Dioscorea* resources is underway in NJBG. It has been shown that among about 60 species of this genus there are 17 species, 1 subspecies and 1 variety which contain diosgenin for medical usage in China. Using domestic resources in China more than 50 medicines from the raw material of *Dioscorea* rhizomes have been produced. Three species, *Dioscorea zingiberensis, D. nipponica* and *D. parviflora*, are the main resources for industrial use. The individual plants with the highest content of diosgenin occurred in species

D. zingiberensis (Ting *et al.*, 1983). Following indications from ancient herbals, active constituents have been extracted in NJBG from *Trachycarpus fortunei* to produce a new medicine to stopping bleeding in gynecological illnesses (Dong, 1981); from *Angelica dahurica* for skin disease; from *Bupleurum scorzonerifolium* for pneumonia in children; and from *Euphorbia pekinensis* for mental illness. In LSBG (1978), research on *Liquidambar formosana* has been conducted and a new medicine for stopping external bleeding was generated. Research on *Gardenia jasminoides* showed that the plant contains effective compounds for birth control (Zhang *et al.*, 1986). It is valuable in making new medicines. In YTBG (1975) 23 species of *Maytenus* have been examined and 12 species have proved to have active constituents, such as, maytansine, for anticancer (Pei, 1979). In GXBG it has been found that the contents of flavonoides in *Ginkgo* leaves are higher in autumn than those collected at other times, and leaves of fruiting branches have a higher content than those from vegetative shoots (Chen, 1988). It indicates that the harvesting time and the parts of organs are critical for *Ginkgo* leaves. In XABG and North-west Botanical Institute research on *Picrasma quassioides* showed that the bitter tasting compounds of this plant possesses an effective function in lowering blood pressure (Sun *et al.*, 1986). It provided new information in searching for new resources for medicines.

In conclusion, during the past 30 years Chinese botanical gardens have done a lot in the investigation, collection, and utilization of medicinal plant resources, but not so much toward their conservation and protection. We now need to address the point that more effort should be made toward the conservation of medicinal plants.

References

Beijing Botanical Garden (1976). Cultivation and propagation of *Andrographis paniculata* in Beijing. *News Letter of Beijing Bot. Gard.*, 11-18.

Chen, G.X. (1988). Contents of flavonoides of different growth periods in *Ginkgo* leaves. *Guangxi Plants*, 8(4), 363.

Cheo, T.Y. & Ting, C.T. (1956). *Traditional herbal medicines in Nanjing*. Science Press.

Dong, Y.F. (1981). Phytochemical research on medicine for stopping bleeding from *Trachycarpa fortunei*. *Bull. Nanjing Bot. Gard. Mem. Sun Yat-Sen*, 130-131.

Dou, F.P. (1981). Introduction and cultivation of *Coptis chinensis* in Nanjing. *Bull. Nanjing Bot. Gard. Mem. Sun Yat-Sen*, 111-116.

Hu, S.T. (1988). Domestication of *Nervilia fordii*. *Guangxi Plants*, 2(3), 263-267.

Huang, S.B. & Chen, Z.M. (1988). *Bencaology*. Southeastern University Press.

Introduction Laboratory (1975). Adaptability of tropical plants introduced from different provenances. *Tropical Bot. Res.*, 32-36.

Ling, Z.S. (1986). Isolation and identification of compounds for lowering blood pressure in *Picrasma quassioides*. *North-west Bot. Res.*, 6(2), 138-140.

Liu, Q.R. (1983). Introduction and observation on medicinal plants from Shengnonjia region. *Wuhan Bot. Res.*, 1, 85-90.

Lushan Bot. Gard. (1978). Research on the function of drug for stopping bleeding from *Liquidambar formosana* leaves. *Material of Bot. Res.*, 1, 71-83.

Nanjing College of Pharmacy & Nanjing Botanical Institute (1965). *A Flora of Chinese Drugs in Jiangsu province*. Jiangsu People's Press.

Pei, C. & Cheo, T.Y. (1951-1985). *Icones of Chinese medicinal plants*. Volume 1-9. Science Press.

Pei, S.J. (1979). Exploration and utilization of tropical plant resources in Yunnan province. *Trop. Bot. Res.*, 12, 1-14.

Shen, L.X. (1980). Vegetative propagation of *Polygonum multifolium* and its protection from pests and diseases. *News Letters of Hangzhou Bot. Gard.*, 18-19.

Ting, Z.Z. (1983). *The steroidal hormone plants*. Science Press.

Xu, Z.F. (1986). A discussion on problems of conservation of plant resources in Xishuanbanna. *Tropical Bot. Res.*, 18, 10-15.

Yunnan Tropical Bot. Inst. (1978). Study on the anticancer function of *Maytenus* Molina. *Tropical Bot. Res.*, 1-9

Zhang, B.X. (1986). Research on constituents from *Gardenia jasminoides* for birth control. *Bulletin of Plant Res.*, 2, 1-7.

Zhang, Z.Q. (1986). Research on induction of callus and reproduction of *Iphigenia indica*. *Wuhan Bot. Res.*, 4(1), 55-57.

Policies to Conserve Medicinal Plants

Policies and Organisation for Medicinal Plant Conservation in Sri Lanka

W.J.M. Lokubandara
Minister of Indigenous Medicine, Government of Sri Lanka *

The conservation of nature and natural resources in Sri Lanka is as old as her culture which goes back 2500 years. The pastoral way of life centering round an agricultural economy, enriched by the Buddhist aversion to the taking of any form of life, gave rise to a unique society which lived in perfect harmony with the environment.

Along with the original settlers from India, who brought their rudimentary agricultural practices, came also the ancient Science of Ayurveda, which means "the science of living". The Lord Buddha himself received Ayurveda treatment and exhorted the value of it. This example was taken up readily and seriously by the clergy, who became excellent exponents of this science. In turn, through the respect the people had for the clergy, they placed their trust in this form of medicine.

The temples became the respositories of the ancient texts on Ayurveda – the "Samhitas" and "Niganduwas" – which contributed in large measure to their preservation and subsequent survival during 400 years of foreign rule.

During those 400 years, the conservation of nature fell into disarray. Only during the past few decades can it be said that there has been a serious attempt at revival.

In this respect the present Government in Sri Lanka has taken numerous progressive steps towards inculcating in the minds of the people a strong pride in their natural heritage. His Excellency the President himself has led the field in doing so. Indeed the responsibility of the people has been enshrined in the present Constitution of Sri Lanka thus: "It is the duty of every person in Sri Lanka to protect nature and conserve its riches".

The fauna and flora of Sri Lanka enjoys a high level of protection, with over four hundred reserves set apart for their conservation. The area of jurisdiction of the Forest Department and the Department of Wildlife Conservation covers over 25% of the land area of this comparatively small island.

---

* *Now* Minister of Education, Cultural Affairs and Information.

Stringent laws apply in these reserves and the government progressively improves both the infrastructure and funding for their implementation. Recent legislation such as the National Heritage Act and the establishment of Mangrove and Biosphere Reserves are putting Sri Lanka back on the road to active conservation again.

But due to foreign rule and to the introduction of Western Medicine to Sri Lanka, Ayurveda was not afforded any opportunity to develop or evolve to suit the changing life styles of the people, i.e. the more urban population. Nevertheless, Ayurveda did survive very strongly in rural Sri Lanka, but had been gradually relegated to the status of "Traditional Medicine".

After independence in 1948, successive governments strove to reinstate Ayurveda by encouraging its practice in towns by setting up hospitals. In doing so, there were also concurrent programmes for the propagation of medicinal plants and for giving government appointments to Ayurvedic Practitioners to man Ayurveda Hospitals and Dispensaries.

However, it was left to the present Government to give autonomy to Ayurveda by separating it from the Health Ministry to form the Ministry of Indigenous Medicine in 1980 with a full time Minister in charge. Sri Lanka is the first country to have such a separate Ministry. Since then it can be said that no pains have been spared in promoting Indigenous Medicine and in developing it to a level which is acknowledged as an acceptable alternative to Western Medicine.

With the spread of Indigenous Medicine, people returned to Ayurvedic Practitioners, particularly for their primary health care. As the number of Registered Practitioners increased, and as the Ministry set up more and more hospitals, the demand for medicinal herbs for the production of Ayurvedic drugs sharply increased. The few scattered nurseries were grossly inadequate to meet the supply and large sums of money had to be spent on importing the drugs primarily from India.

At the same time there was a growing tendency to exploit wild stocks with scant respect for the adverse effects such random extraction would have on natural populations. Going hand in hand was the continuing denudation of forests which further reduced the resource base and tended to threaten species diversity.

There was no question that an aggressive policy of *in situ* conservation was needed to save valuable species from extinction. We turned to IUCN's and WWF's Plants Campaign for guidance and help. At this point a project proposal was submitted to the World Wide Fund for Nature for conservation of medicinal plants *in situ*. This proposal had easy passage because the aims and objectives fell in line with those of the

Plants Campaign and in 1986 WWF Project 3320 "The Conservation of the Medicinal Plants of Sri Lanka" was approved and suitably funded.

My Ministry is of the view that any attempt at conservation should involve the people for they are the ultimate beneficiaries of such action. In pursuance of such thinking the Department of Ayurveda and the Bandaranaike Memorial Ayurvedic Research Institute (BMARI) which come under my purview have been charged with the task of establishing nurseries where medicinal herbs are correctly identified and, also, produced in bulk. The plants are distributed at very nominal prices from dozens of sales points in several districts throughout the island. Handsome provision in the votes of my Minstry makes it possible to disseminate simple literature among school children who are also coopted in growing plants in school compounds, Sunday Schools and the like.

In my view this is a necessary form of *ex situ* conservation, which had actually been instilled in our people for centuries. Our ancestors practised the art of growing herbs in simple containers like flower-pots (pot herbs) or in their gardens (herb gardens) through which whole prescriptions could be dispensed in times of illness. Through this we hope to revive the age-old situation where the mother in a family was a "pharmacist" who dispensed decoctions from her garden "pharmacy".

Of the nearly 3000 species of plants (flora) in Sri Lanka, about 500 are used in Ayurveda. These are mostly identified from the ancient texts which have their origin in India. Sri Lanka and India have about 2000 plant species in common and it is probable that more than the 550 mentioned above can have medicinal properties. In addition Sri Lanka has 600 species of plants which are endemic to the country. A sizeable number of these are used in herbal medicines.

My Ministry recognizes the importance of preserving this vast species diversity for the generations to come. This is all the more necessary when we see how research in plant pharmacology is discovering drugs in plants hitherto unused in Medicine, whether Ayurveda or Western. Very soon, therefore, my Ministry will be declaring special reserves for the protection of medicinal trees and herbs and appointing officials to look after them. These officials will, of course, work in close collaboration with the Buddhist clergy and Ayurvedic Practitioners of the area, who will be knowledgeable in identifying the plants of medicinal value. They will also work closely with the Government agencies in the district, who will assist in enforcing laws.

I am also preparing legislation to prevent the over-exploitation of medicinal plants by unscrupulous collectors who ravage natural forests and collect plants far in excess of the demand. It has been brought to my notice that places famous for their wealth of herbs have been denuded.

One such example is Dolukanda in the District of Kurunegala in Sri Lanka, which had a legendary richness of medicinal plants. But a recent survey carried out under WWF Project 3320 revealed to our surprise that not more than about 10 species of common medicinal plants were left.

This legislation will, therefore, aim at controlling the wasteful collection of wild plants so that the genetic diversity will not be diminished in any way. The legislation will also make exports of certain plants possible on a permit to be issued under the authority of the Commissioner of Ayurveda. One condition of the permit will be that the exporter should prove that the herbs were from a recognized herbal garden and not from a reserve.

The propagation of medicinal plants is now being actively carried out in herbal gardens and nurseries throughout the country. The World Health Organisation is actively participating in this programme with funds, expert advice and equipment. Herbal gardens are being established in school compounds, in the premises of all the Ayurvedic Hospitals spread over the country, in most of the electorates and in public parks and gardens. There is a large nursery even in the premises of my Ministry.

At these herbal gardens are sales points where plants are made available to the public at give-away prices, to encourage the pot-herbs and herb gardens of the people. The sales point at my Ministry is a wayside shop from which passers-by can pick up their plants by placing the money in a till. There as no salesmen or saleswomen. The turnover here is very encouraging for it is in the heart of the city.

The nurseries supply thousands of plants to Government institutions that have taken on the establishment of nurseries themselves. A good example is the 200 acre plantation at Kataragama in southeast Sri Lanka, which is very efficiently run by the government's Janata Estate Development Board. The Ministry of Lands and Land Development has extensive nurseries in many districts where the Land Development Department maintains horticultural nurseries as well. The Forest Department, too, has established nurseries primarily to restock Forest Reserves that have been denuded of these medicinal plants. I am indeed grateful to the UNDP and the WHO for the wonderful encouragement they give in the implementation of this programme.

The propagation of medicinal plants is accompanied by the publication of leaflets and booklets aimed at giving the would-be grower not only an idea of how to maintain the garden but also a thorough knowledge of the uses of the plants he wishes to grow. In this way the message of Ayurveda is being spread at a very encouraging rate.

My officials at the Bandaranaike Memorial Ayurvedic Research

Institute (BMARI) at Nawinna, 15 kilometres from Colombo, are actively engaged in ascertaining the curative properties of the wide range of selected trees and herbs. The UNDP and the WHO have here too very kindly funded a modern Research Laboratory, which has given a great impetus to this institute which lacked the funds to set up one on its own. There is no doubt that with the facilities now provided to the BMARI, Sri Lanka will soon be discovering new plants with medicinal properties especially in the endemic species hitherto not used in Ayurveda.

The organization needed for the Conservation of Medicinal Plants *in situ* is being funded by Project 3320 of the World Wide Fund for Nature. As I said earlier on, the importance of protecting plants in their wild habitats cannot be over-emphasised. At the present rate of forest destruction not a day can be lost in affording such protection. That is why I consider WWF's Project 3320 so essential and timely.

The Project has the following aims and objectives in the conservation of Indigenous Plants:

1. To advise the Government on the need for the protection of representative habitats of medicinal plants through declaration of nature reserves to ensure genetic continuity;
2. To conduct surveys of selected geographical zones and prepare checklists of plants, identifying them scientifically to ensure correct usage in prescriptions;
3. To identify threatened species in order to propagate them using modern methods of propagation (e.g. tissue culture) and reintroduce them into the wild;
4. To assist the Government in framing appropriate laws to ensure protection of medicinal plants and to control their exploitation;
5. To assist the Government in preparing educational and publicity programmes for long-term conservation.

It was obvious that for these objectives to be achieved, the best available expertise would have to be tapped. I therefore sought the assistance and cooperation of a number of institutions, both governmental and non-governmental, who either had a number of senior scientists already engaged in the study of Ayurveda in one or more of their branches or were already involved in protecting fauna and flora in nature. These included the Universities, the Medical Research Institute (where phyto-pharmacological studies are being developed), the Natural Resources Energy and Science Authority (NARESA), the Departments of Forestry and of Wildlife Conservation, the Royal Botanic Gardens and so on. Ayurvedic Practitioners of repute both from the clergy and the laiety and officials from my own Ministry complemented this very representative team.

Since the declaration of natural reserves would take much time to pass through the various legal processes, this team recommended that special areas be demarcated within the 400-500 existing protected areas. The Forestry and Wildlife Conservation Departments have begun work on this aspect and I expect it to be completed within the next 12 months.

Mention has been made in the chronicles and texts of certain areas where concentrations of medicinal plants exist. Three good examples are Dolukanda (already mentioned), Rumassala on the Southern coast and Bibile-Nilgala in the Uva Basin of Sri Lanka. Dolukanda has already been surveyed and action taken to reforest it and also reintroduce medicinal plants which are said to have been on this hill in profusion. Rumassala was also cursorily inventorised and found to have a good population of useful plants, but a team of botanists, local practitioners and enthusiasts under NARESA scientists will be carrying out a three month survey with a view to declaring this as the first nature reserve specific to the protection of the medicinal plants. This will be a historic achievement. The Bibile-Nilgala area is already well protected by the Department of Wildlife Conservation for it forms the western boundary of the Gal Oya National Park.

A list of threatened plant species has already been prepared, also under the aegis of NARESA, by Professor B.A. Abeywickrema, who is an authority on the Flora of Sri Lanka. Plants of medicinal and economic value have been picked out from this list and are being collected for propagation at the nursery at Meegoda (near Colombo) which has been set up with funds from Project 3320. The laboratories at Nawinna and at the Royal Botanic Gardens, Peradeniya, (near Kandy) have already succeeeded in unravelling the mystery of propagation of difficult species such as *Coscinum fenestratum* (Weni-wel-geta), which is of very wide usage, and of *Caesalpinia bonduc* (Kumburu). The first, Weni-wel-geta, also comes into the category of over-exploited species because it is being ruthlessly cut down to meet the wide demand.

The preparation of literature and publicity material has already begun and with funds now available we should be able not only to keep to our target of spreading the message of Ayurveda to the far corners of the country but also to strive to convince those who rely solely on Western Medicine that here is an efficacious, complete and cheap way to leading a healthy and fuller life.

Finally, I wish to draw the attention of the concerned international organisations and national governments to the many advantages that can be derived from sharing the knowledge and experience of different communities in the conservation of plant resources and their efficacious use in simple or home-made remedies. Strategies of both global and regional nature should be developed for this purpose, preferably in consultation with other agencies in allied fields. As a first step, it may be possible to have a programme of this nature to cover the whole or part of the regions of South and South-East Asia. In any such endeavour, we in Sri Lanka would indeed be happy to make available to the rest of the world the fruits of our research and labour in this field and, if necessary, even to serve as the focal point for a selected area or region. I trust that this matter will receive the urgent attention of the authorities conceived.

I wish to conclude this paper with what Lord Gauthama the Buddha said of the forest long years before the present awakening in us to the need to preserve nature's gifts to mankind. Said the Enlightened one:

> "The forest is a peculiar organism of unlimited kindness and benevolence that makes no demand for its sustenance and extends generously the products of its life activity; it affords protection to all beings, offering shade even to the axeman who destroys it."

Experience in the Conservation of Medicinal Plants in Sri Lanka

Lyn de Alwis,
Mt Lavinia, Sri Lanka

At the commencement of Project 3320 in June 1986 (see previous paper), there was a bias towards *in situ* conservation. But as work got underway it became clear that there was a need for a simultaneous *ex situ* conservation as well. The main reason for this was the comparatively recent resurgence of Ayurveda (Indigenous Medicine) in Sri Lanka and its developments as an acceptable form of Alternative Medicine.

As the Sri Lanka Co-ordinator of the Project I saw the relevance of establishing a rapport between the *in situ* conservation authorities, namely the Forest Department and the Department of Wildlife Conservation (DWLC), and those involved in the cultivation of herbs in nurseries. The latter included the Ministry of Indigenous Medicine, which has several WHO-aided Nurseries scattered over the island, the Bandaranaika Memorial Ayurvedic Research Institute (BMARI) near Colombo, the Royal Botanic Gardens (RBG) in Peradeniya and several private growers. This rapport was established by forming at ministerial level a representative Committee to include all concerned.

I have worked closely with the different institutions named above and visited several private herbaria and would like to set down my findings briefly.

Identification of Plants

The main constraint to either form of conservation is the confusion regarding the correct identity of the plant. This has been caused by practitioners resorting to the use of localized vernacular names in prescriptions. These are often misleading as some plants have several synonyms, e.g. *Rubia cordiflora* has 4 sinhala names and 49 sanskrit names. Quite often a synonym of one plant may become the common name of another. This confusion has led to substitutions, and more dangerously, to adulteration. The damage to the medical practice is obvious. However, in certain cases, by consensus, certain substitutes have been accepted, particularly if they have the same chemistry. A good example for Sri Lanka is the substitution of *Coscinum fenestratum* (Weni-wel, sinhala) for *Berberis aristata* (Daruharidra, sinhala), both of which contain berberin. The reason for the substitutions was that *Berberis* occurs in relatively small populations in the Hill Zone, while *Coscinum* occurs more widely in the Low Country Wet Zone.

However, the laboratories at the BMARI are fighting the problem of adulteration by checking samples of herbs available in the market against the correctly identified plant.

The problem of identification is being further pursued at the BMARI for two other important reasons, to compile a list of frequently used plants and to cultivate them in selected areas.

As a first step, 100 ola-leaf manuscripts containing prescriptions and recipes going back some 300 years have been read by competent physicians. Over 1000 plants have been mentioned, but only about 200 could be considered in common use. (Nearly 600 indigenous plants are used in Ayurveda). With the help of plant taxonomists these 200 plants are being identified to their scientific name and then classified according to their distribution in the different climatic zones of the country.

Cultivation in Nurseries

Presently the nurseries grow only a limited number of species, which are inadequate to dispense even simple prescriptions. This is an area which has to be strengthened. Nurseries must have a much better representation of plants than at present. Also there is an undesirable tendency for nurseries to duplicate the same few species, which are invariably fairly plentiful in the wild, e.g. *Azadirachta, Phyllanthus, Terminalia.* One drawback was the unavailability of planting material. Both the BMARI and the RBG have, I am happy to report, succeeded in propagating several species through laboratory and field trials and by tissue culture. Among these are *Rubia cordiflora* (with the help of 2-nodal cuttings), *Woodfordia fruticosa, Caesalpinia bonduc* (by tissue culturing embryos), and *Entada scandens* (by mechanically exposing the embryo and parts of the cotyledons). Others have propagated *Phyllanthus embilicus* by applying heat on the dried fruit to cause it to dehisce, while the seed of *Coscinum fenestratum*, if placed over a pile of earth fired from below, have been found to germinate in 48 days. These are valuable contributions to science. It is intended to encourage more work on propagation by involving the universities too.

There are "zonal" nurseries established with WHO assistance in the Hill-country (e.g. at Haputale – 1500 m elevation) and in the mid-country (at Haldummulla – 900 m elevation). More are under consideration by the Ministry.

In the meantime Project 3320 has begun a 3 1/2 acre nursery in the Low Country Wet Zone, just 20 miles from Colombo. This nursery will contain a minimum of 50 species as listed by the BMARI scientists and they will be classified according to their usage, e.g. *Casia auriculata, Barleria prionitis, Rauwolfia serpentina* for snake-bite (and other ailments, too), *Bauhinia racemosa* for iodine deficiency, *Cissus quadrangularis* for ulcers (leaf), fractures (stem), *Solanum xanthocarpum* for bronchial coughs.

There is a compelling need to establish special nurseries for the cultivation of those plants whose products have now to be imported, e.g. *Piper longum, Glychyrrhiza glabra*, which grow very easily in Sri Lanka. This will save the country a good deal in foreign exchange. If propagation practices increase and improve, an attempt can be made at reintroduction into the wild habitats from where, through over-exploitation of denudation, the species has been endangered.

Botanical Surveys

Project 3320 has set in motion a series of botanical surveys in selected eco-geographic zones to ascertain the availability of medicinal plants. Four surveys have been completed by the staff of the Departments under the Ministry. As Co-ordinator, I obtained the assistance of the RBG, the Natural Resources Energy and Science Authority (NARESA) and University of Kelaniya to undertake saurveys of selected habitats in the Kandy District, Galle (Low Country Wet Zone) and Sigiriya (Dry Zone) with the same objective.

In situ Conservation

There are over 400 Forest Reserves in Sri Lanka and over 50 Protected Areas under the jurisdiction of DWLC. These are being inventorized for compiling check-lists of medicinal plants, after which localized "Medicinal Plant Reservations" will be declared by the Ministry of Indigenous Medicine.

Social and Agro-forestry

It is heartening to see the great interest being taken by local government authorities to establish Herbal Gardens, sales outlets for both plants and drugs and to popularize Ayurveda in towns and suburbs. Project 3320 envisages that through this exercise, there will be a return to home garden cultivations and the pot-herbs of former times. These will be singular achievements, ensuring long-term conservation by the people themselves.

The Forest Department too, plays an active role in the spread of Ayurveda by including the cultivation of medicinal plants in their agro-forestry schemes. A notable success is the emphasis given to medicinal plants in their reforestation programmes among the new settlers in the gigantic Accelerated Mahaweli (River Basin) Development Programme. More recently, medicinal plants have been included in the buffer-zone rehabilitation programme around the Sinharaja Man and Biosphere Reserve, the country's only low country wet zone rain forest.

Project 3320 can do much for the Conservation of Medicinal Plants in Sri Lanka through its involvement with the above activities, which it now monitors.

The Conservation of Medicinal Plants Used in Primary Health Care in Thailand

The late Dr Pricha Desawadi
Formerly Deputy Director-General, Department of Medical Sciences, Ministry of Public Health, Bangkok, Thailand

In Thailand, people are close to nature. They know how to use herbs and plants both as food and medicine and teach this to their children. Over many centuries of experience they have developed a form of traditional medicine, suitable for their own health care and well accepted by them. It has certain advantages over imported systems of medicine in our setting because, as an integral part of the people's culture, it is particularly effective in solving certain cultural health problems. It can and does freely contribute to scientific and universal medicine. Its recognition, promotion, and development would secure due respect for our people's culture and heritage.

The provision of traditional remedies and medicinal plants in a health care system is not second-rate medicine but is part of the national scene with its well-established health care system for the people. It is like a traditional art which has high potential in the qualities and properties of the skills and drugs used. "Health for All" is regarded in Thailand as a broad concept of steady and widely diffused improvement of society as a whole, based on health as a primary need for happiness and well-being. It is consistent with the cherished values and traditions of our nation, the monarchy, and our religious institutions, but emphasizes the urgent needs and rights of the vast majority of the rural poor and other disadvantaged groups.

To help these poor people to stand on their own feet, a pilot project on primary health care was initiated in one district of Chiang Mai Province in 1969. The health services delivery system at that time consisted of curative care, maternal and child health and family planning, communicable disease control, nutrition, school health, health education, sanitation, laboratory services, health statistics and the distribution of government-produced household remedies, under the supervision of health officers. These services achieved only very low population coverage. Analysis of the problem revealed that one of the most important causes was the lack of people's participation in the health delivery system. The

Ministry of Public Health therefore decided to have some people selected from the community and trained as health volunteers and health communicators. Evaluation of the pilot project revealed increased coverage of population with basic services.

Primary Health Care

The experience gained from these studies led to the development of a nationwide programme of primary health care in 1977, with the following objectives:

- To expand coverage of the health services, particularly among the under-served rural population and to help the people to become self-sufficient;
- To utilize community resources and to encourage community participation in order to solve individual health problems and eventually to establish self-help programmes at the village level;
- To promote the dissemination of health information to local people, as well as to assemble and make use of all data concerning the health status and needs of the communities;
- To make basic health services available, accessible and acceptable to the people; and
- To promote the health status of the people who live in the rural areas and increase their own awareness of health problems and the means of solving them.

The primary health care programme in Thailand was set in motion under the Fourth Five Year Plan (1977-1981), was modified and expanded under the Fifth Plan (1982-1986) and continued in the Sixth Plan (1987-1991). Village health volunteer and communicator training will be continued and these volunteers will be key community health resources. However, primary health care activities in the community are being broadened, making use of other groups such as mothers groups for village nutrition activities, village craftsmen for water and sanitation, etc.

The essential elements of primary health care in the Thai national programme are:

- Education concerning prevailing health problems and the methods of preventing and controlling them;
- Promotion of food supply and proper nutrition;
- Maternal and child health care, including family planning;

- Adequacy of safe water supply and basic sanitation;
- Immunization against major infectious diseases;
- Prevention and control of locally endemic diseases;
- Appropriate treatment of common diseases and injuries;
- Provision of essential basic household drugs for the community;
- Provision of dental health care; and
- Provision of mental health care.

The government has laid down a clear-cut policy of providing health services to all the people. To implement this policy primary health care development appears to be the best means and the Ministry of Public Health has been very active in this venture since 1977. Work in primary health care has been seriously pursued, with definite guidelines for the development of basic health and hygience of the people.

Among the 10 elements for the development of primary health care in Thailand, one is the provision of essential drugs to all villagers. The provision of these drugs to villages has met with problems due to shortages, the three main causes of which are:

- Transportation problems: many remote villages are accessible only by bullock carts and bicycles, and sometimes not at all during the rainy season. It is therefore very difficult for villagers to go to the nearest sub-district health centre for any medicines needed;
- Lack of drug supply outlets: drugstores cannot survive in villages because the purchasing power of the people is low and establishing more health stations is not feasible, as villages are small and the government does not have sufficient funds; and
- Poverty: most villagers cannot afford to buy any kind of medicine and as a result, common conditions are left unattended until they deteriorate and become serious.

Through primary health care, villagers can be taught and trained to deal effectively with their basic health needs. They will then be in a position to help both themselves and their own communities. To accomplish this goal the community health volunteer project was set up. The volunteers are selected from the villages in which they live. There are two distinct categories – village health volunteers and village health communicators. The primary health care units supply village health volunteers with simple non-prescription medicines classified as household medicines. These comprise over 60 items, are less expensive than other drugs and are mostly used for the treatment of common illnesses.

Medicinal Plants in Primary Health Care

Apart from the household medicines, which are largely based on western formularies, there is another category of drugs which are equally important since they are efficacious and still popular among the Thai people. These are herbal or traditional medicines and they should be properly studied, evaluated and brought to the public attention. The primary health care programme is trying to promote herbal medication as an important tool for village health care which can help to overcome the shortage of drugs as well.

To solve the problem of shortage of drugs in the villages three approaches are being used:

- A Drug and Medical Product Fund – called "drug cooperatives" – has been established. At present, more than 45,000 villages are participating in this scheme.
- A project for development of traditional medicine in primary health care has been set up. The objective is to promote the use of traditional herbal medicine in rural communities and to encourage extensive cultivation of medicinal plants by villagers, for their own consumption and as a source of income. Forty-eight medicinal plants of well-known action have been classified into the therapeutic categories, e.g. laxatives, expectorants and anthelmintics. A project for the cultivation of medicinal plants for primary health care programmes has been initiated in 1,000 villages.
- Traditional household remedies, with improved formulae and presented as compressed tablets packed in aluminum foil, were distributed to the drug cooperatives and community hospitals in 25 provinces. Seven community hospitals are interested in using traditional medicine and medicinal plants in the treatment of common diseases.

As a result of this project, physicians in the seven hospitals are enthusiastic about making clinical trials of medicinal plants and are collaborating in research work. Young research workers are now being increasingly involved in studies on medicinal plants in various aspects of scientific research in pharmacology, toxicity and phytochemistry, etc.

National Policy on Medicinal Plants

Close collaboration between interested groups in the different institutes in Thailand prevents duplication of effort and unnecessary expenditure. A National Committee on Medicinal Plants has been set up with a multi-disciplinary team consisting of a physician, botanist, chemist, pharmacologist, scientist, private medicinal plants exporter and other supporting staff. This Committee will develop a national policy on medicinal plants which will include: setting priorities for clinical and scientific research; drug trials on animals for testing safety and efficacy; ethno-medical and botanical surveys; the development of an information system and data-base; the manufacture and export of traditional medicines based on medicinal plants; and the cultivation and conservation of medicinal plants on a national basis.

Conservation of Medicinal Plants Used in Primary Health Care

Deforestation through slash and burn cultivation, excessive logging or forest fires inevitably affects the moisture in the atmosphere, and hence the pattern of rainfall. In Thailand, and also all over South East Asia, medicinal plants usually come from the forest. Deforestation therefore results in irreplaceable losses of plants already known to be of medicinal value as well as those the therapeutic uses of which have yet to be discovered.

The Thai Government is well aware of the importance of natural resources especially medicinal plants – and that it is crucial that proper resource management and development in this region be accorded priority by policy makers.

Through the Department of Forestry, the Ministry of Agriculture issued a regulation in 1975 (revised in 1987) for the conservation of 158 important items from more than 200 species of trees, as well as the special conservation of 13 rare species of trees. These numbers include at least 61 species of medicinal plants and the parts derived from those trees which are strictly regulated.

To mark the celebration of Sixtieth Royal Birthday Anniversary, the Department of Medical Sciences will implement a project for the distribution of 150,000 seedlings of medicinal plants used in primary health care to all provinces, through the six regional laboratory centres.

The main purpose of this project is the maintenance of the medicinal plant resources of this country, both for their conservation as well as for the people's own basic health care.

A Green Revolution in Northeast Thailand

The national policy on medicinal plants will be to promote the proper management of these resources by the distribution of medicinal plant seeds and seedlings all over the country. Under the Sixth Plan it is intended to make the Northeast green with trees and medicinal plants, and to establish conservation areas all over the 17 provinces of Northeast Thailand, in addition to other projects for irrigation, fresh water supplies, and the cultivation of fruit trees and fast-growing trees. The Government seriously hopes to make the Northeast completely green by 1991, under the motivating title of "ISAAN Khieo".

Medicinal Plants and the Law

Cyril de Klemm
Paris, France

Introduction

Medicinal plants, as many other products of nature, have been collected from time immemorial by traditional communities all over the world. Natural resources were considered to be the common or collective property of the group. Collection was regulated by unwritten rules based on tradition and often on religion. Stiff penalties could usually be imposed when the rules were broken. As traditional societies were often strongly territorial, wild plants and animals were protected from harvesting by outsiders by well-defined boundaries that members of other groups were not allowed to cross.

These traditional closed systems have now almost completely disappeared. A few, however, still seem to survive. In some districts of the Indian Himalayas, the collection of certain plants continues to be restricted to certain days of the year, during religious festivals (Gadgil, 1985).

The opening-up of many traditional systems brought about profound changes in the legal status of wild animals and plants. Animals became *res nullius*, that is to say things which cannot be the subject of ownership, even by the owner of the land on which they occur. Ownership may, however, be obtained by "occupancy", in other words by actually killing or capturing the animal. This concept inherited from ancient roman law is now embodied, with some variations, in the legislation of a very large number of countries. A few, however, have recently enacted laws conferring on wild animals the status of public property. Whether this new status will result in more effective conservation remains to be seen.

Wild plants, on the other hand, being attached to the land, are immovables and the property of the landowner. In practice, however, with the exception of trees, which have a recognized economic importance, plants usually continue to be considered by the public as a product of nature, which may be freely collected by anyone, even on private land. The reason for this is that the taking of plants of little monetary value, or at least of their fruit, foliage and flowers, is generally regarded as part of the ordinary enjoyment of the countryside (Bonyhady, 1987). In certain countries, for instance Norway and Switzerland, this customary freedom has been translated into legislative provisions establishing a right for members of the public to gather wild plants anywhere. In one case, in the German *Land* of Bavaria, this right is even enshrined in the Constitution.

For all practical purposes, therefore, property rights on wild plants must be considered as purely nominal. In fact, if not in law, plants must also be considered as *res mullius*.

This situation is, however, slowly changing. When plants have a recognized monetary value, their collection on private land without the owner's consent is now sometimes considered by the courts to be theft. Recent decisions by French and Italian courts concerning the mass collection of mushrooms are typical of this new trend. Statutes making it a criminal offence to collect wild plants for profit on the land of another are still, however, very rare. In the United Kingdom, the Theft Act 1968 provides that a person who picks mushrooms, or wild flowers, fruit or foliage for reward, sale or other commercial purpose, is guilty of theft. The uprooting of any wild plant, whether or not for profit, is also regarded as theft by that act (Bonyhady, 1987).

It is doubtful, however, that the potential imposition of criminal penalties for the theft of wild plants will be sufficient to deter collectors, as few landowners will normally be prepared to warden their property and to institute proceedings against thieves unless the plants concerned have a high market value (as in the case, for instance, of truffles, *Tuber* spp.).

But even if wild plants were adequately protected on private land from collection by outsiders, they would not be protected against the action of the landowner himself, who, as a consequence of his property rights, is always free to exploit them or destroy them at will.

Finally, on public lands, although wild plants are usually nominally the property of the State or of a public agency, they will almost always be considered by the public as ownerless.

To sum up, wild plants have the worst of two worlds. As private property they can be destroyed by their owners. As a free product of nature they can be collected by anyone almost anywhere. Any legislation aiming at the conservation of wild plants should always be viewed against this background.

Legislation Protecting Wild Plants

With an increasing number of wild plant species threatened with extinction, States have begun to adopt protective legislation. This legislation is, however, generally recent and exists only in a relatively small number of countries, including all European countries, the United States, Australia and South Africa. In the rest of the world, there is either no legislation to protect plants at all, or, in the very few cases where laws have been enacted, the number of species covered is usually very small.

The main purpose of most of the laws which have been adopted so far is to control collection and to limit the demand for wild plants by the imposition of trade prohibitions or restrictions.

The destruction of protected wild plants in the course of legitimate activites such as agriculture, forestry or construction remains, however, generally authorised. Habitat protection requirements are most often ignored.

a) Collection

Collection controls aim at remedying the problems arising from the ambiguous legal status of wild plants. Controls applicable to persons other than owners or occupiers of land are legally easy to institute because they are merely a manifestation of the polical powers of the State. Controls imposed to landowners are more difficult to establish as they imply restrictions to property rights. Many countries, therefore, hesitate to prohibit the taking of wild plants by the owners of the land on which they grow. This is particularly true of common law countries. This situation is, however, gradually changing and many of the most recent laws now restrict or prohibit collection by any person on any land. In both cases, however, collection conrols are notoriously difficult to enforce.

Protection from collection may be total or partial. Total protection consists in an absolute prohibition to collect plants or parts of plants of certain listed species. Exceptions may, however, be made, subject to a permit, for the purpose of scientific research. For partially protected plants, there is usually a prohibition to uproot them or to dig out their subterranean parts but the collection of a small number of aerial parts remains authorised.

Plants which are not listed as totally or partially protected may usually be collected freely. In a small number of jurisdictions there is a general ban on the collection of wild plants for whatever purposes. In the United Kingdom, for instance, under the Wildlife and Countryside Act of 1981, the uprooting of any species of plant by persons other than landowners or persons authorised by them is prohibited. In several Italian regions, Austrian *Länder* and Swiss cantons, not only is the uprooting or the collection of the subterranean parts of plants prohibited but there are also restrictions on the gathering of their aerial parts. The number of flowering stems or branches which may be picked varies from five to twenty according to jurisdictions. In some cases the maximum quantity allowed is that which may be held in the hand.

Certain countries control the collection for commercial purposes of otherwise unprotected plants. This is achieve by requiring collectors to apply for commercial collection permits.

In some cases these controls only apply to certain listed species. In France, for instance, a permit from the Ministry of the Environment is required for the commercial collection of 31 taxa, including many medicinal plants. In Zaïre, the collection of *Rauwolfia* spp. is prohibited except under a permit.

In some countries, the permit requirement applies to the collection for commercial purposes of any wild plant. In the state of Western Australia commercial collection permits are required for all species of vascular plants, even when the collection is carried out by the landowner on his land. In Switzerland, under federal legislation, a permit must be issued for the collection of any unprotected species of wild plant except for mushrooms, berries and medicinal herbs, provided these are gathered in conformity with local usages. Many Swiss cantons have, however, adopted stricter measures and require a permit for the commercial collection of any wild plant, without exceptions.

In a very small number of jurisdictions the commercial collection of certain species is completely prohibited even under a permit. In Belgium, seven taxa are so protected. Their taking for private use remains, however, authorised. In some Italian regions, as for instance Val d'Aosta, no commercial collection at all is allowed for any wild plant except, under a permit, for a small number of medicinal species.

When permits have been issued conditions are generally attached specifying the name of the species concerned, the quantities that may be collected, the dates and places of collection and sometimes the collection methods that must be used.

Enforcement is relatively easy as the possession of quantities exceeding those authorised by the permit is generally sufficient to establish that an offence has been committed.

Offences are usually punished by fines and sometimes even by imprisonment. The most effective penalty, however, is the withdrawal of the permit and the prohibition to apply for another permit for a certain period of time, for instance, one or two years. In the case of repeated offences, the permit may be revoked permanently. In addition, plants illegally collected as well as the tools and vehicles which have been used to commit the offence may usually be confiscated.

b) Trade

1) Domestic trade

Trade in fully or partially protected plant species is generally prohibited. In addition, the possession and transport of specimens of these species, or of a number of specimens exceeding the authorised daily collection limit for partially protected species, are also usually prohibited. These rules are primarily designed to eliminate the demand for protected species and to facilitate enforcement of collection restrictions. When the possession of a protected plant is in itself an offence, there is no need to prove that the plant has been unlawfully collected.

Trade in unprotected species is generally unrestricted. However, where commercial collection is prohibited or otherwise regulated, trade in the species concerned is generally also forbidden or restricted. To be effective, however, a collection permit system must necessarily be supplemented by controls applied to all the other links in the trade chain. These include the licensing of traders and of plant processing industries, the obligation for licensees to keep records of all their transactions, and a system of tags that must accompany specimens in trade as far as their final destination. Furthermore, there must be a prohibition for traders to buy or sell plants obtained from other persons than licensed sellers or growers.

Examples of these elaborate trade controls can be found in several Australian states, e.g. New South Wales and Western Australia and in the Cape Province in South Africa. It was necessary to institute these controls in these jurisdictions because of the large volume of trade in cut flowers. The tagging system exists, in particular, in certain states of the United States such as California and Texas, mainly to control trade in desert plants such as cacti.

There is also a need to control the commercial growers of plants that are subject to collection restrictions. This is important to ensure that growers are not obtaining their propagation material from specimens unlawfully removed from the wild, and also that they do not sell wild specimens under the guise of artificially propagated ones. Certain jurisdictions, for instance several Italian regions, require certificates of origin as a proof of artificial propagation. Others, like the Cape Province in South Africa, have instituted a licensing system for horticulturists and require them to keep books where all their transactions must be recorded. In France, tags must accompany artificially propagated specimens of protected species. Specimens without tags are deemed to have been unlawfully collected from the wild.

To enforce these controls, inspections must be carried out by specialised personnel on the premises of traders and growers. As this is expensive, it is doubtful that it is often practised. The need for such inspections has, however, been clearly emphasised in a resolution of the Fifth Meeting of the Conference of the Parties to the Convention on International Trade in Endangered Species of Flora and Fauna (CITES) in 1985 (Resolution 5.15).

2) International trade

Where domestic trade is controlled, prohibitions or restrictions often extend to the import and export of protected species. Import controls are necessary to ensure that illegally collected and exported specimens are not subsequently imported as if they were of foreign origin. They may also serve to ensure that only plants that have been lawfully collected and exported from their country of origin are imported into the country of destination. Import controls generally include the requirement for an import permit and/or the presentation of an export permit or a certificate of origin from the exporting country. The law may also require that only licensed traders be authorised to import specimens of plant species which are subject to these specific controls. In France, a tag must accompany imported specimens of species which are protected under national legislation when they are put on the market.

Export controls are necessary to prevent the export of unlawfully collected specimens. These generally consist in a requirement for export permits and in the establishment of controls at points of exit to verify the nature and quantities of the exported goods. As an example, the law of Zaïre which institutes a commercial collection permit for *Rauwolfia* spp. also provides that at the time of export the Customs shall check that the quantities exported do not exceed those mentioned on the permit.

The Convention on International Trade in Endangered Species of Wild Fauna and Flora (CITES), signed at Washington in 1973 and to which some 103 states are now a party, is the basic instrument regulating international trade. Its main purpose is to establish co-operative links between importing and exporting countries to ensure that illegally obtained wildlife belonging to species listed on the Convention appendices cannot be freely traded at international level.

> Trade in species listed on Appendix I to the Convention, when carried out for commercial purposes, is prohibited. Transactions for non-commercial purposes require both an import and an export permit. Trade in Appendix II species is subject only to the presentation of export permits. Importing countries must not accept exports from other countries if they are not accompanied by such permits. The listing of a plant on Appendix I makes it, therefore, legally impossible to trade in wild specimens of the species. Artificially propagated specimens of Appendix I plants are, however, considered by the Convention as if they were listed on Appendix II. They may, as a result, be traded, provided exports permits have been issued.

As long as it does not provide for a better preservation against destruction and for habitat conservation measures, legislation protecting wild plants will, however, never be sufficient in itself to ensure the conservation of endangered plants. This does not mean that it is useless. Indeed, the very fact that a plant is protected by legislation may have important effects even though these will generally not be legal requirements. Thus, the presence of protected plants will often be taken into consideration when environmental impact assessments are prepared or zoning plans developed. Permit to undertake certain activities that would destroy protected plants are sometimes refused. Nature reserves, are, in some cases, established only for the sake of preserving some of these plants.

Legislation Applicable to Medicinal Plants

Legislation protecting wild plants is primarily designed to preserve endangered or rare species. Lists of protected species generally include some medicinal plants but usually not many. Commercial collection restrictions sometimes apply to particular medicinal plants. When they are applicable to all wild plants, medicinal plants are of course covered. Controls have been instituted in some countries on the export of certain species. Examples are Zaïre, where, as already mentioned above, the export of *Rauwolfia* spp. is subject to a permit; Botswana, where the export of *Harpagophytum procumbens* is regulated; India, which completely prohibits the export of certain medicinal species, e.g. *Atropa acuminata, Dioscorea deltoidea, Rauwolfia serpentina* and *Saussurea lappa*. For other species, such as *Dioscorea prazeri* and *Rauwolfia canescens*, only artificially propagated specimens may be exported, under a permit. Plants listed so far on the CITES Appendices are mostly ornamentals such as orchids and cacti. There are, however, a few medicinal species such as *Saussurea lappa* on Appendix I and *Dioscorea deltoidea* and *Panax quinquefolius* on Appendix II.

Few countries have laws dealing specifially with medicinal plants. In Poland there is a list of such plants, and they cannot be collected without a permit. In Italy a law was enacted in 1931 to regulate the collection of medicinal plants. Permits to collect listed species may only be issued to holders of a herbalist degree from a school of pharmacy. Small quantities may, however, be collected by members of the public for personal or household use. The quantities allowed are specified in the legislation. Some Italian regions, as they are now empowered to do under the Constitution, have also enacted regional laws on the collection of medicinal plants, under stricter rules than those in national legislation. The act of 31 March 1977 of the region of Val d'Aosta is a particularly good example of these new laws. An analysis of its main provisions is, therefore, worthwhile.

This act, which deals with the protection of the alpine flora in general, lays down the rule that all wild herbaceous and shrubby plants are protected. It is prohibited to collect and to trade in any such wild plants. For certain endangered species collection is totally prohibited except, under a permit, for scientific research. For another group of species, the act establishes a daily collection limit to six flowering stems. For all other species the daily limit is twenty flowering stems. The collection or destruction of the subterranean parts of all species is also prohibited. Trade in all species is forbidden, even under a permit. There is, however, one exception to this last rule which concerns a certain number of medicinal plants which are listed on an appendix.

Those plants can be collected for commercial purposes provided a permit has been issued. The permit application must mention the name of the species for which the permit is requested, the place of collection and the quantities which the applicant proposes to collect. The written permission of the owner of the land where collection is to take place is also required. Landowners or occupiers have, however, the right to collect, but only on their own land, small specified quantities of medicinal plants for their personal use. They must obtain a permit for any quantity exceeding these limits.

The permit is personal and cannot be transferred to third parties. It lays down collection conditions which must be complied with. It may only be granted if the proposed collection will not result in the depletion of the species concerned at the place of collection. The quantities specified in the permit may be subsequently reduced in the event of unfavourable climatic or man-caused conditions.

Only certain implements, such as sickles, may be used. The digging up of the subterranean parts of medicinal plants may be authorised by the permit where these parts are the ones which have medicinal properties. The survival of the species and the stability of the soil at the place of collection must, however, not be impaired. If the soil has been damaged in the process it must be restored by the collector. To ensure that he complies with this condition he may be required to deposit a bond.

Prospects For Future Legislation

Medicinal plants are different from other plants in that they are important for human health. Their extinction, or serious depletion, should, therefore, be avoided as a matter of priority and their sustainable exploitation organised on a sound basis. Legislation protecting endangered or rare species is of course essential to prevent extinctions. It is, however generally inadequate to ensure sustainable exploitation.

The very few laws that deal specifically with the conservation and exploitation of medicinal plants, such as the law of the Val d'Aosta region, provide a useful basis for the management of wild populations of these plants but do not take into consideration all the relevant factors, in particular the necessity to maintain the resource base and to protect important habitats.

Ideally, legislation on the exploitation and management of medicinal plants should address the following points:

a) Prevention of over-collection

The prevention of over-exploitation is essential. Medicinal plants may be particulary at risk because the newly discovered medicinal property of a plant, or even a sudden influx of demand for a well-known plant may result in the collection of very large quantities to supply the pharmaceutical industry. By the time adequate controls have been instituted the species may be seriously depleted. The solution is a permit requirement for the collection of any wild plant for commercial purposes.

b) Permits as management instruments

The advantage of a permit system is that it allows for considerable flexibility, at the discretion of the issuing authorities, on the way collection is to proceed. There is no need, therefore, to adopt special regulations governing the collection of each species. To close an area to collection, either permanently or on a rotational basis, all that is needed is to refrain from issuing permits authorising collection in that area. Similarly, the quantities that may be collected may vary from one permit to another, according to areas or dates; the total number of permits may also be limited. Collection pressure may, therefore, be adjusted to the optimum sustainable yield of the species concerned or to the relative abundance of the plants in any given region. Conditions attached to permits may also impose collection methods causing the least possible damage to the plants concerned. They may, for instance, prohibit uprooting if only the aerial parts of the plants need to be collected, or the use of certain implements or methods which may unnecessarily damage the plants or the ecosystem.

The permit system may be an extremely refined management tool, provided it is based on good scientific data and that enforcement is reasonably effective.

c) Management plans

The permit system will, however, only operate satisfactorily if there is a management plan for the species concerned (or where the species is seriously depleted, a recovery plan). Any management plan will have to be based on adequate knowledge of the distribution, conservation status and ecology of the species. Adequate staff should, therefore, be provided to operate the system. Appropriate penalties for violations of permit conditions are also essential. The most effective type of penalty is probably the revocation of the permit and the prohibition to apply for a new permit for a certain period of time. For serious or multiple offences, the revocation of the permit for life can be envisaged.

d) Trade controls

Trade in medicinal plants should always be subject to a permit. Transactions between private persons and small scale local trade will, however, generally be difficult to control. In most cases, however, such trade will not have serious consequences. Prohibitions are, however, important as a deterrent. Dealers, growers, industries and laboratories, importers and exporters are easier to control provided that certain basic rules such as licensing and the obligation to hold registers where all transactions are recorded, are provided and enforced. Inspections will often be necessary and adequate staff should, therefore, be available. Penalties such as the revocation of licences can be extremely effective.

e) Licence fees

The system will be expensive to operate. Without adequate staff, however, it will be ineffective. A solution could be to collect a fee on commercial permits and licences. The fees could be proportional to the quantities and value of the plants concerned.

f) Habitat protection

There is clearly a need to prevent the destruction of medicinal plants and their habitats in the course of activities unconnected with collection, such as agriculture, forestry or construction. Legislation prohibiting the destruction of wild plants and natural habitats is, however, notoriously ineffective unless it can be supplemented by adequate land-use controls. In most cases, there will be a need to establish special reserves as a means to safeguard the gene pool of the species concerned and as a source of propagation material for artificially propagated plants. In these reserves, it should be prohibited to disturb the natural vegetation, unless specific management measures are required to enhance the chances of survival of the medicinal plants concerned. Usually, there will also be a need for a management plan. Activities which do not affect the vegetation, such as hunting, may not necessarily be prohibited.

It is particularly important to ensure that reserves provide lasting protection for the plant species they contain. Obviously, government ownership of the land, if need be by acquisition or expropriation, will usually be the answer to this problem. This may not, however, always be possible and other forms of protection, such as the imposition of land-use controls by administrative order, may often provide an acceptable substitute to State ownership. Landowners may, in these cases, be entitled to compensation.

In any event, it will always be of major importance to tailor conservation regulations to the requirements of the species. In particular, the biological processes which govern the reproduction of the plant species which the reserve purports to preserve will also have to be preserved. This could mean that the pollinators and/or seed dispersers of the species, and perhaps also their habitat, in the reserve and sometimes outside it, may need to be protected by regulations.

Where plants requiring protection are thinly scattered over large areas, such as in tropical rain forests, the reserve concept is manifestly inadequate to ensure their survival. Ideally, a network of national parks encompassing large areas of all natural habitat types would be the answer to the problem. This may not, however, always be possible and other means will have to be found. The establishment of wilderness or roadless areas could be a solution in some cases. Other means include prescribed forestry practices in certain areas or the impositon of controls prohibiting certain activities. The recent (1988) National Heritage Wilderness Areas Act of Sri Lanka constitutes an interesting and perhaps unique example of Wilderness Area legislation which specifically refers to medicinal plants. The act provides that the establishment of wilderness areas will be carried out in consultation with the Minister of Indigenous Medicine and that management plans for the areas so established shall be developed with the assistance of the Ayurveda Department.

g) Artificial propagation

Legislation should provide suitable incentives for the development of artificial propagation of medicinal plants. Special permits may have to be instituted to allow the collection from the wild of the propagation material that growers may require. In addition, adequate controls must be imposed to ensure that permit conditions are complied with and that growers are not supplied with unlawfully obtained specimens or do not sell wild specimens under the pretence that they have been artificially propagated. The licensing of growers and an obligation to keep a record of all their transactions is, therefore, necessary.

Incentives to artificial propagation may include special government subsidies. Several Italian regional laws provide for this possibility. An example is the law of the region Basilicate of 16 January 1978, which provides for the conclusion of contracts between the Region and universities and other public research bodies for the carrying out of experimental studies on the cultivation of medicinal plants and the protection of natural populations of these plants.

References

Bonyhady, T. (1987). *The Law of the Countryside*. London: Professional Books.

Gadgil, M. (1985). Social Restraints on Resource Utilization: The Indian Experience. In *Culture and Conservation*: *The Human Dimension in Environmental Planning*, ed. J.McNeely & D. Pitt, pp. 135-154. IUCN. London: Croom Helm.

Cited Legislation

Australia: New South Wales: Native Plants Protection Regulations, 1975 (as amended).

Australia: Western Australia: Wildlife Conservation Act, no. 77 of 1950 (as amended).

Belgium: Arrêté royal relatif aux mesures de protection en faveur de certaines espèces végétales croissant à l'état sauvage, 16.2.1976.

Botswana: Regulations Concerning the Control of Exporting the Grapple Plant (*Harpagophytum procumbens*) (no date available).

France: Arrêté relatif à la liste des espèces végétales protégées sur l'ensemble du territoire national, 20.2.1982.

France: Arrêté relatif à la production, à l'importation et à la commercialisation d'espèces végétales protégées, 12.10.1987.

India: Exports (Control) Order, 1977 (as amended).

Italy: Legge n°99. Disciplina della coltivazione, raccolta e commercio delle piante officinali, 6.1.1931.

Italy: *Basilicate*: Legge regionale n°4. Interventi per la ricerca, la coltivazione e l'incremento delle piante officinali, 16.1.1978.

Italy: *Val d'Aosta*: Legge regionale n°17. Protezione della flora alpina, 31.3.1977.

Poland: W sprawie wprowadzenia gatunkowej ochrony roslin, 30.4.1983.

South Africa: *Cape Province*: Nature Conservation Regulations, 29.8.1975.

Sri Lanka: National Heritage Wilderness Areas Act, n°3 of 1988.

Switzerland: Loi fédérale sur la protection de la nature et du paysage, 1.7.1966.

United Kingdom: Wildlife and Countryside Act, 1981.

United States: *California*: Desert Native Plants Act, 1981.

United States: *Texas*: Parks and Wildlife Laws, ch. 88, as amended in 1985.

Zaïre: Arrêté départemental portant réglementation de l'exploitation du *Rauwolfia*, 12.5.1977.

Let's Stop Talking to Ourselves: the Need for Public Awareness

Paul Spencer Wachtel
Head, Creative Services, WWF International, Gland, Switzerland

Many of us, I suspect, spend most of our time speaking with people who are interested in the same things we are interested in. For example, I am in daily contact with colleagues who explain the importance of traditional medicine and related conservation issues.

Whenever possible, though, I escape from the ivory tower of conservation and meet people who are lawyers and carpenters and auto-mechanics and cooks. By doing so I learn a sobering fact. Their priorities are not necessarily the same as ours.

This shouldn't surprise me, but it does. We're not getting our message across. You can see this for yourself. Look at the best-selling magazines on any news-stand in your country. I would bet that not one of the top ten publications is even vaguely related to the conservation of medicinal plants and traditional medicine. You'd be hard-pressed to find one out of the top 100 magazines which deals directly with the issue we will discuss all week, and to which many of us have devoted our lives.

Psychologists and sociologists tell us that the most important things in a person's life are the themes represented by these magazines – love, shelter, health, adventure, sex, food. The conservation of medicinal plants is nowhere on the list.

Does this mean the conservation of medicinal plants is unimportant? Of course not. What it does mean is that we spend too much of our time talking to ourselves.

Because the rest of the world doesn't know about our concerns they cannot support our programmes, they cannot give us money, they cannot vote for legislators who encourage conservation work, they cannot teach their children about the importance of saving plants. We, and by "we" I refer to many conservationists and health-care officials, are guilty of the most basic mistake any supplier of a product or service can make.

We are talking, not communicating. In commercial terms, we are selling, not marketing. What does this mean? I don't propose to go into a lecture on marketing techniques, but ask you to consider this.

Selling is inward-directed. It means giving people what *we want them to have.* Let's say that I think the world is ready for a glossy new monthly magazine called "The Exotic World of Euphorbiaceae". I'm doing what I want to do. But there's no reason to think that anybody except my relatives will buy the publication. People are people. They already have enough problems, whether it's getting enough water or enough firewood, dealing with sick goats or neighbours who steal their pigs. They might be burdened by broken plumbing or a nagging spouse or funding their child's college education or deciding which colour Mercedes to buy.

People don't have time for our problem or our dreams. We have to make time for *their* problems.

Marketing is the opposite of selling. Marketing is positioning your product or service is such a way so that it *meets an individual's needs, desires, goals, ambitions and dreams.* Look at the way Unilever markets soap in many countries. Now soap is soap – it keeps you clean. But that's not the way Unilever markets the product. They know that women cherish the image of beauty and soft skin, hence the marketing of Lux soap by featuring movie-stars in what marketing experts call a "beauty platform". Some mothers want a reliable, inexpensive, so-called healthy soap for the whole family, and so you get ads for Lifebuoy soap showing lots of happy kids splashing around in a bath together – a "health platform". Other people are concerned about personal hygiene and preventing body odour; the result is a deodorant soap called Safeguard. This is marketing. There really isn't a great deal of difference between these bars of soap, but there *is* a significant difference in the way they are marketed to different groups of people who have differing needs and desires.

Let's consider conservation of medicinal plants as a business. What is the product we have to offer? I suggest that our product is the vast genetic library of wild plants and animals that contribute to good health.

And what is the benefit we are offering? To most people, I suggest the benefit is well-being, freedom from illness, peace of mind. But it might also be low-cost medication, it could be safety of treatment, it might be respect and continuation of traditional social values, it could be adventure through reading about the real-life dramas that occur to field investigators.

How do we choose? How do we communicate? To greatly simplify things I suggest we look at two essential elements of communication.

Choosing the right Message. And selecting the right Medium. Let's look at the *message* first. What simple idea do we want to convey?

The message can best be defined by understanding the consumer. There is no such creature as a universal message. I wish there was, it would make our life much easier, but one of the facts of life is that people in different countries, in different societies, in different communities and age groups, want different things.

Now you may say that people are people, regardless of what language they speak and whether they live in wooden shacks or palaces. Well, yes and no. Certainly we are all part of the family of man, but surely the message that will work with the Prime Minister of Thailand is different to the message that will be effective in communicating with the head of a pharmaceutical company in Bangkok and different again to the message that will be well-received by the Karen slash and burn agriculturalist.

Which target groups are you most interested in influencing? Do you want to reach members of parliament to get them to enact new legislation to protect medicinal plants? Do you want to reach the middle class to get them to contribute money to plant conservation? Are you trying to convince rural farmers to take a more active role in conserving the plants on which they rely?

Basically my advice is to use professional communicators to communicate the message. These communicators might be advertising agencies, or commercial firms, or football stars, or dance troupes or producers of puppet plays.

The other essential point to consider is the *medium*. Think of the dozens of ways people get information in their lives. Conservationists, particularly in the north, often get bogged down by evaluating our communication success by whether we get coverage in newspapers or on television. But there are other media that are sometimes more effective, which work within the existing social patterns.

Don't think you can create a new medium. You can't. The best you can hope for is to make good use of an existing medium. But choose the right one.

How has this combination of the right medium with the right message achieved conservation?

In India and Bangladesh, European demand for frogs legs led to over-harvesting of the animals, which were natural predators of insects that ate padi and spread disease. The villagers engaged in cruel and excessive frog collection. Although the people earned some ready cash, other problems arose shortly after, mainly the multiplication of insect pests. For the first time chemical pesticides had to be used, but the cost of the chemicals exceeded the value of the income from the frogs – and pesticides, of course, are harmful to humans and wildlife. Also because there were too few frogs to eat many mosquitoes, diseases such as malaria increased.

The conservation objective was to convince villagers to stop catching frogs. The message was that frogs are helpful, and that by killing frogs serious problems will arise. The communicator chosen by WWF in India was a local artist, Ranjit. His medium: he prepared traditional scroll paintings and went from village to village singing his story about rice, amphibians, snakes, chemicals and greedy Europeans. He sang in a local language and used local idioms and easily understood examples.

In another example, the Penans in Malaysia's Borneo state of Sarawak wanted to convey a message to the State government that their traditional homeland – the tropical rainforest – was being destroyed by timber operators. The Penans' life style was threatened and they were angry.

The medium chosen by these forest-dwelling people was very direct – they blockaded timber camps, prevented lorries from transporting timber or supplies, and generally disrupted, in a peaceable way, timber activities. The result: worldwide press coverage which resulted in pressure on government officials to change their policies.

The Kuna Indians in Panama chose a different means of communicating. They recognize that the future of their society depends on maintaining the rain forest, in which botanists have already identified 35 new species of plants. Panamanian officials wanted to build a road through their territory, which would have, in effect, destroyed the sanctity of their environment. The message the Kuna wanted to give to the government was twofold: first, we rely on this land, and second we want to manage it ourselves. I deliberately use the world manage, not preserve. The Kuna understand that the best chance they have of succeeding is to bring in tourists, allow scientists to work in the forest, encourage school groups to visit, but all *on their own terms*. The medium they chose was political lobbying, involving Kuna leader Rafael Harris. He tells the story of an occasion in 1980, when Panama's former leader, General Omar Torrijos, visited a number of Indian leaders at a Kuna village. Torrijos, astounded at the vast rainforest, asked, "Why do you Kuna need so much land? You don't do anything with it. You don't use it. And if anyone else cuts down so much as a single tree, you shout and scream." To which Kuna leader Harris responded: "Suppose I got to Panama City and stand in front of a pharmacy, and because I need medicine but have none of your money, I pick up a rock and break the window. You will take me away and put me in jail. For me, the forest is my pharmacy. If I have sores on my legs, I go to the forest and get the medicine I need to cure them. We Kuna need the forest, and we use it and take much from it. But we take only what we need, without having to destroy everything as your people do." Arguments like these, coupled with $100,000 of Kuna cash to get some of these projects in motion, have kept the rest of Panama out of Kuna country.

Let us say you have a central programme that you want to communicate widely. How do you reach rural target audiences with various conservation messages?

In Indonesia folk singer Ully Sigar volunteers her services to sing at village gatherings throughout the vast archipelago, and she trains and encourages some 10,000 young entertainers to do the same. These young singers and dancers return to the rural areas with new confidence, prepared to work with environmental NGOs to alert people to the importance of protecting nature and advising them of steps they can take to ensure that their children will have a naturally productive future.

Appropriate communication, of course, can occur anywhere. In Italy WWF supporters wanted to convey a simple message to Italy's hunters – stop killing wildlife. But instead of creating a head-on conflict, the conservationists used what I call lateral communication. Several weeks before the start of the hunting season they announced that on the opening day they would be out in force in the forests and fields, wearing bright clothes, playing music, singing, and in general disrupting the hunters, but doing so in a gentle way. When dawn broke on the first day of the hunting season thousands of conservationists throughout Italy got out their trumpets, sang Verdi, and offered breakfast rolls and coffee to the bewildered, and generally bemused, hunters.

I lived for many years in Indonesia, and worked for an advertising agency that helped market basic consumer products. It was interesting to go into isolated villages, at the end of dirt roads, far from the nearest school or health clinic. Who had the best communication network in these isolated areas? It wasn't the Ministry of Health, nor the Ministry of Agriculture, nor the Ministry of Education. The best and most effective communication between the wealthy industrial centres and the rural backwaters was created by the commercial companies and their advertising agencies who marketed detergent, Aji-No-Moto brand monosodium glutimate, and Eveready batteries One type of product which was found literally everywhere was *jamu*, the generic term for Indonesian medicinal herbs.

The use of jamu goes back thousands of years in Indonesia, but it wasn't until the last two generations that commercial jamu manufacturers trained pharmaceutical chemists, built modern factories, hired sophisticated package designers, created extensive distribution networks and hired experienced advertising agencies to produce TV commercials with the most popular Indonesian entertainers. The *jamu* manufacturers and their advertising agencies understand marketing – they understand that the message must be simple and repeated countless times, and they understand the medium – how best to communicate to various target audiences. Their distribution is so effective that the Indonesian family planning board uses jamu distribution networks to distribute condoms.

And do not think that this communication is limited to rural folk. The influence of the jamu industry is so strong it reaches the highest level. Indonesian President Soeharto was invited to open a new multi-million dollar factory for the pharmaceutical company Ciba Geigy in Jakarta a few years ago. The President was so convinced of the value of traditional medicine that he infuriated his Swiss hosts by speaking, not about the wonders of western pharmaceutical products, but about the importance of herbal medicine to the country's well-being.

These are examples of the right message, and the right medium. There will be different in each region. In Africa the strongest marketing presence in rural areas is often that of the breweries. Work with them. Don't try to do it yourself. Instead of trying to reach everybody through the big national circulation newspapers or television networks perhaps a more targeted approach through fishing societies, or gardening clubs, or scouts associations, or automobile touring groups, might pay off.

Use the conservation organizations which do have effective communications. There are many, but the one I am most familiar with is WWF. We have helped make people worldwide aware of the importance of tropical rainforests, and the contribution plants make to our lives.

In March 1989 WWF launched an ambitious campaign to raise funds to save the world's biological diversity. The first thrust of this five year campaign is on the relationship between nature conservation and health. This is a relationship that all people can understand because it affects everyone. We're all in the same boat. Plants help people. And conservationists can help health care professionals.

That is an important point to remember when dealing with partners of all types, but particularly commercial sponsors. Help each other. Whatever you do should be of mutual benefit. Why should a commercial company help you? Not simply because you are a nice person or because you represent a worthy cause. You should be prepared to understand your partner's needs and help him achieve his specific marketing goals. This is not always easy, but it will pay off in the long run. One example of what we call "mutual-benefit opportunities".

Gevalia coffee is a premium brand marketed by General Foods Denmark. They chose to support nature conservation in Brazil's rainforest, and re-designed their pack to explain the importance of conservation to Scandinavian coffee drinkers. Every time you bought a box of coffee the company made a donation to save the rainforest. Let us be under no illusions about why they did this. It was not love of nature, nor was it due to the marketing director's wife being a member of WWF.

Gevalia found that supporting nature conservation gave them a marketing edge, it gave consumers a reason to choose Gevalia over other coffee brands. The result: their sales increased 10% during the promotion period.

I would like to close with another example of appropiate cultural communication which I hope will stir our creativity. In India and Sri Lanka some of the richest sites for medicinal plants used in Ayurvedic medicine are the sacred groves that are scattered throughout the countryside. In predominantly Buddhist Sri Lanka these areas are often under the control of the local Buddhist monastery; in predominantly Hindu India the sacred groves are generally under the protection of a naga (serpent) deity, which represents fertility, or Parvati, the consort of Siva who represents the Earth Mother. The origins are thought to date back to an episode in the Ramayana, the classic Hindu epic which is, together with the Mahabarata, the most influential and well-known fable of South and Southeast Asia.

Near the end of the story, Rama and his brother Laksmana are injured in their fight with the villain Rawana. Rama's doctor needs medicinal herbs to cure the warriors, and sends Rama's faithful friend, Hanuman, who is a flying white monkey general, to the Himalaya to fetch the special plants. Hanuman, being an impatient and scatterbrained fellow, gets to the Himalaya and realizes he forgot what plants he was supposed to collect. So he decides to take back the entire mountain. As he flies over India back to Sri Lanka, where Rama and Laksmana were dying, clumps of the mountain fell to the ground. The spots where this rich earth, full of medicinal plants, landed became sacred groves, which have stood for thousands of years. Today these areas are protected by local people, who, in some cases, are allowed to pluck a limited amount of medicinal plants for their own use.

How can these forested areas remain intact when they are surrounded by rice fields, coconut plantations and villages? Simply because they are part of the society's cultural traditions. They are local "insurance reserves" which survive, not because the central government has authorized them, but because the people themselves want them. Could we not build on this tradition, and develop new stories and characterizations of Rama and Hanuman, of naga and Parvati, to actually create new reserves? It has been done, in south India, where just ten years ago several industrialists, who were concerned about environmental destruction near Bangalore, created a "traditional" sacred forest.

We – the conservationists and the health care professionals – have messages to convey which are of great importance. But I urge this group to spend less energy talking to ourselves about these issues and instead devote more creative energy to communicate with the rest of the world.

Germplasm, Genetic Erosion and the Conservation of Indonesian Medicinal Plants

Mien A. Rifai
Herbarium Bogoriense, Puslitbang Biologi-LIPI, Bogor, Indonesia, and
Kuswata Kartawinata
Unesco Regional Office for Science and Technology for Southeast Asia and the Pacific, Jakarta, Indonesia

Introduction

We are currently facing four great problems which are interrelated to one another, and our success to tackle them will determine the survival and the existence of *Homo sapiens*. The earth is presently already densely populated and this is attributed to the improvement of the standard of living which in turn resulted in the population explosion. The population explosion is the first main problem which at the same time is the root of other problems. There is no other choice but that we have to satisfy all basic needs of human beings, which comprise food, cloth, shelter, education and health. The supply of the basic needs in sufficient quantity constitutes the second problem, which, because it has reached a critical point, has to be handled through international cooperation. Large-scale exploitation of terrestrial and marine biological resources, agricultural intensification (including the use of high yielding varieties, synthetic fertilizers, pesticides, irrigation and mechanized equipment) and agricultural expansion (such as the conversion of tidal swamps and forests in arable lands) as well as the exploitation of other natural resources (such as oil and gas mining and utilization of solar energy) are examples of human activities in an effort to satisfy the basic needs. These activities give rise to the third problem, i.e. environmental degradation and the exhaustion of the biological resources. This ecological crisis leads to the fourth problem, that is how to conserve the natural resources in such a manner so as to guarantee the survival of human beings.

As a part of one of the components of the basic needs of Indonesian people, *jamu* (herb medicine) and other traditional medicines are directly linked with the above-mentioned serious problems. The rapid population increase along with the educational improvement have stepped up the awareness of people towards the need for a better health. This has in turn stimulated the rapid development of the *jamu* industries, ranging from cottage industries to large factories with high capital inputs. As a consequence a large amount of raw plant material has to be harvested from the wild to cater these industries. The process of exploitation of medicinal plants may lead to the extinction of many species, since so far it is not complemented by any effort to conserve them. It is well known that traditionally a concoction of a *jamu* requires a variety of wild plant species. Hence, to some people it is perhaps strange that wild plants would have to be domesticated, but such a domestication will be inevitable if the traditional use of *jamu* is to be maintained.

Warnings concerning the danger that some species were becoming rare and even extinct were spelled out some time ago in a forum that brought various parties interested and involved in traditional medicines (Rifai, 1976). Furthermore, factors affecting processes that lead to the rarity and extinction of plant resources have been also amply indicated (Rifai, 1979). Suggestions concerning the basic strategy to slow down the rate of exploitation of the main raw material for *jamu* have been put forward also. Various activities to conserve medicinal plants have been initiated, but by far such efforts are still without clear direction, uncoordinated, not holistic in approach and are relatively insignificant compared with the magnitude of the threat. Hence, as yet there is no significant and long-term impact. The problem is so great that it needs a strong political will and clear, well-planned and integrated actions.

Basic Concept of Germplasms

The flora of the Malesian region (which covers Indonesia, Malaysia, Philippines and Papua New Guinea) contains about 40,000 plant species, ranging from algae, mosses, ferns and seed plants. This implies that more than 10 % of the world flora occur in Malesia, hence it is not exaggerated to say that this region is extremely rich in plant resources, and has a high species diversity. There is no doubt also that many species, genera and families are endemic in Malesia. Thus this region has a high species diversity and high genetic variation and is the most important centre of genetic variability of many taxa, such as Apocynaceae, Euphorbiaceae, Moraceae, Myrtaceae, Piperaceae, Rubiaceae, Simarubaceae, and Zingiberaceae.

Every plant population has a set of characters and special characteristics that are controlled by a genetic system. Thus, every individual of a population contains germplasm, i.e. a substance that regulates its inherited behaviour that makes one population differ from the others.

Hence, a population of *pasak bumi* (*Eurycoma longifolia*) occurring on the eastern slope of Meratus Mountains in South Kalimantan differs in genetical make up from that found on Bukit Lawang in North Sumatra. These differences may be expressed in their resistance against diseases, the size of the root systems, chemical contents, ability to grow on poor soils, etc. If one of the above-mentioned *pasak bumi* populations becomes extinct, the whole set of germplasm with unknown properties will disappear also.

Fortunately many plant species consist of thousands of populations, which all together form a gene pool in which more or less free exchange of genes takes place. This is the feature utilized by plant breeders in creating new high-yielding varieties through breeding. Through well-planned and systematic crossings of individuals from various populations (or biotype, forma, cultivar, variety, clone, etc.), a set of desirable characters can be accumulated in a new cultivar. As yet no such selection and breeding efforts involving medicinal plants in Indonesia have been carried out. To date no one has thought of applying breeding practices to obtain high quality medicinal plants having, for example, high medicinal properties, high productivity, low ash content, etc. It is possible that one day when this stage is reached in the *jamu* industry it is already too late, because then the germplasm having desirable properties will be already extinct.

It was at one time thought that the sources of plant genetic variability for breeding purposes is unlimited. However, plant genetic resources at the centres of diversity are not continually available at any time when we need them. A variety of events have taken place that resulted in the extinction of populations, along with the germplasm they contain. Thus the high genetic variability is continually degrading, leading to the occurrence of genetic erosion, whose degree of intensity differs from plant species to another. The degree of *genetic erosion* may be measured by system of scoring 0-5 or by other methods. Genetic erosion may also be expressed according to criteria adopted by *The IUCN Plant Red Data Book* (Lucas & Synge, 1978) in which 5 types of category of rareness are recognized:

- **Extinct**: applies to plant species known or considered extinct; degree of erosion is usually scored 5.
- **Endangered**: species threatened by extinction and will not survive without strict protection; *purwoceng* (*Pimpinella pruatjan*) belongs to this category, usually scored 4.
- **Vulnerable**: a category for species which are not immediately threatened by extinction, but occur in small numbers and because of

their continual exploitation they need protection; *ki kuning* (*Arcangelisia flava*) is an example of a medicinal plant belonging to this category with score of 3.

- **Rare:** species with large population but occurring locally or have a wide distribution but rarely found and have undergone severe erosion; *pulai pandak* (*Rauvolfia serpentina*) is an example of this category with score of 2.
- **Indeterminate:** species that are definitely undergoing a process of becoming rare but information concerning the actual situation is not adequate; most of medicinal plants considered rare are in this category with score of 1.

Various conservation efforts have been undertaken to maintain the variability of germplasm, hence the survival of plants. The establishment of nature reserves is an effort to exercise *in situ* conservation in order to maintain the natural distribution of species. This is further strengthened by *ex situ* conservation, that is to preserve species through cultivation and maintenance of medicinal plants in botanical gardens, arboreta and other sites specially set aside for this purpose. Conservation of germplasm can be also implemented through storage of seeds, propagules, tissue culture or other parts of plants under "minimum survival" condition. By applying the latest conservation technology, the longevity of parts of plants may be extended to several years. Cells of *Atropa belladona*, for instance, are able to live up to two years if they are kept in tissue culture under the temperature of -196° C, while seeds of *Amaranthus tricolor* have been calculated to be able to survive for 955 years if they are stored under the temperature of -6°C and humidity of 5%.

Exploitation and Rareness

The use of rhizomes of *alang-alang* (*Imperata cylindrica*) as medicine to cure fever will not cause genetic erosion because only few people use them. Moreover, the ability of this plant to grow and spread makes it possible for this species to maintain a large number of individuals, hence exploitation of this plant will not in any way affect its survival and existence. This holds true also for plant species used in the preparation of home-made *jamu*. Generally, the raw material needed is sufficiently available and usually cultivated in large numbers. The frequency of use is usually not so high, hence it will not result in their genetic erosion and endanger their existence. Although the preparation of *jamu* for home consumption uses a great deal of wild plants, it will not lead to a crisis of plant resources in view of the fact that the magnitude of plant material exploited from the wild is within the limit of the ability of these resources to maintain their existence. In the past, the principles of sustainable harvesting of plants from forests was traditionally exercised also. When harvesting the rhizomes of *temu putih* (*Curcuma zedoaria*), for instance, it was assured that some of them should remain in the soil to enable wild plants to regenerate, so that the sustainability of plants is guaranteed. Therefore, people in the past knew exactly where to find plants ready for harvest; at present it is hard to find such people.

To a certain extent utilization of medicinal plants in small-scale *jamu* home industries will not cause genetic erosion. Most of the raw material used in home-made *jamu* are products of common plants and are easily available in the market or shops specially selling them. In recent years, there is a tendency of the *jamu* home industries to increase significantly. In Bogor, West Java, for instance, the *jamu* vendors tend to increase in number and sell various kinds of *jamu* that were not known before. Co-operatives and even organized syndicates seem to have developed and control this traditional *jamu* business. If this trend continues, those medicinal plants that are already becoming rare will suffer most and are threatened by extinction.

Meanwhile the modern *jamu* industries and big pharmaceutical companies continually need and use plant material in large quantity and this has resulted in genetic erosion of medicinal plants that have not been cultivated. This situation is aggravated by the fact that the demand for the traditional *jamu* is steadily increasing. Seventeen years ago, for instance, several species of *Curcuma* were growing abundantly in the teak forest around Randublatung (East Java), and people in the area (especially at Saradan) produced large quantity of *Curcuma* powder, but it is not so at present. The change is attributed to the fact that the area is one of the principal suppliers of this medicinal plants for large-scale modern *jamu* industries whose capacity of production increased drastically during the last 15 years. The increasing market demand, increasingly better price, excessive stimuli and bonus provided, strong competition, and the desire to get quick profits, have all destroyed the principles of sustainable harvesting practiced by people in the past. At present raw material, which were previously abundantly available in Java, have to be imported from the other islands, and already some years ago *kencur* (*Kaempferia galanga*) was imported from abroad.

Kedawung (*Parkia roxburghii*) is one of many species used in jamu preparation. It grows in the wild and no attempt has been made to cultivate it. In some areas of East Java, it has become rare because of excessive and indiscriminate harvesting of fruits by felling the trees. If this trend continues, in no time the species will undoubtedly become extinct. Similarly, in West Java *Parkia intermedia*, the closest relative of this species , which is presumed to be the hybrid of *P. roxburghii X P. speciosa*, is almost extinct, as only few trees are now still living. The rapid disappearance of *ki kuning* (*Arcangelisia flava*), which is already rare, is due to the fact that the whole stem is harvested and used in *jamu* preparation and also its natural population is very small because it rarely produces seeds.

Export of medicinal plants has been going on for many years, and the demand in the international market keeps increasing. One big Swiss pharmaceutical company, for example, has requested eight tons of seeds of *Voacanga grandifolia* (Apocynaceae) and is willing to pay a high price. This species is rare and has very light seeds. To satisfy the above request, all available seeds in the forest will perhaps have to be harvested, leaving nothing for its regeneration. Similarly, five tons of the rhizomes of a rare species of *Curcuma* (*temu badur*) has been sought by a West German pharmaceutical company, and 100 kg/year of *pili cibotii* (fine hairs of *Cibotium barometz*) are demanded by a French firm. It can be imagined how many plants of these species will have to be destroyed should such requests be satisfied.

In view of the fact that no attempt has been made to cultivate naturally growing medicinal plants, a recent request for 200 tons of *daun tempuyung* (*Sonchus arvensis*) from the Netherlands could not be fulfilled by the exporters. Such plants as *pulai pandak* (*Rauvolfia* spp.), *temu-temuan* (*Curcuma* spp.), *kumis kucing* (*Orthosiphon aristatus*) and other medicinal plant species constitute important export commodities which would certainly provide foreign exchange earning should they already be in large-scale cultivation.

It should be noted that the threat to the survival of medicinal plants is not only attributed to their utilization in *jamu* and other pharmaceutical industries but also to activities in other sectors. In an effort to increase food production to satisfy ever increasing demand, agricultural lands have been expanded. The expansion has involved conversion of various natural ecosystems (in particular dryland and swamp forests) into food-producing lands, resulting in the destruction of natural habitats of many potential and known medicinal plant species. The expansion of agricultural lands is often accompanied by establishment of new human settlements that involved also clearing of forests and other ecosystems. Many medicinal plants are sensitive to such changes, and would not be able to survive in such severely disturbed habitats. It is well known that most of the medicinal plants are herbs, shrubs and climbers growing on the forest floor under the shade of closed forest canopy. Such indiscriminate exploitation of medicinal plants along with deforestation and timber extraction that lead to destruction of plant habitats will undoubtedly make it difficult for the plants to sustain themselves. Such forest species as *tabat barito* (*Ficus deltoidea*), *krangean* (*Litsea cubeba*), *kayu angin* (*Usnea misaminensis*) and *simbar* (*Lycopodium carinatum*) and many other medicinal plants need serious protection because of the above-mentioned forest destruction.

Owing to the unavailability of certain jamu plants, there is a tendency now to substitute certain plants presumed to have similar medicinal value for another known one. Five to six species of *Cinnamomum*, for instance, have been currently used to substitute *sintok* (*Cinnamomum sintoc*) and *massoi* (*Cryptocarya aromatica*), *kesambi* (*Schleichera oleosa*) for *pulai pandak* (*Rauvolfia* spp.), and *kumis ucing* (*Cleome gynandra*) for *kumis kucing* (*Orthosiphon aristatus*), in most cases because of difficulty in obtaining the material substituted.

Considerations in Conserving Medicinal Plants

To ensure the continuous supply of the raw material in the future, the existence of medicinal plants in the wild should be maintained. Not only their survival that has to be safeguarded, but also their genetic diversity in order to make it possible to develop them and to improve the quality of the *jamu* in the future. The importance of conserving plants and their germplasm is indicated by the establishment of an international organization that specifically handles the plant germplasm conservation. The Government of Indonesia has also set up the National Commission for Conservation of Plant Germplasm.

Various activities to conserve medicinal plants have been undertaken. Inventory of species, exploration, and collection of samples of population, and preparation of legislation to protect them have been initiated. However, more effective preventive actions are yet to be developed, since mere legislative gestures are not sufficient to stop the process of exploitation as has been going up to now.

Not all medicinal plants are threatened by extinction and not all methods of harvesting jamu raw material are endangering the survival of medicinal plants (Rifai, 1979). If the *jamu* industries only rely on annual plants such as *Orthosiphon aristatus* which have been both intensively and extensively cultivated, there will be no problem of medicinal plants becoming rare and extinct. However, the current practices of utilization of plants for *jamu* and the ways of the *jamu* industries securing supplies of raw material are frequently against the principles of sustainable management and utilization of renewable natural resources. Consequently such species as *pasak bumi* (*Eurycoma longifolia*), *pulai pandak* (*Rauvolfia serpentina*), *purwoceng* (*Pimpinella pruatjan*) and other wild medicinal plant species, whose roots are extracted for various medicinal uses, do not seem to have a bright future since the method of harvesting used is by killing them. The reproductive and regenerative capacity of *sintok* (*Cinnamomum sintoc*), *kayu rapat* (*Parameria laevigata*), *massoi* (*Cryptocarya aromatica*) and *pulai* (*Alstonia scholaris*) is reduced because the whole trees are felled in order to harvest their bark.

The foregoing account and the fact that a large number of species is utilized should be taken into consideration in setting up priorities for conserving medicinal plants. Therefore, wild plants with small populations should have greater priority over plants that are already in cultivation. Among wild plants greater priority should be given to perennials over annuals. Furthermore, plants that are extracted for their roots and barks should receive special attention. Meanwhile the future of medicinal plants that are harvested for their leaves, flowers and fruits is to a certain extent relatively safe, and they become endangered if they are sensitive to habitat disturbances and grow only in forests that may be clear-cut indiscriminately at any time.

Data on the status of rareness of medicinal plants are required to establish a national priority list. Efforts that have been initiated by Wiriadinata (1976) should be continued, where he used the scoring system adopted by *The IUCN Plant Red Data Book* (Lucas & Synge, 1978).

Implementation of Conservation

To ensure the availability of raw material for *jamu* industries and to explore the possibility of future development, sustainability of medicinal plants and preservation of the variability of germplasm are absolutely necessary. Although not specifically designed for conserving medicinal plants, the Government of Indonesia has implemented the conservation efforts by designating various types of natural ecosystems as nature reserves, wildlife sanctuary, national parks, biosphere reserves, protection forests, recreational forests and other types of protected areas. *In situ* conservation is at present the only significant action undertaken by the government, although it is not well established in the sense that the management of protected areas is not as yet effective because of the vastness and widely scattered distribution of the protected areas. Stricter control in and better management of protected areas still have to be developed further.

In situ conservation of medicinal plants should be further strengthened by *ex situ* conservation through cultivation and maintenance in botanical gardens, arboreta, parks, and in any other possible sites. At present Indonesia has two general living plant collections (Botanical Gardens and Gardens of the Centre for Research and Development of Industrial Plants) and medicinal plant gardens (Hortus Medicus Tawangmanguensis and one private garden), which may be considered significant but are still far from sufficient to accommodate the need of conserving medicinal plants. In the Botanical Gardens and Industrial Plant Gardens, the medicinal plant collection is only a complement to the collection of other plants, while the Tawangmangu garden accommodates only medicinal plants growing at high altitudes. Those gardens, therefore, have not been properly functioning as sites to conserve endangered medicinal plants in a real sense and the plants cultivated there have not been selected according to the priority indicated above.

It has been suggested that to complement the Hortus Botanicus Tawangmanguensis, new gardens at different locations with different climatic conditions should be established and careful planning regarding the species selection for cultivation is exercised. It should be noted that establishment of such a garden will require substantial funding to cover the costs of site planning, securing plants for planting and in particular for its maintenance. It would be more advantageous if such a garden is so designed as to have multiple functions, in the sense that it is not only devoted to a *basic collection* aimed at germplasm conservation, but it should constitute also an active *working collection* which can be utilized for a variety of activities. The plants contained therein will provide benefits and usefulness in improving the quality of *jamu* through the provision and development of better quality propagules of medicinal plants. In this respect it is necessary to distribute samples to parties interested for further evaluation of their genetic properties, variability, chemical contents, cultivation techniques, economic potentials, etc. In so doing in-depth and thorough research and development of the use of the plants for medicinal purposes could be undertaken. Research results and all the plant collections should be well documented so that their utilization for various purposes could justify the substantial funds expended for their conservation. This set of activities should be incorporated into the planning package of the establishment of new gardens or utilization of the existing ones to make the gardens more attractive to funding agencies.

Moreover the Department of Health should also appeal to other agencies (such as Department of Agriculture and Department of Public

Works) to assist or to be involved in the conservation of medicinal plants. Possibilities should be explored to make use of roadsides, school gardens, city parks, public graveyards, green belts, hospital gardens, open sites designated for regreening and reforestation programmes and any other suitable sites to be planted with trees having medicinal values, such as *ganitri* (*Elaeocarpus sphaericus*), *ki sariawan* (*Symplocos odoratissima*), and *nagasari* (*Mesua ferrea*). The Department of Agriculture, Department of Public Works and other agencies should be approached and convinced of the usefulness of medicinal plants for various purposes. With such an effort species that are becoming rare and endangered could be accommodated in these places which would then function also as reserves capable of providing supply of seeds, propagules and material for other purposes.

The Department of Health's country-wide programme on 'living pharmacy' (home gardens planted with medicinal plants for home use) could be also used for conservation purposes by including local rare species in the list of medicinal plants to be planted. By so doing it is expected that local people will become gradually aware of the importance of conservation and that by cultivating rare species they can gain benefits through the continuous supply of plant material which would be otherwise not be possible to obtain. Furthermore possibilities should be explored also to make use of the *taungya* agricultural system which has been practiced by farmers in Java to cultivate forest lands. Agronomic research should be encouraged to support the success of the above schemes, which in the long range would be able to supply *jamu* and other pharmaceutical industries with plant material needed.

It has been indicated elsewhere (Rifai, 1976, 1979) that the main agent responsible for the disappearance of medicinal plants is the utilization of these plants in *jamu* industries and the export of raw plant material. They should, therefore, share the responsibility and burden of conserving medicinal plants, in view of the fact that their survival is depending heavily on the availability of the plant material in large quantities. Efforts should be made to convince the *jamu* industries that it is wise and desirable that they should have nurseries and in cooperation with other agencies they should develop plantations of medicinal plants. The Forest Department controls a large area of unproductive lands (such as *Imperata* grassland and secondary forests), which can be made economically productive by cultivating them with medicinal plants. Establishment of such plantation will involve the use of a large area of lands, manpower, and substantial capital. The Nucleus Estate Scheme (*Perkebunan Inti Rakyat*), which has been successfully implemented in some agricultural enterprises, seems to be one of the possible means to develop medicinal plant plantations. In this scheme private companies and small-scale farmers are working together in a mutually beneficial fashion. At any rate, conservation of medicinal plants will be successful if the government has the political will and in this case it should be made mandatory for large *jamu* and other pharmaceutical industries to produce and manage their own supplies of raw plant material without endangering the survival and existence of medicinal plants growing in the wild.

Participation of People in Conservation

Conservation of medicinal plants is concerned mainly with activities to protect them against human disturbance, hence public participation is mandatory. Exploitation of medicinal plants that leads to their extinction is not directly done by *jamu* industries and exporters but by rural people with capital provided by private creditors. People who collect plants in the field are not aware that they are annihilating resources that nurture them. They are in general lowly educated and poor people who continuously exploit the resources in order to get as much cash as possible for their subsistence without thinking of the future. Provision of information concerning sustainable means of harvesting resources so as not to endanger both plants and people and on other alternative ways of earning living is extremely important in this respect.

Various non-governmental organizations concerned with environment and nature conservation and such organization as boy scouts may be urged to participate in the conservation of medicinal plants. They could be involved in regreening and reforestation of degraded land areas by planting them with medicinal plants, monitoring rare plants, and exercising social control on conservation. It would be desirable if scientists and other parties directly involved in dealing with medicinal plants could get together in a working group and in cooperation with government sector actively work to formulate ways and means of conserving medicinal plants. Professional associations can also have a big share in this activity because of their independence from government policy enable them to provide objective evaluations

It would also be desirable if the Department of Health could take the initiative to promote the involvement of *jamu* industries and exporters of plant material in all activities related to conservation of medicinal plants. The existing medicinal plant gardens should function more than just an exhibit of plant collections but they could also become centres for propagating medicinal plants and suppliers of propagules needed for large-scale cultivation.

Only with profound awareness and direct commitment along with sincere willingness to sacrifice time, energy and funds of all parties concerned that efforts in conservation of medicinal plants will produce significant results. If the efforts are successful, not only the knowledge on *jamu* but also the medicinal plants themselves can be inherited by future generations. We have yet to prove that we are a responsible generation.

References

Lucas, G. & Synge, H. (1978). *The IUCN Plant Red Data Book.* Morges: IUCN.

Rifai, M.A. (1976). Genetic erosion and conservation of medicinal plants in Indonesia. Paper presented at the *Symposium on Traditional Medicine I,* Semarang. (In Indonesian).

Rifai, M.A. (1979). The process of decimation of medicinal plants. Paper presented at the *Seminar on Supervision of the Application of Traditional Medicine,* Jakarta. (In Indonesian).

Wiriadinata, H. (1976). The future survival of some medicinal plants. Paper presented at the *Symposium on Traditional Medicine I,* Semarang. (In Indonesian).

Experiences from Programmes to Conserve Medicinal Plants

Medicinal Plants in India: Approaches to Exploitation and Conservation

S.K. Alok
Joint Secretary, Ministry of Health and Family Welfare,
New Delhi, India

In India, medicinal plants are widely used by all sections of the population, whether directly as folk remedies or the medicaments of the different indigenous systems of medicine, or indirectly in the pharmaceutical preparation of modern medicine. The country is richly endowed with a wide variety of plants of medicinal value which represent a great national resource.

Traditionally, practitioners of the Indian systems of medicine – Ayurveda, Unani, Siddha – made up their own prescriptions for their patients but, nowadays, most of their remedies are manufactured products. The increasing demands of the pharmaceutical industry have created problems of supply and one of the major difficulties being experienced by the Indian systems of medicine is that of obtaining sufficient quantities of medicinal plants for the manufacture of genuine remedies. In the absence of standards for crude drugs, adulteration and substitution have become common.

To correct this situation, measures are needed to promote the cultivation of medicinal plants, to improve methods of collection, to ensure effective quality control and to regulate commerce so as to protect both the producer and the consumer.

There is also a need to create greater general awareness amongst the population as a whole, government officials (particularly those in agriculture and forestry), farmers and scientists of the medicinal and economic value of these plants, so that this heritage may be wisely used and exploited, and at the same time conserved for future generations.

With this in mind, the Ministry of Health and Family Welfare sponsored four regional seminars on medicinal plants in 1986. These were held at Gunagadh (western region), Manali (northern region), Gauhati (northeastern region) and at Coimbatore (southern region). The seminars were interdisciplinary and brought together a wide range of expertise representative of the various interests concerned, at Central Government, State and Union Territory levels.

Representation in the field of health included participants from the Central and State Ministries of Health and Family Welfare, the Indian Council for Medical Research, the Central Council of Indian Medicine, the Central Councils for Research in Ayurveda, Siddha, Unani Medicine and Homoeopathy, the Central Drug Research Institute and the Central Indian Pharmacopoeial Laboratory. At regional level, participants came from the State health services, university medical faculties and research centres, and included local practitioners of Indian systems of medicine and homeopathy.

Government and university departments of agriculture, botany, forestry and horticulture also had an important contribution to make. Their participation included staff of the Indian Forestry Service, the Forest Research Institute, the Botanical Survey of India, the National Bureau of Plant Genetic Resources, the Council of Scientific and Industrial Research, and a State Department of Tribal Welfare. Other disciplines represented included life sciences, physics, zoology and folklore research. From the private sector, there were directors of pharmaceutical companies, marketing executives, owners of pharmacies, experts on medicinal plants, cultivators and others.

Thus, these regional seminars constituted a series of broadly-based consultations which, through discussions and working papers, made it possible to share experiences, present different points of view and arrive at a consensus of opinion on the situation in the country as a whole.

Reviewing the Situation

The same general approach was used at each seminar, participants being divided into working groups to consider the various aspects of the subject.

For example, for the northern region, three working roups were given common terms of reference but applied them to different States (Group 1 – Himachal Pradesh; group 2 – Jammu and Kashmir; group 3 – Haryana and Punjab). Each group was provided with a list of about 50 medicinal plants found in the area concerned and was invited to:

- Assess their availability in the area;
- Estimate the annual requirements for each (indicating the parts of the plants used);
- Identify shortages due to export, over-exploitation and collection of plants by untrained and unskilled labour;
- Draw up plans for cultivation, collection and conservation;
- Suggest measures for improved storage and preservation; and
- Explore possibilities for developing new drugs and substitutes.

In the seminar for the north-eastern region, a list of 112 medicinal plants readily available in the area was provided as a basis for discussions. One group was to consider their availability (abundant, common, scanty), distribution in the region and whether they were imported or exported. A second group was concerned with cultivation, collection and conservation. A third group was to make recommendations regarding supply and demand, marketing strategy, promotion of commonly-used medicinal plants as natural remedies, and the training of unskilled labourers, collectors and cultivators.

The seminar for the Southern Region followed a similar pattern, participants being given a list of 90 medicinal plants available in the area.

Assessing Medicinal Plant Resources

Examining the lists of medicinal plants commonly used or found in their respective regions, the working groups found that they could make only rough estimates of their availability, generally classifying them as abundant, common or scarce. However, even this was of value in focussing attention on species which were or might soon become endangered and on the need for their cultivation, conservation or substitution. Omissions were repaired, regional lists of medicinal plants duly completed and a number of exotic species suitable for local cultivation included.

Reliable data on supply and demand were not available and indeed, the multiplicity of government departments and agencies dealing with the various aspects of medicinal plants made the collection of such data extremely difficult. Nevertheless, a start could be made in putting data collection on a regular basis. For example, drug licensing authorities should require pharmaceutical manufacturers to supply estimates of annual requirements of medicinal plants as well as returns of the quantities used. State industries should provide similar estimates, as should the Expert Promotion Councils of Bombay and Calcutta. The monthly statistics on the "Foreign Trade of India" were another source. Data on medicinal plant production and utilization might well be collected by students of Indian systems of medicine and homoeopathy as part of their practical training.

Coordination and Support

It was generally agreed at the regional seminars that some form of overall authority was needed to coordinate the various activities relating to medicinal plants. A central agency, such as an Indigenous Drug Development Corporation, with subsidiary agencies at State level, could serve this purpose. It would draw on expertise from the different disciplines involved to formulate policy and would provide a mechanism or keeping market conditions, prices, supply and demand, internal trade, exports, cultivation, utilization and conservation under regular and comprehensive review.

Within the Minstry of Health and Family Welfare, a central monitoring cell, with corresponding cells at State Government level, would maintain close liaison with the drug licensing authorities and would monitor the use of medicinal plants and raw drugs by pharmaceutical and other industries in the public and private sectors.

These government bodies would bring out periodical bulletins giving details of cultivation, availability, requirements, import and export potential, etc., of other different medicinal plants, the desirability of using substitutes and the need for conservation.

Cultivation

Considerable expansion is required in the cultivation of medicinal plants, not only to meet the requirements of the health sector and commerce but also to counteract the harmful effects of over-exploitation of species in short supply.

At State level, large-scale cultivation should be undertaken of a variety of the most commonly used medicinal plants, both for demonstration purposes and as a source of supply. Similarly, medicinal plant gardens should be set up at district level to serve as demonstration- cum-training centres and also as nurseries. Even at village level, small medicinal plant gardens would be of practical and educational value and could be developed simultaneously with the social forestry programme.

Agricultural universities and other research organizations have a major role to play in establishing and maintaining model medicinal plant gardens, in carrying out research, in serving as reference centres, in providing technical guidance, in laying down agronomic practices for farmers and in studying the economics of medicinal plant production. For plants in short supply, they may use tissue culture techniques to provide large numbers of plantlets for supply to cultivators while, at the same time, assessing the content of active principles of the plants obtained by such means.

The cultivation of medicinal plants may also be encouraged under the Social Forestry scheme, by using abandoned land in shifting cultivation areas and also by roadside planting of suitable trees.

Collection from the Forests

A great deal of India's wealth of medicinal plants lies in her forests. In forested areas, the collection of medicinal plants for sale in the markets forms an important part of the livelihood of the local inhabitants, particularly of tribal people, or of the traders who send their own workers into the forests to collect these plants.

Much of the collection of medicinal plants from the forests, therefore, is done by people from the economically less favoured sections of the population who have not been trained for the purpose. As a result, collection methods are wasteful and there is over-exploitation with no attempt at regeneration. It is not surprising, therefore, that some plants, traditionally collected from the forests for medicinal purposes, are now in short supply. Finally, the disappearance of certain species is an inevitable result of the increasing destruction of India's forests.

This combination of factors poses a major threat to the survival of many plant species and obviously requires urgent action if measures for conservation are to be effective and to have a change of success.

Market surveys of medicinal plants and the drugs derived from them are needed so that projections of likely demand may be made by States and suitable programmes developed to ensure adequate supplies. Such survey data, collected by the Central Government and made available to other States, would facilitate rational planning.

Where medicinal plants are cultivated, establishing cooperatives with an appropriate network of drug depots has obvious advantages in making a proper marketing, storage and grading infrastructure available and trade more viable. This has been done by the Government of Tamil Nadu, in the southern region.

However, in areas where medicinal plants are not cultivated but are collected from the wild by forest dwellers, tribal people and others, cooperatives may not be easy to organize. In such cases, it may be preferable to entrust the setting-up of collection and storage centres and the organization of marketing to a forest or tribal development corporation. Alternatively, collection and storage may be left to private enterprise, individuals or companies being registered with the State for this purpose.

Storage

Adequate storage facilities need to be established as near as possible to collection centres and provision made for sorting and preserving the plant material (leaves, fruit, bark, roots, gum, resins, etc.). This may require special studies under local conditions to determine the best methods of drying and avoiding deteroration. By keeping regular records, these centres may help in providing reliable data on the production of medicinal plants.

In forest areas, these plant should be identified and listed as "minor forest produce" so that their collection, storage and marketing may be promoted by State Forest Departments.

Processing and Manufacture

Concern has been expressed about the quality of the dried medicinal plants entering the market. This underlines the importance of proper identification, sorting, grading and storage of raw material, the need for quality control and the observance of recognized standards at all stages of processing.

Dealers, manufacturers and exporters should have their material duly identified, tested and certified for purity by laboratories approved by the State and Central drug control authorities. Experience and data acquired by the Research Councils of Ayurveda, Siddha, Unani, Homoeopathy and those of modern medicine may be of value in this connection.

Periodical workshops on the technology of processing medicinal plants would help to improve manufacturing practices and product quality and could be organized jointly by industry, research institutions and medical colleges.

In areas producing large quantities of medicinal plants (for example, in the north-east region of India) encouragement should be given to setting up manufacturing units. This would not only increase local interest in the collection and cultivation of medicinal plants but would also generate much-needed employment. By producing crude preparations of reliable quality for sale to the trade, the decline in the prices of raw materials may be arrested.

This presents an opportunity for state Governments to exercise a beneficial influence over commerce in medicinal plants and their derivatives and to ensure fair support prices for collectors and cultivators. There may also be a role for cottage industries in the partial processing of certain raw materials and this needs to be explored.

Incentives

Promotion of the cultivation, processing and utilization of medicinal plants may require financial and other incentives. Individuals and private institutions wishing to engage in these fields on a commercial scale should be eligible for government grants, interest-free or low-interest loans and short- or medium-term credit facilities. The State itself may become involved in marketing or production, either by holding share capital in private enterprise or by assuming direct responsibility as may be necessary in the economically under-developed tribal or forest areas.

Conservation and Research

Sustainable utilization and conservation go hand in hand. The various activities and approaches described above – review, assessment, cultivation, collection, marketing, storage, processing, manufacture and incentives – are all essential to any balanced plan to make full use of the nation's wealth of medicinal plants while avoiding over-exploitation. Space does not permit a more detailed discussion.

The working groups identified certain specific plants growing in their regions for which protective measures seemed to be urgently needed. They recommended that "custodian institutes" be designated, each being made responsible for the *ex situ* conservation and propagation of a few selected endangered medicinal plants.

The establishment of herbaria and medicinal plant gardens by universities and regional stations was considered to be a priority for all States. Used not only for cultivation and propagation but also for educational purposes, they would meet an important need since it was felt that the herbaria maintained by the Botanical Survey of India were too large and too technical to be approached by the general public.

In situ conservation requires the creation of protected areas and forest reserves in conjunction with afforestation programmes, a ban on trade in endangered species, the identification of suitable substitute plants where they exist, ethnobotanical studies and the provision of incentives to tribal people to conserve these species.

Restrictions or a ban on exports of certain plants may be essential, particularly as a short-term measure. Such a decision would be a matter of the Central government, which would take into account the overall supply position, something which individual States could not do, and would keep the situation under regular review.

The conservation of germplasm would form part of the responsibilities of a network of institutions which would also engage in a co-ordinated programme of research to develop high-yielding disease-resistant strains. They would employ the most modern techniques, including tissue culture and genetic engineering. The proposed National Centre for Research on Medicinal Plants would have a valuable role in coordinating the different aspects of research in this field.

A major element in the conservation of medicinal plants will be the extent to which people can be brought to have a better and fuller understanding of the part these plants play in their lives and the need to protect them. Devising suitable information programmes for the different sections of the public concerned will call for great skill.

Schoolchildren are probably the most receptive audience and the easiest to reach. Such instruction can be introduced at an early stage in their education and can serve as a means of teaching them about the conservation of plant and animal life as a whole. Similarly, the subject should be included in the general education of college and university students, as part of an orientation to human ecology, and in adult education programmes.

Forest dwellers, tribal people, collectors, cultivators, unskilled labourers and others who make a living from medicinal plants would greatly benefit from suitable training in identification, methods of collection, preservation and storage. This would help them to avoid overexploitation and wastage and, complemented by better arrangements for marketing, would allow them to make the best use of the resources available.

In this work, colleges of Indian systems of medicine, regional research centres, universities and other educational institutions have an obvious place in the development of courses and teaching material, the training of staff and the creation of facilities for practical demonstrations and field work.

Conclusion

India has a vast reservoir of nearly 400,000 practitioners of Ayurveda, Siddha, Unani medicine or homoepathy who, unfortunately, have not so far been adequately utilized in the health care delivery systems. National health policy stresses the need for preventive and promotive medicine which these systems are well-placed to deliver and, at some stage, there has to be some sort of integration of all these systems if the people are to be properly served.

About 80% of the raw materials for drugs used in the Indian systems of medicine and homoeopathy are based on plant products. The credibility of these systems of medicine depends, therefore, on having available authentic raw material in sufficient quantities. With some 46,000 licensed pharmacies manufacturing the traditional remedies of these medical systems, it is necessary to plan for large-scale cultivation of medicinal plants and to ensure that they are accurately identified, properly processsed, free of adulterants and of acceptable quality.

The regional seminars, briefly described, have explored many aspects of the subject and it is clear that progress in the cultivation, utilization and conservation of medicinal plants depends on the willingness to work together of numerous government departments, agencies and private enterprises. The work of the Department of Environment in bringing out a list of endangered species deserves special mention. There is, naturally, always an interval between the making of recommendations and their implementation. It is, therefore, of interest to record that steps have already been taken to set up a Coordinating and Monitoring Cell on Medicinal Plants in the Ministry of Health and Family Welfare and that similar cells will be set up in each State. A start has been made and awareness has been created of the importance of medicinal plants. We now have to secure the effective cooperation of the Departments of Environment, Forest, Agriculture, Rural Development and other appropriate organizations so that we may pool our resources and work together.

The Chinese Approach to Medicinal Plants–Their Utilization and Conservation

Xiao Pei-gen
Institute of Medicinal Plant Development (IMPLAD), Chinese Academy of Medicinal Sciences, WHO Collaborating Centre for Traditional Medicine, Beijing, People's Republic of China

The Present Situation

In China, medicinal plants enjoy an inherent and prominent role in the general health service and are by no means secondary in functions to synthetic drugs and antibiotics. This fact is well reflected by the Chinese Pharmacopoeia (in 2 volumes, 1985 ed.), of which the first volume is totally devoted to traditional Chinese crude drugs (consisting nearly exclusively of medicinal plants) and their varied preparations, with a total of 713 items, including 385 items of crude drugs (excluding their preparations), of medicinal plant origin (Yang, 1985, Xiao, 1989); the second volume lists all the chemical, antibiotic and biological drugs along with their preparations, totalling 776 items (Yuan & Zhang, 1985). The importance of medicinal plants is also evidenced from the statistic that the market for traditional Chinese drugs accounted for an average to 40% of the total consumption of medicaments in recent years (Xiao, 1983).

Unfortunately, due to the destruction of forests, overgrazing of meadows, expansion of industry and urbanization, as well as excessive collection in the wild of rare and endangered plants, the natural resources of medicinal plants are being reduced day by day. There has, therefore, been an urgent need to draw up the necessary plans for medicinal plant utilization and conservation by our State Government.

The Strategies Adopted

Investigation and Systematization

As the first stage of a long-term programme on the utilization and conservation of Chinese medicinal plants, a nation-wide survey project on this subject, involving several years of hard work, has been conducted, with the items recorded including their origin, distribution, ecological features, resources and their therapeutic effectiveness. Total number of plant species involved in this survey (Xiao, 1987) is given in Table 1.

Table 1. Chinese medicinal plants identified to date

Origin	Number of species
Thallophytes	281
Bryophytes	39
Pteridophytes	395
Gymnosperms	55
Angiosperms	
Dicotyledons	3690
Monocotyledons	676
Total	5136

Subsequent comprehensive studies of the data collected, supplemented by information from the relevant literature provided the basis for the compilation of several series of monographs:

1. **Chinese Materia Medica** (1979-). This is regarded as the leading, up to date compilation in this field. The old edition contained 994 items of crude drugs in four volumes. The new edition will be in six volumes of which the first four (1979, 1982, 1985 and 1988) have already been published. The chief editor was the Institute of Materia Medica, and is now the Institute of Medicinal Plant Development (IMPLAD), Chinese Academy of Medical Sciences.
2. **Encyclopaedia of Chinese Materia Medica** (1977). This was edited by the Nanjing College of Traditional Chinese Medicine and includes 5767 drug items (botanical, animals, and mineral origins). It is regarded as the most comprehensive compilation of traditional Chinese drugs and Chinese medicinal plants.
3. **Colour Atlas of Chinese Herbal Drugs** (1982-). Edited by the People's Medical Publishing House and the Xiong Hun Publishing House, Japan. It will be in 25 volumes and will include a total of 5,000 drug items along with colour paintings, most of them of medicinal plants. The first eight volumes have already been published.
4. **New Compendium of Chinese Materia Medica**, with Professors Wu Zheng-Yi, Zhou Tai-Yan and Xiao Pei-Gen as the Chief Editors. It will be in three volumes, including some 6,000 medicinal plants. Two volumes appeared in 1988 and 1989 respectively.
5. **Colour Album of Chinese Herbal Medicines** (1988-), with Professor Xiao Pei-Gen as Chief Editor. It includes 5,000 colour photographs of Chinese herbal medicines in ten volumes, of which vols. I, II (1988) and vol. III, IV, V, VI (1989) have already been published.

In addition, 35,000 items of ethnopharmacological data on Chinese medicinal plants have also been collected and subsequently systematized by electronic computer (Xiao *et al.*, 1986, 1989).

Conservation and Implementation

Though abundant and rich, the natural resources of medicinal plants are not unlimited and must be well-protected by law and by the people.

Collection of any medicinal plant should be guided by precise knowledge of the species, including its locality, time of its maturation, the part(s) to be collected, and its conservation needs being taken into consideration at the same time. Steps should be taken to avoid over-exploitation and excessive collection. The gathering of rare and endangered species, such as *Panax ginseng, Coptis chinensis, Gastrodia elata* and *Paris polyphylla* etc. should be prohibited.

Genebanks of medicinal plants should be established. Up to the present, quite a number of important Chinese medicinal plants have already been preserved in genebanks under the auspices of several agricultural institutions.

Introduction and Acclimatization

Not a few of the plants listed in the ancient Chinese Materia Medica were of foreign origin and even today some of the crude drugs prescribed have to be imported still, for example, American Ginseng from the United States and Canada, *Amomum kravanh* from Thailand etc. In order to meet the market demand several botanical gardens in China initiated the task of introducing these plants for acclimatization, aimed at eventual plantations in farms. The introduction policy included also those non-officinal wild origin plant crude drugs which have been used by people for practical medication. Some examples of these two categories are as follows:

Of foreign origin: *Panax quinquefolium, Amomum kravanh, Strychnos nux-vomica, Rauwolfia vomitoria, Syzygium aromaticum, Cassia acutifolia* and *Vinca minor* etc.

Indigenous species of wild origin: *Schisandra chinensise, Gentiana scabra, Cistanche deserticola, Asarum sieboldii, Scutellaria baicalensis, Bupleurum chinense, Atractylodes lancea, Anemarrhena asphodeloides, Fritillaria unibracteata* and *Alpinia oxyphylla* etc.

Three major botanical gardens are currently carrying out studies on the medicinal plants to be introduced:

- Institute of Medicinal Plant Development (IMPLAD), Chinese Academy of Medical Sciences (Beijing) and its two branch institutes (Hainan and Yunnan);
- Guangxi Botanical Garden of Medicinal Plant (Nanjing);
- Nanchuan Institute of Cultivation of Chinese Materia Medica (Sichuan, Nanchuan).

Several other botanical gardens, located in Beijing, Guangzhou, Nanjing, Shanghai, Kunming, Lushan, Hangzhou and Xian, also participated in the work.

The total number of medicinal plants under study is estimated to be 4,500 species.

Cultivation

With the increasing use of medicinal plants in China, it is necessary to cultivate the most commonly used ones to guarantee supplies.

About 100 species of medicinal plants are under cultivation (Table 2) covering some 460,000 hectares.

The most important cultivated medicinal plants are: *Panax ginseng, P. notoginseng, Astragalus mongholicus, Angelica sinensis, Coptis chinensis, Codonopsis pilosula, Rehmannia glutinosa, Paeonia suffruticosa, Cinnamomum cassia, Amomum villosum* and *Atractylodes macrocephala*; they are all commonly used traditional Chinese drugs.

At the same time, several wild growing medicinal plants which are needed in vast quantities are now being introduced and cultivated, e.g., *Glycyrrhiza uralensis, Rheum palmatum, Cistanche deserticola, Poria cocos* and *Dioscorea nipponica* etc.

In addition, modern biological technology has also been used: for instance, tissue culture is used for propagating *Lithospermum erythrorhizon, Panax quinquefolium, Corydalis yanhuosu, Scopolia tangutica* and some others.

Utilization

The road which led to the recognition of the importance of medicinal plant in its justified position has been long and bumpy. Not until 1949, when the meaningful resurgence of Traditional Chinese Medicine effected under the direction of the Chinese Government, did Chinese scientists go all out for a joint multidisciplinary research for the better utilization of medicinal plants for people's health care. Several papers can be cited here for further information: for a general introduction

(Xiao, 1981a, 1983, 1986); for pharmacologically active principles (Xu *et al.*, 1985, Liang *et al.*, 1986, Xiao & Fu, 1987, Xiao, 1987, Zhu, 1987); for clinical trials (Xiao & Chen, 1987, 1988); for new drugs developoment (Xiao, 1981b).

Table 2. List of cultivated medicinal plants in China

Achyranthes bidentata (rt), *Aconitum carmichaeli* (rhz), *Alisma orientale* (rhz), *Allium tuberosum* (sd), *Amomum villosum* (fr), *Angelica dahurica* (rt), *A. sinensis* (rt), *Andrographis paniculata* (pl), *Areca catechu* (sd), *Artemisia argyi* (l), *Asparagus cochin-chinensis* (rt), *Aster tataricus* (rt), *Astragalus mongholicus* (rt), *Atractylodes macrocephala* (rhz), *Aucklandia lappa* (rt), *Biota orientale* (sd), *Brassica juncea* (sd), *Carthamus tinctorius* (fl), *Cassia acutifolia* (l), *C. obtusifolia* (sd), *Celosia cristata* (fl), *Chaenomeles speciosa* (fr), *C. grandis* (exocarp), *Citrus aurantium* (fr), *C. reticulata* (exocarp), *Codonopsis pilosula* (rt), *Coix lacryma-jobi* var. *ma-yuan* (sd), *Coptis chinensis* (rhz), *C. deltoidea* (rhz), *Cornus officinalis* (fr), *Corydalis yanhuosu* (rhz), *Crataegus pinnatifida* (fr), *Curcuma aromatica* (rhz), *C. domestica* (rhz), *C. zedoaria* (rhz), *Datura metal* (fl), *D. innoxia* (fl), *Dendrobium nobile* (pl), *Dioscorea opposita* (rhz), *Dolichos lablab* (sd), *Eriobotrya japonica* (1), *Eucommia ulmoides* (bk), *Euphorbia longan* (aril), *Eupatorium fortunei* (1), *Euryale ferox* (sd), *Evodia rutaecarpa* (fr), *Gardenia jasminoides* (fr), *Fritillaria thunbergii* (bulb), *Gardenia jasminoides* (fr), *Gastrodia elata* (rhz), *Ginkgo biloba* (l, sd), *Gleditsia sinensis* (fr), *Glehnia littoralis* (rt), *Illicium verum* (fr), *Impatiens balsaminea* (sd), *Isatis indigotica* (l, rt), *Lilium brownii var. viridulum* (bulb), *Lonicera japonica* (fl-bud), *Lycium barbarum* (fr), *Magnolia officinalis* (bk), *M. biondii* (fl), *Melia toosandan* (fr, rt-bk), *Mentha haplocalyx* (pl), *Morus alba* (bk, l), *Nelumbo nucifera* (l), *Ophiopogon japonicus* (rt), *Paeonia lactiflora* (rt), *P. suffruticosa* (rt-bk), *Panax ginseng* (rt), *P. notoginseng* (rt), *Perilla fructicosa* (l, sd), *Phellodendron amurensis* (bk), *Piper nigrum* (fr), *Pogostemon cablin* (l), *Polygonatum cyrtonema* (rhz), *Poria cocos* (pl), *Prunus armeniaca* (sd), *P. mume* (fr), *P. persica* (sd), *Pseudostellaria heterophylla* (rt), *Psoralea corydifolia* (fr), *Raphanus sativus* (sd), *Rehmannia glutinosa* (rhz), *Ricinus communis* (sd), *Salvia multiorrhiza* (rt), *Schizonepeta tenuifolia* (pl), *Scrophularia ningpoensis* (rt), *Sesamum indicum* (sd), *Sinapis alba* (sd), *Sophora japonica* (fl-bud), *Stephania tetrandra* (rt), *Trigonella foenum-graecum* (sd), *Tussilago farfara* (fl-bud), *Zingiber officinale* (rhz), *Ziziphus jujuba* (fr)

Key: bk = bark, fl = flower, fr = fruit, l = leaf, pl = plant, rt = root, sd = seed rhz = rhizome

In China, the quantity of medicinal plants used for direct decoction in the traditional Chinese doctor's prescription and as ingredients in the officinal medicine is huge and the current annual demand is reportedly to be 700,000 tons. This figure excludes traditional usage as mentioned above and also the production of several minor industrialized commodities.

1. **As raw materials**: *Dioscorea nipponica, D. zingiberensis* (for diosgenin), *Rauwolfia verticillata* (for reserpine or total alkaloids), *Scopolia tangutica* (for hyoscyamine, anisodamine, anisodine), *Datura metel, D. innoxia* (for scopolamine), *Ephedra sinica* (for ephedrine), *Berberis poiretii* (for berberine, berbamine), *Cephalotaxus mannii* (*C. hainanensis*) (for harringtonine, homoharringtonine), *Sophora japonica* (for rutin), *Thevetia peruviana* (for cardiotonic glucoside).
2. **As sweeteners:** *Glycyrrhiza uralensis, G. glabra; Rubus suavissimus* (similar to stevioside); *Siraitia grosvenorii; Stevia rebaudiana* (stevioside); *Lithocarpus polystachys.*
3. **As bitterness agents**: *Humulus lupulus, Gentiana scabra, Swertia japonica, Citrus aurantium, Taraxacum mongolicum.*
4. **As health drinks**: *Actinidia chinensis* (fr), *Rosa roxburgii* (fr), *R. dahurica* (fr), *Tamarindus indica* (fr), *Psidium guajava* (fr), *Crataegus pinnatifidus* (fr), *Hibiscus sabdariffa* (calyx), *Chrysanthemum morifolium* (inflor), *Vaccinium uliginosum* (fr).
5. **As general tonics or health foods**: *Panax ginseng* (rt), *Astragalus mongholicus* (rt), *Cordyceps sinensis* (dead caterpillar), *Ganoderma lucidum* (fruitbody), *Zizyphus jujuba* var. *inermus* (fr), *Dioscorea opposita* (rhz) *Lycium barbarum* (fr), *Tremella fusiformis* (pl), *Acanthopanax senticosus* (rt).
6. **As natural colourings**: *Hippophae rhamnoides* (orange), *Curcuma longa* (yellow), *Gardenia jasminoides* (yellow), *Lithospermum erythrorhizon* (violet), *Capsicum annum* (red), *Carthamus tinctorius* (red, orange).
7. **For essential oils**: *Cinnamomum cassia* (bk), *Illicium verum* (fr), *Pogostemon cablin* (pl), *Mentha haplocalyx* (pl), *Thymus serphyllum* (pl), *Eucalyptus globulus* (l), *Jasminum officinalis* (fl), *Michelia alba* (fl), *Magnolia liliflora* (fl), *Vanilla planifolia* (fr), *Foeniculum vulgare* (fr), *Artabotrys hexapetalus* (fl), *Nardostachys chinensis* (rt), *Rosa rugosa* (fl), *Origanum vulgare* (pl), *Amomum villosum* (fr), *Zingiber officinale* (rhz), *Cymbopogon citratus* (pl), *Cinnamomum camphora* (1), *Litsea cubeba* (fr), *Ocimum gratissimum* (pl).

For key see Table 2.

Task of IMPLAD

The Institute of Medicinal Plant Development (IMPLAD), a WHO Collaborating Centre of Traditional Medicine in Beijing, with its two branches in Yunnan province and Hainan Island, is an organization specializing in the research of medicinal plants, under the auspices of Chinese Academy of Medical Sciences (CAMS).

IMPLAD is committed to a three-fold developing programme aimed at the better utilization of Chinese medicinal plants, particularly those commonly used as traditional Chinese medicaments. This involves, firstly, to enlarge the resources and improve the quality of the medicinal plants, either of foreign origin or wild-growing ones, which have been had to be imported or are in short supply in the market from natural resources only, together with resolving problems of plant diseases and insect pests affecting the medicinal plants studied including: *Panax ginseng, P. quinquefolium, Coptis chinensis, Rehmannia glutinosa, Amomum villosum, A. kravanh, Lycium chinense, Astragalus mongholicus, Lonicera japonica, Glehnia littoralis, Fritillaria thunbergii, F. pallidiflora, F. unibracteata, Gastrodia elata, Cistanche deserticola, Corydalis yanhuosu, Ganoderma lucidum, Vinca minor, Polyporus umbellatus* and *Areca catechu*. Secondly, to develop new preparations or other products from medicinal plants, for example, a health drink derived from the fresh juice of *Hipphophae rhamnoides*, "Ji-Sheng injections" from the spore of *Ganoderma lucidum*, "Zhen-Qi capsules" from the extracts of *Astragalus mongholicus* (rt) and *Ligustrum lucidum* (fr), and a potion "American Ginseng Royal Jelly" from *Panax quinquefolium* and *Schisandra chinensis* etc. And thirdly, searching for new drugs derived from medicinal plants, especially drugs that are effective in the fields of immunology, the cardiovascular and nervous systems; several new drugs have already been found, such as: anisodine and anisodamine from *Scopolia tangutica*, dimer-anthocynidin from *Fangopyrum dibotrys* etc.,

In order to maintain a natural genebank of medicinal plants, an enlarged botanic garden is being established in Beijing IMPLAD headquarters. This project is financed by the Chinese Government and occupies a total area of more than 143,000 m^2. It has various sections with appropriate cultivation, research areas and sections for the breeding and conservation of seeds, together with several greenhouses with available area of over 1800 m^2. It has at present living collections of 1300 species of medicinal plants.

Conclusions

As a Member State of WHO, with one fifth of the world's population, we in China rely heavily on "natural medicine" of plant origin, and wish to recommend this idea to other countries, especially those of the "third world", to share our experience in utilizing medicinal plants and hence their conservation and their research and development for the welfare of the whole world.

Acknowledgement

My grateful thanks are due to Professor Wei-Liang Sung of IMPLAD for his valuable help in preparing this manuscript.

References

Liang Xiao-Tian, Yu De-Quan & Fang Qi-Cheng (1986). Structural Chemistry of Natural Products in Chinese Herbal Medicine. In *Advances in Science of China, Chemistry*, Vol. 1, ed. Tang You-Qi, pp. 1-39. Beijing: Science Press; and New York: John Wiley & Sons.

Xiao Pei-Gen (1981a). Some experience on the utilization of medicinal plants in China. *Fitoterapia*, 52, 65-73.

Xiao Pei-Gen (1981b). Traditional experience of Chinese herb medicine, its application in drug research and new drug searching. In *Natural Products as Medicinal Agents*, eds J. Beal & E. Reinhard, pp. 351-394. Stuttgart: Hippokrates Verlag.

Xiao Pei-Gen (1983). Recent developments on medicinal plant in China. *J. Ethnopharm.*, 7, 95-109.

Xiao Pei-Gen (1986). Basic research on herbal medicine in China. *Herba Polonica*, 32(3-4), 235-245.

Xiao Pei-Gen (1987). Ethnopharmacologic investigation on pharmacologically active principles from Chinese medicinal plants. In *Proceeding of the International Symposium on Traditional Medicines and Modern Pharmacology*, eds H.P. Lei, Z.Y. Song, J.T. Zhang, G.Z. Liu & H.Z. Tian, pp. 106-115. Beijing.

Xiao Pei-Gen (1989). Medicinal plants – expert from the Chinese Pharmacopoiea. In *Herbs, Spices and Medicinal Plants*, eds L.E. Craker & J.E. Simon, vol.4, in press.

Xiao Pei-Gen & Chen Ke-Ji (1987). Recent advances in clinical studies of Chinese medicinal herbs 1: Drugs affecting the cardiovascular system. *Phytotherapy Research*, 1(2), 53-57.

Xiao Pei-Gen & Chen Ke-Ji (1988). Recent advances in clinical studies of Chinese medicinal herbs 2: Clinical trials of Chinese herbs in a number of chronic conditions. *Phytotherapy Research*, 2(2), 55-62.

Xiao Pei-Gen & Fu Shan-Lin (1987). Pharmacologically active substances of Chinese traditional and herb medicines. In *Herbs, Spices and Medicinal Plants*, eds L.E. Craker & J.E. Simon, vol. 2, pp. 1-55.

Xiao Pei-Gen, Wang Li-Wei, Lu Shuang-Jin, Qiu Gui-Sheng & Sun Jing (1986). Statistical analysis of the ethnopharmacologic data based on Chinese medicinal plants by electronic computer (I) Magnoliidae. *Chin. J. Intregrated Trad. Western Med.*, 6(4), 253-256.

Xiao Pei-Gen, Wang Li-Wei, Qiu Gui-Sheng & Sun Jing (1989). Statistical analysis of the ethnopharmacologic data based on Chinese medicinal plants by electronic computer (II) Hamamelidae and Caryophyllidae. *Chin. J. Integrated Trad. Western Med.*, 9(7), 429-432.

Xu Ren-Shen, Zhu Qiao-Zhen & Xie Yu-Yuan (1985). Recent advances in studies on Chinese medicinal herbs with physiological activity. *J. Ethnopharm.*, 14, 223-253.

Yang Dai (1985). An introduction to the Chinese Pharmacopoiea (1985 ed.) volume I. *Chin. Trad. Herb. Drugs*, 16(11), 34-36.

Yuan Shi-Cheng and Zhang Qing-Xi (1985). General aspects of the Pharmacopoeia of People's Republic of China. *Chin. Pharm. Bull.*, 20(10), 582-585.

Zhu Da-Yuan (1987). Recent advances on the active components in Chinese medicines. *Abstract of Chinese Medicine*, 1(2), 251-286.

Conservation of Medicinal Plants in Kenya

J.O. Kokwaro
Department of Botany, University of Nairobi, Kenya

Introduction

Kenya is located on the equator in East Africa, and has a total land area of 582,644 sq. km. The land rises gradually from the sea in the south-east to about 1,800 m some 500 km inland where the Great Rift Valley divides it from north to south. The valley is about 70 km wide and 300 m deep with precipitous walls in some parts. The land slopes gradually down to the Lake Victoria shore at 1070 m on the west of the Rift Valley. There are two main mountain masses of Mt Kenya (5199 m) and Aberdare Mts (3964 m) to the east of the Rift Valley, and the Mau Range (3097 m) and Mt Elgon (4321 m) to the west.

More than half of Kenya in the north and north-eastern parts and also in the southern section of the Rift Valley floor is composed of arid or semi-arid land. Most western part of the country, and the higher land from 1370m to 2740m in the central part receives a good rainfall and are fertile. The mountain slopes and these areas of high rainfall had in the past been characterised with dense natural high forests.

The need to conserve the individual medicinal plants used is brought about by the rapid increase of population and more demand for land for both agriculture and human settlement. The estimated population in 1960 was 6.5 million people, and the present (1987) estimate is three times that figure. A number of the medicinal plants widely exploited are from the natural forests. The total area of closed forests in Kenya by 1963 when Kenya attained her independence was about 772,000 ha and this represented only 2.7% of the total land area. This is a very small percentage of forest land in a country where the population growth is so high as in Kenya.

The Kenya Government had no formal forest policy until 1957 when White Paper No. 85 of 1957 was published. The Forest Department similarly had no written policy but followed the recommendations of visiting experts from Britain. The White Paper No. 1 of 1957 clarified the interpretation of the Ordinance and enabled three categories of local district council forest reserves to be recognized. The Forest Department of the central government manages all three kinds:

1. In the case of productive forests, it pays net revenue to the local district council.
2. In the case of under-developed forests, it meets expenses from the Central Government funds and pays revenue if any, to the local district council who may be called upon to contribute to costs by deduction from revenue.
3. In non-productive forests, it meets expenses from Central Government funds without the accumulating debit balance being charged to the local district council.

Provisions are made for revising the declared status of a forest from time to time. These favourable financial terms for reservation of forests in private local areas had greatly facilitated reservation and conservation, and some local district councils have converted their by-law forests to forest reserves. Reservation and conservation of important catchment areas under local district councils has also progressed rapidly since such terms were introduced.

Despite the rapidly rising population the Forest Department has continued to preserve the major portion of the closed forest, and many valuable and denuded hill catchments which it has also been afforesting or protecting from further destruction. But despite all these efforts, the total area of forest reserves represents only 2.7% of the country (which is 12% of the fertile and densely populated part) and this is far from being adequate. The more forest reserves we have the more valuable medicinal plant species we can conserve. Whereas it is desirable to have more forest reserves to serve as a reservoir for medicinal plants, this desire however continues against mounting population pressure for land and mounting political resistance from politicians.

The threatened wild animal species are generally protected in part in the national parks and partly in the forest reserves. But the threatened wild plant species have no definite place for their protection because even the 2.7% of forest reserves continue to be reduced in size daily. The Forest Department has an obligation to produce enough forest products to meet the country's needs. On the other hand, we know that indigenous forests are not fast-growing, and have to be replaced with the fast-growing softwood plantations. In forest reserves the big game normally conflicts with reafforestation policy.

Major Ecosystems of Kenya

High and medium potential land is about 17% of the country, and these areas support 90% of the population, 75% of which is rural. Plant communities in high potential areas such as the highland forests, Kakamega Forest and the coastal forests are usually the most threatened by over-utilization. Most land in Kenya, about 75%, is composed of woodlands, bushlands and grasslands, and is mainly found in the arid and semi-arid areas. The inland water bodies cover about 2.1% of the surface areas. Although the forested land represents only about 2.1% (1,475,000 ha) of the total land area, they provide some of the essential needs for Kenyans. The depletion rates of the forest resource including herbal remedies are therefore very high. For example, Kakamega, North and South Nandi Forests which occur in high potential areas, are being cleared at the rate of 245, 295 and 490 ha per year respectively.

The woodlands, bushlands and savannas are situated mainly in the arid and semi-arid zones of Kenya, and cover approximately 75% of the total land area. They support the greater part of the national livestock and much of the wildlife. Degradation in these areas tend to be a much more gradual process than in the forest areas.

Current Problems in Conserving Medicinal Plants

We can observe the need for a more forceful reservation of both the nature reserves and the national parks with more immunity from excisions of temporary expediencey. Just like in most African countries, medicinal plants play a vital role in the health services of large areas of the rural Kenya, and yet there are no areas for their conservation. The Government is quite aware that the major function of the national parks is to conserve wild animals as the main tourist attraction. The same Government is aware that the main function of the Forest Department is to produce enough timber and other forest products from both natural forests and plantations for Kenya's needs. We are all aware that most wild game in the national parks feed on wild plants within the parks and nature reserves. We are also aware that many Kenyans continue to use herbal remedies the majority of which are found in both forest and national reserves. But the most important issue today is that there is no national policy on either conservation (protection) or propagation of the indigenous medicinal plants in the country. It is only through vigorous conservation that a representative sample of the natural herbal plants be preserved for future prosperity.

Most endangered medicinal plants occur in threatened plant communities of the country. Such areas have direct economic importance to the Kenyan economy for agriculture, industry and tourism. All these threatened plant communities are intimately linked to the survival and livelihood of Kenyan families in general.

Some General Conservation Recommendations

Using Kenya as an example, the following suggested recommendations may be applied.

A. For Recognised Individual Plant Species:

1. Forest Department, National Universities, Botanic Gardens, research institutions connected with medicinal plants and herbal practitioners must be encouraged to propagate medicinal plants in adequate quantities.
2. Conservation of germplasm should be undertaken by collecting seeds of medicinal plants.
3. Propagation methods should be devised to produce fast-growing medicinal plants, especially indigenous trees.
4. Herbal practitioners should be encouraged to grow their own medicinal plants in their own private gardens.
5. The value of public environmental education as an essential and effective method of action in protecting individual medicinal plant species should be recognized.

B. For Threatened Plant Communities

1. Up-to-date inventories of the existing plant communities should be carried out and the data made available to the public. A floristic inventory of the plant communities in the country should be made for planning its conservation and use.
2. Vegetation mapping is also an essential method of conserving the threatened plant communities since it also reveals what exists.
3. The value of public environmental education is an essential and effective method of action in protecting plant communities.
4. Collaborative research programmes by national institutions such as Universities, Research Institutes and international bodies is another effective method of approaching the conservation of plant communities.

5. A properly integrated land use policy and planning is a vital tool in promoting conservation of plant communities in the country. In this context, all land management plans at district level should incorporate the information on environmental impact studies. Such studies include the identification of endangered plant communities which require urgent protective management.
6. To protect indigenous forests from overexploitation, the Forest Department should plant more fast growing softwoods for local needs such as timber, charcoal, paper, etc.

Conclusions

There are many endangered individual medicinal plants in Kenya, but their conservation will depend largely on the overall conservation of the endangered ecosystems in which they occur. Kenya, like most tropical countries, has a majority of its medicinal species in the forests. Every year known and unknown potential medicinal plants become extinct. One reasonable suggestion to this problem is to cultivate such trees as crops. However, this is only a partial solution to the problem since some of these species require similar shade conditions existing in the tropical moist forests in order to grow. There is, therefore, a need for silvicultural requirement studies in order to bring useful germplasm into cultivation, and the full intraspecific variation of the species is conserved to meet changing needs in environmental conditions. In summary, conservation methods must be based on and supported by genecological exploration and evaluation of existing germplasm, as a method towards sound utilization of the medicinal plants for the benefit of mankind.

Complexity and Conservation of Medicinal Plants: Anthropological Cases from Peru and Indonesia

Christine Padoch
Institute of Economic Botany, New York Botanical Garden, USA.
Timothy C. Jessup[1]
Department of Human Ecology, Rutgers University, New Brunswick, USA.
Herwasono Soedjito
Herbarium Bogoriense, Puslitbang Biologi - LIPI, Bogor, Indonesia, and
Kuswata Kartawinata
Unesco Regional Office for Science and Technology for Southeast Asia, Jakarta, Indonesia

Introduction

The continued existence of many medicinal plants and of the knowledge of how to use them is threatened by two worldwide, inexorable processes: the rapid disappearance of tropical forests and other natural vegetation types, and the destruction or rapid change of tribal cultures and local medical traditions. While acknowledging the seriousness of these problems, in this paper we focus instead on some aspects of medicinal plant use that have received less attention. These are: the prevalence of users other than tribal people, the diversity of sources they exploit (including sources other than primary forests), and the varying effects of commercial exploitation on medicinal species. We briefly review the general problem before turning to specific cases.

Deforestation, especially of old growth or primary forests in the humid tropics, is endangering many medicinally useful plant species, although the precise number is not known. Species are threatened both by general destruction of their habitats and by overexploitation of particular products.

The rapidity with which medicinal plant knowledge is being lost is even more difficult to quantify. We cannot know what the destruction of native cultures, which often has entailed the actual physical destruction

---

1. *Now* World Wide Fund for Nature, Bogor, Indonesia.

of entire groups, signifies in terms of knowledge lost. The erosion of medicinal traditions through progressive deculturation of tribal peoples is perhaps more significant world-wide. Increased reliance upon mass-marketed chemical cures and abandonment of local medical traditions are occurring in many areas.

These processes of loss of plants and knowledge have been mentioned by many others concerned with the conservation of medicinal plants. While they are of enormous importance, we would like to add several refinements and cautions to these generally accepted views.

Primary forests are indeed the sources of many medicinal plants. However, people also use plants from many other habitats. Secondary forests, managed swidden fallows, agricultural fields, and house gardens are all places where people gather, manage, or grow important useful plants. In recent years there are also increasing efforts at cultivating medicinal plants. Traditional agricultural systems very commonly exhibit great plant diversity, with medicinals, food plants, and other useful species growing side by side. Change in such systems can affect the availability of medicinal species, and research on plants used for medicinal purposes should not overlook human-influenced habitats.

While medicinal plants grow in all sorts of environments so also all sorts of people know and use medicinal plants. Tribal peoples have traditionally been the focus of most anthropological and ethnobotanical studies. The medicinal plant lore of non-tribal peoples, especially rural peasants and urban dwellers has, in many areas, received little or no attention. The participation of such populations in markets where commercially produced chemical medicines are available is widely assumed to preclude their continued use of naturally occurring medicinals. However, while urbanization and greater involvement in national cultures is often correctly assumed to lead to the loss of medicinal plant knowledge, such results are not always predictable.

Similarly, marketing of plant products for medicinal uses will not necessarily affect the continued availability of any particular plant in an easily predicted fashion. While commercial collecting of wild forest plants may deplete or even eliminate those resources (Rifai, 1983), market demand for a plant medicine may alternatively lead to an increase in its availability because of cultivation or other management.

We agree that, as forests are being lost together with the rich traditions of forest-dwelling peoples, research and conservation efforts must concentrate on attempts to understand and limit these losses. However, some of the less commonly discussed aspects of medicinal plant use should not be overlooked, as they often have been in the past, by anthropologists and other researchers. The following examples of medicinal plants use in Peru and Indonesia are presented to correct some possible misconceptions concerning the use of medicinal plants and to encourage a broadening of research and conservation efforts.

Plant Use among the Ribereños of the Peruvian Amazon

It is often acknowledged that hunters and gatherers and small groups of shifting cultivators who live deep within Amazonian forests are keepers of profound knowledge about plants, including medicinal plants. However, when such tribal peoples are incorporated in the wider society, adopt the language of the state, embrace a world religion, and begin to participate in a market economy, they are often reclassified as peasants, as "people with no culture". It is assumed, often erroneously, that even their environmental knowledge disappears with their abandoned rituals.

In the Peruvian Amazon, the overwhelming majority of rural dwellers are such non-tribal people. Known as *ribereños*, "people of the river-banks", many of them only recently lost their tribal identities. Others are the descendants of unions between immigrants – who often came during the Rubber Boom at the turn of the century – and native Amazonians. Culturally their daily lives are an amalgam of European and Amazonian traditions.

While these people converse in Spanish, sell some of their produce, and often resort to patent and prescription medicines when their health fails them, they still know and expect much of medicinal plants. Most *ribereños*, regardless of sex and age, have a surprising knowledge of plants that they rely on to soothe aches, stop diarrhoea, relieve fevers, and cure snakebite. A number of plants that also serve to forecast the future, or to succeed in business or love are often known to all but the very young. Ideas on the correct preparation of these substances, especially powerful hallucinogens or other magic preparations, are however, more often the domain of a specialist, a *curandera/o* or *vegetalista*. Many small villages, as well as the larger towns and cities that are now centers of immigration, boast one or more well known curers. Their healing ceremonies, which include both Christian and Amazonian elements, make frequent use of hallucinogens and other plants long used by tribal groups.

Both the pharmacopoeia of the specialist and that of the village dweller includes local plants as well as exotics. Many frequently used plants, such as yerba luisa (*Cymbopogon citratus*), and *albahaca* (*Ocimum basilicum*) are of Old World origin, while numerous others have doubtless been brought in from other parts of the Amazon and of South

America. Since very little is known of aboriginal methods of curing, little can be said with certainty about the blending of diagnostic or curative procedures of various groups. However, *ribereños* in many villages are descended from numerous tribal groups (Padoch & de Jong, in press). Since we know that their agricultural and hunting technologies reflect a diverse tribal heritage, we also assume that different peoples have contributed to a combined medicinal lore. Thus, while detribalization processes may have resulted in the loss of much valuable knowledge from any individual culture, it should not be forgotten that new combinations of knowledge have doubtless also resulted in new cures and new medicinal techniques.

Rural-based curers in the Peruvian Amazon frequently maintain large gardens, where medicinal plants are cultivated. In one village along the lower Ucayali River in Peru, one of us (Padoch) encountered a village *curandera* with 78 species of plants in her home garden. All but a few of these were used for curative, diagnostic, or magical purposes. The curer had transplanted many of these from the forest for her own convenience. In addition to tending these garden plants, she knew the location of many other trees, vines, herbs, and shrubs that grew in the vicinity and could be gathered when necessary.

While we do not question the urgency of recording the plant knowledge and environmental wisdom of tribal groups that are rapidly being acculturated or even destroyed, we do want to suggest that the profound plant knowledge of groups such as Peru's *ribereños* should not be ignored.

Commercial Exploitation of Medicinal Plants in Indonesia

We turn now to another aspect of rural people's use of medicinal plants, that is, their exploitation of certain species for sale. The implications of exploitative activities for the conservation of a species depend on whether exploitation leads to depletion, as if the species were a non-renewable resource, or to a sustained or increased yield under some form of resource management. Attention must be paid to specific biological characteristics of exploited species as these influence and are affected by the activities of collection and management (Jessup & Peluso, 1986).

To illustrate the point, we consider two examples of commercial exploitation of medicinal plants with contrasting effects on the exploited species. The two plants, both trees indigenous to Indonesia, are known there as *gaharu* and *kayu putih*. The first is collected irregularly from wild sources in primary forest, a practice that has led in at least some areas to its depletion. The second is managed or cultivated and has spread in that way far beyond its native range.[2])

Gaharu is a resinous product of *Aquilaria malaccensis* Lamk. and other species in the family Thymelaeaceae. Not all individuals of a species contain the valuable wood, which develops in trees infected by an unknown pathogen. *Gaharu* is used in Chinese and Malay pharmacopoeias to treat a variety of ailments, particularly those associated with pregnancy and childbirth. It has been collected and traded for centuries throughout Southeast Asia and to China and the Middle East.

Many collectors of *gaharu* in Indonesia belong to remote tribal groups living in or near extensive forests. The wood they collect is traded through a series of middlemen until it reaches its point of consumption, often a distant city such as Hong Kong. At times the price of *gaharu* has been quite high – it was about US$ 100 per kg in Indonesian cities in 1980 – but it is also subject to considerable fluctuation. Prices fluctuate apparently because of changes in the location, abundance, and quality of sources as well as uncertainty among traders and collectors.

The trees that yield *gaharu* grow in primary rain forest, where they are "rather widely distributed, but ... nowhere abundant" (Foxworthy, quoted in Burkill, 1935, I: 205). Collectors must therefore cover a wide area. One of us (Jessup) studied the activities of *gaharu* collectors in a remote part of Kalimantan (Indonesian Borneo). He observed that collecting invariably entails felling of the trees, which resprout weakly, if at all. The occurrence of valuable grades of *gaharu* is sporadic, and many trees must be felled to obtain a small amount of the wood.

The combination of scattered, slowly regenerating trees and heavy collecting when prices are high has led repeatedly to the depletion of *gaharu* sources, leaving traders and collectors to seek other sources elsewhere. Nevertheless, we know of no cases of cultivation or even silvicultural management of *gaharu*.

A different story can be told of *kayu putih*, or *Melaleuca leucadendra* (L.) L., a species of tree or large shrub in the family Myrtaceae. Oil obtained from *kayu putih* is used in Indonesia and elsewhere to treat a wide range of complaints. It is sold commercially in Southeast Asia and, at least until the 1930's (Burkill's time), was an ingredient in European medicines as well.

The sources of *kayu putih* are less remote from traders and users than those of *gaharu* for two reasons. First, *kayu putih* is not restricted to primary forests; it thrives in disturbed areas and is also cultivated.

2. **Our sources on both *gaharu* and *kayu putih* are the following: Burkill (1935), Corner (1952), Heyne (1950), Hou (1960), Jessup & Peluso (1986), Ng (1973), Peluso (1983), Perry (1980), and Whitmore (1980).**

Second, the species has been introduced into cultivation as far north as China from its native region, which extends from Tenasserim in India to the Moluccas in Indonesia, where it grows naturally especially in swamp forests.

A notable characteristic of *kayu putih* is its ability to resprout vigorously after natural or anthropogenic disturbance, particularly fire. Under natural or semi-natural conditions, the plants form dense stands in areas subject to periodic burning. Its ability to resprout doubtless contributes to the relative ease with which *kayu putih* can be cultivated. Furthermore, as a concentrated resource it is more amenable to management than are the scattered trees that produce *gaharu*. Thus, the effects of commercial exploitation and the need for conservation efforts are very different for these two medicinal plants.

Conclusions

Our aim has been two broaden the discussion of people's use of medicinal plants in tropical countries by focusing attention on users other than tribal people, on the diversity of sources they exploit, and on the varying effects that commercial exploitation has on the continued abundance of the species. Our conclusions can be briefly stated as follows:

1. People who use medicinal plants are a diverse group, comprising not only remote forest-dwellers but also rural and urban folk living in ethnically heterogeneous communities. Many have a variety of medicinal traditions on which to draw. The plants they use come from a number of different habitats, both natural and altered. Some plants or plant products are obtained through commercial trade. Research and conservation efforts aimed at restricted localities, habitats of a single type, or narrowly defined indigenous cultural traditions are liable to miss much of the diversity of medicinal plant use.
2. The consequences of exploiting a medicinal species depend on biological characteristics of the plants as well as on market forces and the activities of plant collection and management. Conversely, plant biology constrains the ways in which an exploited species can be managed as a renewable resource. Rather than consider human activities and plant responses as parts of separate "anthropological" and "biological" systems, scientists ought to view exploitation as an interactive process in which both people and plants are dynamic agents.

Literature Cited

Burkill, I.H. (1935). *A Dictionary of the Economic Products of the Malay Peninsula,* reprinted in 1966, vol. 1. Kuala Lumpur: Ministry of Agriculture and Cooperatives.

Corner, E.J.H. (1952). *Wayside Trees of Malaya* 2nd edn, vol. 2. Singapore: Government Printing Office.

Hou, Ding (1960). Thymelaeaceae. In *Flora Malesiana* ser. I, vol. 6, ed. C.G.G.J. Van Steenis, pp. 1-48. Groningen: Wolters-Noordhoff.

Heyne, K. (1950). *De Nuttige Planten van Indonesie*, 3rd edn. vol. 1. s'Gravenhage & Bandung: Van Hoeve.

Jessup, T.C. & Peluso, N.L., (1986). Minor forest products as common property resources in East Kalimantan, Indonesia. In *Proceedings of the Conference on Common Property Resource Management*. pp. 505-31. Washington, D.C.: National Academy Press.

Kochumen, K.M. (1978). Myrtaceae. In *Tree Flora of Malaya*, vol. 3, ed. F.S.P. Ng, pp. 169-254. Kuala Lumpur and Singapore: Longman.

Padoch, C. & de Jong, W. (in press). Santa Rosa: the impact of the forest products trade on an Amazonian village. In *Advances in Economic Botany*. New York: New York Botanical Garden.

Peluso, N.L. (1983). Markets and Merchants: The Forest Product Trade of East Kalimantan in Historical Perspective. Unpub. Master's thesis, Cornell University.

Perry, L.M. (1980). *Medicinal Plants of East and Southeast Asia : Attributed Properties and Uses*. Cambridge: MIT Press.

Rifai, M.A. (1983). Germplasm, genetic erosion, and the conservation of Indonesian medicinal plants. *BioIndonesia*, 9, 15-28. (In Indonesian)

Whitmore, T.C.(1978). Thymelaeaceae.In *Tree Flora of Malaya*. vol. 2, ed. T.C. Whitmore, pp. 483-91. Kuala Lumpur and Singapore: Longman.

Utilization of Indigenous Medicinal Plants and their Conservation in Bangladesh

A.S. Islam
Department of Botany, Dhaka University, Dhaka, Bangladesh

Although there is voluminous literature on the medicinal plants of the Indian subcontinent, considerable work is needed to investigate their properties and elucidate their chemical composition. Extracts of most of these plants have not yet been submitted to rigorous clinical trials for the broad range of ailments for which they are used. Besides this, there are a large number of plants which have never been tested for their medicinal properties.

While this is the situation, the list of endangered plants species well-known for their medicinal properties is increasing and unless special efforts are made for their conservation some of them will soon become extinct. For instance, *Tylophora indica* (Burm. f.) Merr. (= *T. astmatica* Wight & Arn.) – a well-known member of the family Asclepiaceae used for its efficacy against asthma is not available in abundance. Wild *Zanonia indica* (Cucurbitaceae), used as purgative, is no longer available around Dhaka and the surrounding areas. Mention may be made here of the recent publication of IUCN in which M.S. Khan of Bangladesh National Herbarium (Davis *et al.*, 1986) indicated that there are some 35 endangered species in Bangladesh. A list of further endangered species in Bangladesh is given below:

Table 1: List of endangered species in Bangladesh

Name	Family
Zanthoxylum rhetsa (Roxb.) DC (*Z. budrunga*)	Rutaceae
Gloriosa superba	Liliaceae
Urginea indica Kunth	Liliaceae
Marsdenia tinctoria (L.) Br.	Asclepiadaceae
Neptunia oleracea Lour.	Leguminosae
Withania somnifera Dunal	Solanaceae
Viscum album	Loranthaceae
Dendrobium formosum Roxb.	Orchidaceae

The reasons that some of the wild and semi-wild species (which also include some introduced species, such as *Adansonia digitata* (Bombacaceae) and *Tabebuia trifolia* (Bignoniaceae)), become scarcer and scarcer with the passage of time are not far to seek. In Bangladesh, side by side with the population explosion, the number of city headquarters with administrative office buildings and residential quarters for the officials has increased many-fold. As against 20 district headquarters and 460 police stations, there are now 60 districts and 460 subdistricts, called in local language, *upazila*. Many forest areas have already been and are being cleared to house the administrative units and other ancillary facilities. As a result of this invasion of natural habitats together with indiscriminate collection of raw material for export to foreign countries, the population of some of the plant species, such as *Rauvolfia serpentina*, which used to be abundant, is declining rapidly.

The process of urbanization has already cost us a great deal in terms of loss of many valuable plant species. On top of this, almost every year large numbers of many plant species are destroyed in the areas invaded by recurrent flood waters. It is therefore suggested that in any future plan of urbanization some land in the area of disturbance be set apart for conservation of wild plants along with manpower and other facilities for their proper maintenance. It may be mentioned here that a pharmaceutical industry, *Ganasastha* (= health for all), has already specimens of some 300 indigenous species of medicinal importance growing in its huge premises at a place called *Savar* some 25 miles from Dhaka city. The management of this company has already taken upon itself the responsibility to conserve as many medicinal plant species as possible, including the endangered ones.

Concerning the need to investigate properly the medicinal properties of the wild plants of Bangladesh, mention may be made of *Euphorbia hirta* commonly called 'ghaupata' which is used to treat shigella dysentery. This has recently been found by the Institute of Nutrition and Food Science at Dhaka University to be very effective against tuberculosis. Some patients failing to respond to standard antibiotics recovered when treated with this herbal medicine.

More recently Azad Chowdhury and his associates (1982, 1984, 1989) have published a number of papers describing some of the medicinal properties of a number of plants not reported earlier. Of these, the antifertility activity of two species, namely *Marsdenia tinctoria* and *Plumbago zeylanica*, have created a good deal of interest in pharmaceutical circles. Their reports of antibacterial properties of flower buds of *Saraca indica* (Azad Chowdhury *et al*., 1990a), bulbs of *Acorus calamus* and cloves of garlic (*Allium sativum*) have been well received although so far their experiments against gram positive and gram negative bacteria, including shigella, have been confined to *in vitro* testing (Azad

Chowdhury *et al.*, 1989) and that in experimental shigellosis in rabbit (Azad Chowdhury *et al.*, 1990b).

The same group of workers has studied the effect of isolated principles from *Xanthium strumarium* on liver and kidney of AID rats (Azad Chowdury *et al.*, 1987). They have also reported that the extracts of *Moringa oleifera* can be used to treat hypertension (Moniruddin *et al.*, 1986).

Using water extract of betel nut (*Areca catechu*) Mannan (personal communication) at the Institute of Post Graduate Medicine and Research, Dhaka has been able to arrest cardiac failure. Unfortunately no research has so far been done to determine the kind of ingredients in the water extract which bring about this remedy against heart ailments.

Establishment of a Research Institute on Herbal Medicines in Bangladesh

There is a need in Bangladesh now to prepare a comprehensive plan for research and development on indigenous medicinal plants. Such a plan should envisage the modernization of the research institute on traditional medicines so that it can deal with various aspects of medicinal plants on the pattern of some of the SAARC countries, such as India, Pakistan, Nepal and Sri Lanka.

A Research Institute on the *Unani* and *Ayurvedi* system was started in Bangladesh a few years back but unfortunately neither have the facilities been developed for research nor qualified manpower appointed to handle the various problems associated with traditional medicines in that Institute.

Until full facilities for research are developed in the above Institute, an interim arrangement should be made by the Ministry of Health to make better use of existing facilities by bringing together the various institutions and individuals into an effective and well-coordinated working relationship.

Need for Developing an Area in the National Botanic Garden at Mirpur, Dhaka for Conservation

Because of its area, size and built-in facilities, a part of the National Botanic Garden at Mirpur, Dhaka can be used for conservation of endangered species, particularly those of medicinal importance. In fact, the Research Institute on *Unani* and *Ayurvedi* medicines and the National

Herbarium now located at a rented house in Dhaka city may be shifted to the Botanic Garden for expansion and broadening of their scope and utility.

Some new areas of research, such as biotechnology, with emphasis on production of secondary metabolites through tissue culture, as well as study of adulterants usually mixed in the preparation, can be initiated in the above Research Institute. If facilities for production of secondary metabolites are developed, *in vitro* multiplication of callus tissue of many species not native to our country can be done and used as ingredients which are not normally available for the preparation of certain drugs. Tissue culture facilities can also be used for conservation of those endangered plant species in which seed dormancy is absent or very brief.

Clinical Trials

Evaluation of the effectiveness of traditional remedies will enable the physicians to demonstrate whether for given clinical conditions these are effective, and if so, whether their effectiveness is of the same order, superior or inferior to modern treatments.

In the Institute, facilities should be developed so that modern methods can be employed to test crude extracts of plants well-known for their medicinal properties as it is being done elsewhere in the subcontinent. Plants belonging to this category include the following (their equivalent Bengali names are shown against each species within brackets): *Ocimum sanctum* (tulsi), *Adhatoda vasica* (vasak), *Bacopa monniori* (brahmi), *Embelica officinalis* (amlaki), *Terminalia chebula* (haritaki), *T. beleriva* (bahera), *Aegle marmelos* (bel), *Eclipta alba* (kesraj), *Zingiber officinale* (ada), *Centela asiatica* (thankuni), *Azadirachta indica* (neem), *Paederia foetida* (gandha bhadal), *Allium sativum* (rasun), *Aloe barbadensis* (ghritakumari), *Curcuma longa* (halud) etc.

Although the clinical constituents of some of the plants have been thoroughly investigated, the task of isolating various components in the majority of the plants known to be medicinally important is far from being complete. Mention has already been made about the effectiveness of betel nut in curing certain types of heart disease but neither we know of its chemical constitutents nor has any work been initiated so far to determine them. It is only after this is done that clinical trials with its chemical constitutents can be carried out both separately with each of its components and in combination.

Medicinal Plant Formulary

Ideally, the best elements of traditional eastern system of medicine should be combined with modern medicines, since no single system is likely to be sufficient. This, however, will have to be a gradual process requiring a critical evaluation of traditional remedies by modern methods prior to their approval for inclusion in a national formulary.

Fortunately, the Ministry of Health, Government of People's Republic of Bangladesh, with the support of World Health Organization, set up some time back a national Formulary Committee to prepare a formulary of traditional medicines. The committee was formed with veteran *hekims*, *kaviraj*, university professors of the department of Pharmacy and Botany. The draft of the formulary in two volumes, one for *Ayurvedic* and the other for *Unani* medicines (each containing about 550 formulae) is ready to go to the press for publication. This formulary has an appendix prepared by two experts in pharmacy; in the appendix the main chemical constituents of some 30 important herbal plants and their medicinal properties are shown.

One very important outcome of this Formulary Committee's meeting is that the experts in pharmacy have suggested that certain preparations in the formulary, for example in the treatment of headache or for deworming, traditional medicines will be a better substitute. This is because unlike the ingredients used in analgesic and anthelminitics used in modern medicines, those in the herbal preparations have hardly any side effects. In fact, these experts would incorporate the above suggestions in the formulary to appear in future on modern medicines.

Following the publication of a national formulary on traditional medicines and the proper functioning of the Research Institute, which at present exists mostly in paper, except for the building and rudimentary service by a few physicians not trained in the modern methods of medicines, the stage will be set for evaluating critically traditional remedies by modern methods. Once this is done and traditional medicines pass the test of scientific assessments, there will be less of a problem to get their approval for inclusion in the national formulary on modern medicines. This process will have to be gradual and cannot be achieved overnight.

Processing and Production

When the therapeutic value of a medicinal plant preparation has been established and has been shown to be at least equal or superior to conventional modern treatments, the various aspects of processing and production would need to be taken up, initially on a pilot basis. Such projects would ensure dependable sources of supply of raw material, type of formulation i.e. whether the preparation is to be dispensed in tablets, capsules, ampoules or any other form, standardization, suitable conditions of storage and studies of the shelf-life of products.

Licensing and Marketing

Once approval has been obtained from the drug regulatory authorities, further work in this direction, namely product development, promotion and marketing can be taken up by semi-government or state approved private agencies. The help of big pharmaceutical companies, already functioning in this country and which have elaborate arrangements for handling all kinds of issues connected with marketing of a good product, may be sought.

Mention may be made here that some well-known institutes of this subcontinent, namely the Central Institute of Medicinal and Aromatic Plants (CIMAP), Lucknow, India; the Department of Medicinal Plants, Khatmandu, Nepal; and the Hamdard Research Institute, Karachi, Pakistan, all manufacture traditional medicines and market them successfully. In fact, the Hamdard Trust has finalized arrangements for setting up a private university in their *Madinatul Hikmah* (science city) complex to deal with all aspects of traditional medicines. The socio-economic conditions of Bangladesh being more or less the same and because of this country's membership in SAARC, it should not be difficult to launch collaborative projects with neighbouring countries on promoting traditional medicines in Bangladesh with the object of ensuring health services for all by the end of this century.

Use of 'Khas' Land for Cultivation and Conservation of Plants of Medicinal Importance

The land owned by the Government is called 'khas' land in Bengali; such land which occupies quite a substantial area in the country can be leased out to landless cultivators on easy terms for the specific purpose of growing rare plant species to conserve them as well as cultivating frequently used medicinal plants to ensure regularity of supplies of the latter group so that the manufacturing of the essential drugs does not suffer for want of ingredients. This system of leasing out of 'khas' land to the landless peasants introduced some time back in Nepal is working there satisfactorily. The supply of raw material in abundance will, on one hand, bring down the price of traditional medicines and, on the other, reduce the chances of adulteration which are quite frequent on account of non-availability of some essential constituents of certain preparations.

There is a tendency all the world over to turn to nature cures and it is important to establish scientific basis of the treatments accorded through traditional medicines. The dream of the teeming millions of the developing countries would be fulfilled if they get at their doorstep facilities for traditional medicines at a cost which is within their financial means. It is the duty of the scientists concerned to accomplish that objective with the support of the Government and other interested international agencies.

The author is grateful to Professor A.K. Azad Chowdhury and Mr Md Shah Jahan of the Department of Pharmacy and Mr A Hasan of the Department of Botany, Dhaka University and Mr Monzur-ul-Kadir Mia of Bangladesh National Herbarium for helping him with some valuable information in the preparation of the manuscript.

References

Azad Chowdhury, A.K., Ahsan M., Sk.N. & Ahmed, Z.U. (1989). *In vitro* antibacterial activity of garlic extract and allicin against multiply drug resistant strains of *Shigella dysenteriae* type I and *S. flexneri*. *Dhaka Univ Studies*, 4(1), 27-32.

Azad Chowdhury, A.K., Satyajit Dey Sarker, Sultana, E. & Islam, Sk.N. (1990a). Antimicrobial activity of the aqueous extract and the flavonoids isolated from the flowers and flower buds of *Saraca indica*. *Bangladesh Pharm. J.*, 9(1), 4-8.

Azad Chowdhury, A.K., Ahsan M., Islam, Sk.N. & Ahmed, Z.U. (1990b). Therapeutic efficacy of aqueous extract of garlic and its active constituent allicin in experimental shigellosis in rabbits. *Indian J. Med. Res.*, in press.

Azad Chowdhury, A.K., Chakder, S.K. & Rahman, S. (1982). Antifertility activity from *Plumbaga zeylanica*. *Indian J. Med. Res.*, 76 (Suppl.), 99-101.

Azad Chowdhury, A.K. & Farouq, A. (1990). Antimicrotrial of *Acorus calamatus*. *Bangladesh Med. Res. Council Bull.*, in press.

Azad Chowdhury, A.K., Khaleque, R.A. & Chakder, S.K. (1984). Oxytocic principle(s) from *Marsdenia tinctoria*. *Dhaka Univ. Studies*, Part B, 32, 85-88.

Azad Chowdhury, A.K. & Mannan, R.H. (1987). Antidiabetic principles from *Xanthium strumarium*. *Dhaka Univ. Studies*, Part B, 35(1), 7-14.

Davis, S., Droop, S.J.M., Gregerson, P., Henson, L., Leon, C.J., Villa-Lobos, J.L., Synge, H. & Zantovska, J. (1986). *Plants in Danger: What do we Know?* Gland, Switzerland: IUCN.

Moniruddin, A., Hossain, S.M.A., Haque, K.H.H.S.S., Rashid, A. & Azad Chowdhury, A.K. (1986). Clinical trial of *Moringa oleifera* for treatment of hypertension. *Chest and Heart Bull.*, 10, 1-14.

[illegible]

[illegible]

[illegible]

[illegible]

[illegible]

[illegible]

[illegible]

[illegible]

[illegible]

[illegible]

[illegible]

[illegible]

Development of a Conservation Policy on Commercially Exploited Medicinal Plants: A Case Study from Southern Africa

A. B. Cunningham
Institute of Natural Resources, University of Natal, South Africa

Introduction

The majority of the population of many developing countries use traditional medicines either because people cannot afford western pharmaceuticals or because traditional medicines are more acceptable. The southern African region is no exception. Recent studies in Botswana (Staugard, 1985), Zimbabwe (Gelfand *et al.*, 1985) and South Africa (Berglund, 1976; Cunningham, 1988; Ngubane, 1977) all document the important role that traditional medicines plays in the lives of people in rural and urban communities in southern Africa. This urban demand for traditional medicines generates a commercial trade in medicinal plants from rural source areas to urban markets and shops.

Commercial exploitation of medicinal plants has important implications for their conservation and management. Using medicinal plants was formerly a specialist activity of traditional practitioners, who had a limited effect on plant resources (Figure 1). Commercial harvesters (who are not trained traditional practitioners, but are rural people otherwise unemployed) are the main suppliers of traditional medicinal plants to urban areas, with disastrous effect on many popular species (Cunningham, 1988). The effects of this trade are most clearly seen in the ring-barking of tree species that are a source of popular medicinal barks (Figures 2 and 3). The trade in herbal medicine is well developed in South Africa due to the large size and rapid growth of urban populations and is on a scale that is cause for concern amongst conservation organisations and rural herbalists (Cooper, 1979; Cunningham, 1984, 1988; Gerstner, 1946; MacDonald, 1984).

Over-exploitation of popular plant species as a result of this trade is not a new problem. Local depletion of the Swamp Forest climber *Mondia whitei* was recorded in Natal province, South Africa, in 1898 (Medley Wood & Evans, 1898). The problem in Natal has steadily grown since then. There are two main reasons for this: firstly, a dramatic decrease in

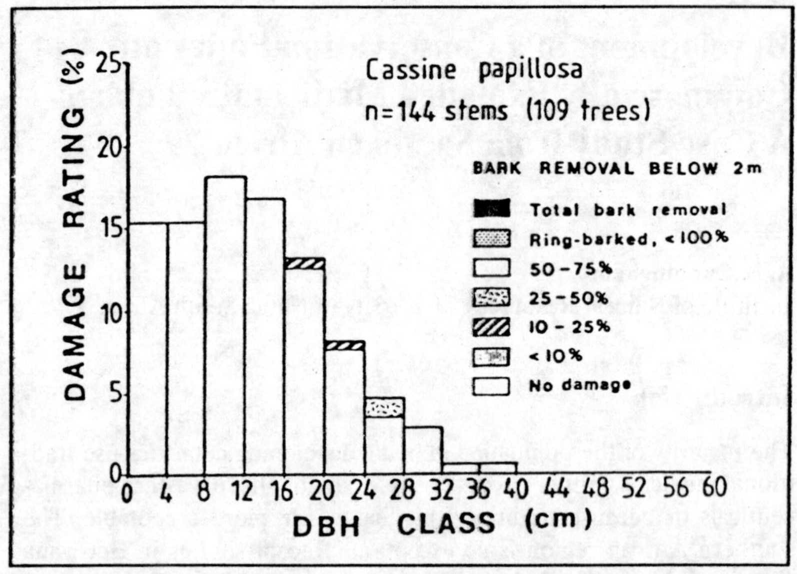

Figure 1. Damage to *Cassine papillosa* trees in sub-tropical coastal forest inaccessible to commercial medicinal plant gatherers. This shows a low level of damage from bark removal by traditional practitioners (from Cunningham, 1988).

area of indigenous vegetation due to the expansion of agriculture, afforestation and urban development with a consequent loss of medicinal plant resources; secondly, a rapid increase in those who use medicinal plants (represented by black South Africans who represent 78% of Natal's population) from 904,000 in 1904 to 4,766,000 in 1980 (Grobbelaar, 1985), with a corresponding increase in demand for herbal medicines. At present, over 400 indigenous species and 20 exotic species are commercially sold to Zulu people as herbal medicines (Cunningham, 1988).

Indigenous species, virtually all of which are gathered from the wild, vary greatly in their geographical distribution, population density, growth rate, growth form and population biology. Demand for common, fast-growing species such as *Artemisia afra* and *Gomphocarpus physocarpa* is easily met from wild stocks. What is of concern is the depletion of wild stocks when demand exceeds supply. Scarce, slow-growing forest species are particularly vulnerable to over-exploitation. Indigenous forests only cover 0.3% of South Africa (Cooper, 1985), but are a source of over 130 commercially exploited traditional medicinal plants (Cunningham, 1988).

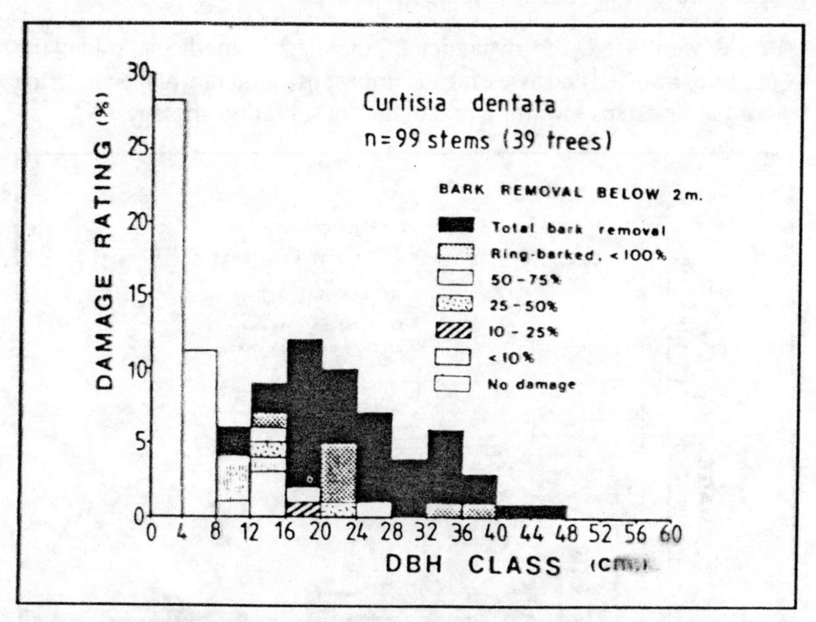

Figure 2. Damage to *Curtisia dentata* trees (Cornaceae) in Afro-montane forest (eMalowe forest, south-eastern Africa) from bark removal by commercial medicinal plant gatherers (from Cunningham, 1988).

The consequences of this have been:

1. The over-exploitation of wild populations of certain popular species;
2. A rapid increase in the prices of species that have been depleted in the wild, with the result that fewer people can afford to buy the medicines;
3. A major loss of income from valuable timber from *Ocotea bullata* due to ring-barking.

Over the past 40 years, law enforcement by forest and game guards has been the dominant strategy in an effort to control commercial exploitation of indigenous medicinal plants. This has not succeeded and more trees are ring-barked in "protected" forests than ever before. New solutions to this old problem were therefore developed at the request of both resource users (herbalists, herb traders) and resource managers (conservation departments) (Cunningham, 1988). The aim of the study was not to find out what the medicinal plants were used for, but how to go about conserving them. Whether a species has physiological action or purely psychological effect is not relevant in this case. If a plant species is popular for either reason it will continue to be used. Determining which species were used, the dynamics of the herbal medicine trade, the quantities involved and assessing the impact of medicinal plant collecting were the first steps towards a pragmatic conservation strategy.

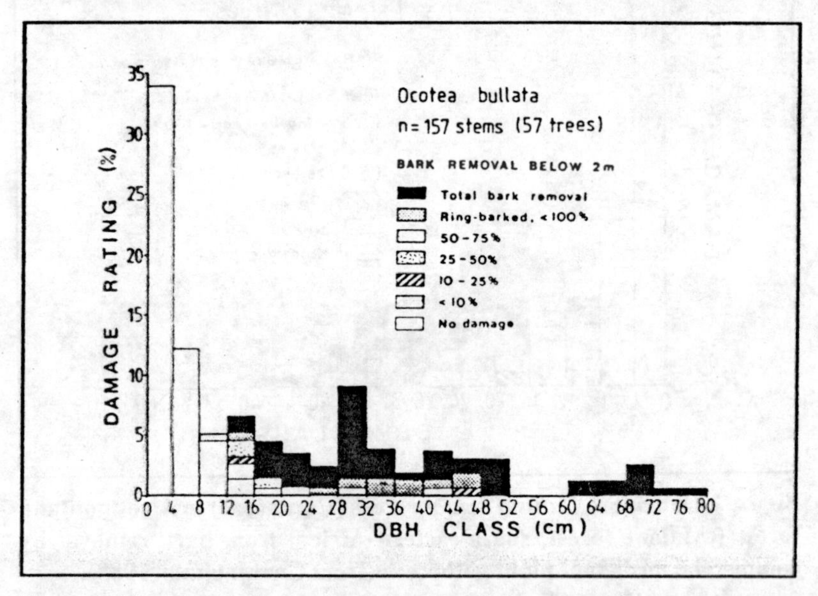

Figure 3. Damage to *Ocotea bullata* trees (Lauraceae) in Afro-montane forest (eMalowe forest, south-eastern Africa) from bark removal by commercial medicinal plant gatherers (from Cunningham, 1988).

Sustainability and Effects of Harvesting

Due to the number of species involved and the paucity of published data on biomass, primary production and demography of indigenous plants in southern Africa (Rutherford, 1978; O'Connor, 1985), no detailed assessment can be made of sustainable offtake of medicinal plant material from natural populations. Unlike the harvesting of low diversity, high biomass production vegetation types (e.g. *Phragmites australis* reeds) or encroaching species (e.g. *Dichrostachys cinerea*) that need to be removed, sustainable use of scarce, slow-growing but popular medicinal plants requires an intensive management effort. This is impractical with limited money and manpower. Even if data on biomass production were available therefore, it is unlikely it would be of much practical value due to narrow margin between sustainable use and over-exploitation for scarce or slow-growing resources.

Despite limited data on population biology of indigenous species, it is generally accepted that a relationship exists between resource stock or population size, and the sustainable rate of harvest. Low stocks are likely to produce only small sustainable yields, particularly if the resource is represented by slow-growing species that take a long time to reach reproductive maturity. Large stocks of species with a high biomass production and short time to reproductive maturity would similarly be expected to produce high sustainable yields, particularly if competitive interaction was reduced by "thinning". There is also a clear relationship between the part of a plant being harvested and the impact on the plant. The response of plants to exploitation and the implications of a declining productivity under a high frequency or intensity of exploitation is critical to developing a conservation policy for particular species.

Life Form and Part Used : the Implications of the Plant's Response to Exploitation for Sustainable Use

Degree of disturbance to the species population and vulnerability to over-exploitation depend on demand, supply, part used and life form. In the absence of detailed information on plant demography, conservation priority species were selected according to four basic criteria:

1. Life form categories represent a useful classification for establishing resource management principles, bridging the gap in knowledge about plant demography and enabling a first approximation of categories of vulnerability of species to commercial exploitation. The life form categories used were those of Raunkiaer (1937) with elaborated sub-sections for vines and lianas, chamaephytes and hemi-cryptophytes (Box, 1981; Mueller-Dombois & Ellenberg, 1974) and the addition of a category for parasitic plants. Life form categories represent a natural sequence from phanerophytes, through chamaephytes and hemi-cryptophytes to therophytes (Figure 4) (Rutherford & Westfall, 1986), in other words from K-selected to r-selected species. Forest trees (phanerophytes) represent the most vulnerable category due to preferential exploitation of thick bark from large (old) plants (Figure 4 a,b,c) which have a long period to reproductive maturity (Figure 4 d), a low ratio of production to biomass (Figure 4 e) and specialised habitat requirements (Figure 4 f, g).

2. Part used (bark, roots, bulb, whole plant, leaves, stems, fruits, sap). The category of greatest concern to resource managers and rural herbalists is that of the slow-growing, popular species with a restricted distribution which are exploited for bark, roots, root tubers,

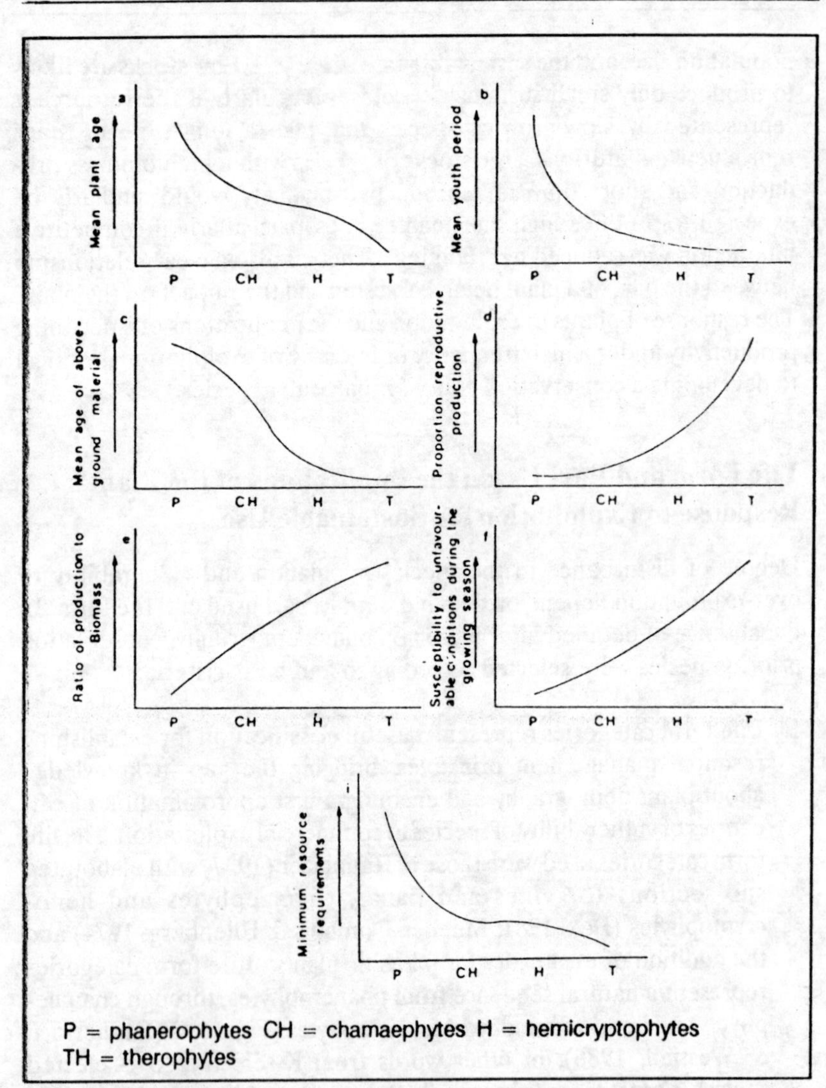

Figure 4. The generalised relationships between life form and plant ecological and demographic properties.
a, mean plant age; b, mean plant youth period; c, mean age of above ground biomass; d, the reproductive portion of production; e, ratio of production to biomass; f, sensitivity to unfavourable growing conditions during the growing period g, minimum resource requirements for individual plants (from Rutherford & Westfall, 1986).

bulbs/corms or where the whole plant is removed. Both *Barringtonia racemosa* and *Warburgia salutaris*, for example, are relatively slow-growing trees (mesophanerophytes) with restricted distributions in Natal. Both are popular sources of herbal medicines. In *Barringtonia*, however, use of the abundant fruits poses little threat to the species, whilst commercial exploitation of *Warburgia salutaris* bark is considered to be the major threat to populations of this plant in South Africa (Cunningham, 1988), Zimbabwe (Mavi, pers comm.) and Tanzania (FAO, 1986).

Whole plants, bark, roots or bulbs represent the main ingredients of Zulu herbal medicine (Cunningham, 1988). The effects of their removal have a more immediate and damaging effect on habitat and species populations than harvesting of leaves or fruits. Although a high frequency and intensity of *Hyphaene petersiana* leaves for commercial purposes is recorded from southern Africa (Cunningham & Milton, 1987), field observation and the oral evidence of resource users suggests that a similar situation does not yet exist for plant species whose leaves are exploited for medicinal purposes. The only known case from the literature is the "severe reduction" in *Artemisia granatensis* populations in Spain due to commercial exploitation of leaves and inflorescences for "Artemisia tea" (Lucas & Synge, 1978).

3. Distribution within the area immediately affected by the commercial trade in medicinal plants (the Natal region), according to five basic categories used by Moll (1981), namely categories for rare (r), uncommon species, subdivided into uncommon but widespread (uw), uncommon and localised (ul) and common species, subdivided into common and widespread (cw) and locally common (cl) species (Figure 5).

4. The demand, based on the estimated quantities (50 kg size maize bags sold/year) of each species sold annually by 54 herb traders. Unpopular species are of less concern to resource managers due to the low frequency and intensity of exploitation compared to popular, commercially sold species.

Coppicing ability and the vulnerability of trees to bark removal are also important attributes which vary with the physiology of different species. At one extreme, *Faurea macnaughtonii* and *Podocarpus henkelii* are the species most sensitive to bark removal. *Warburgia salutaris, Prunus africana* and *Nuxia floribunda* are at the other extreme, where complete bark regrowth can occur after the trunk has been completely ring-barked. Resilience after ring-barking does not enable trees to survive when demand exceeds supply. Gatherers continue to debark *Warburgia*

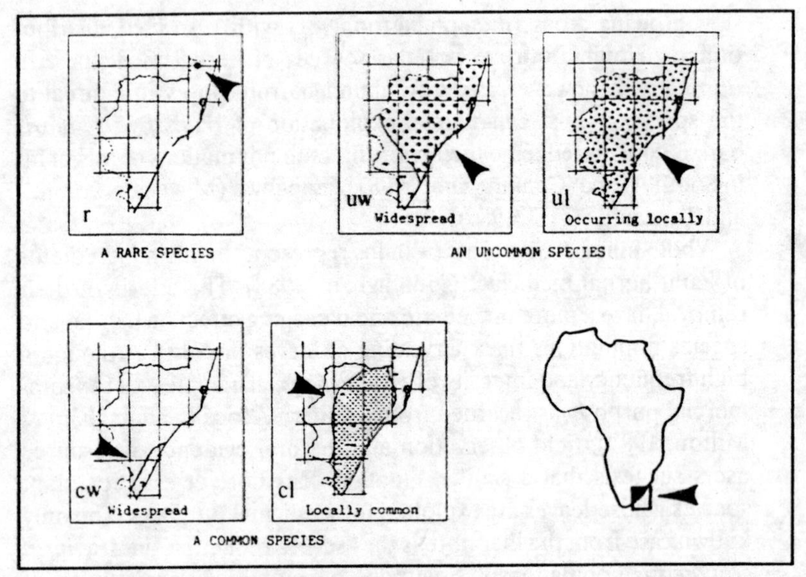

Figure 5. The five-point scale used to rate the scarcity of medicinal plant species in Natal (from Moll, 1981)

salutaris trees when bark is only partially regrown because of the scarcity of alternative sources of supply, finally debarking large roots and killing the tree.

Assessing the Value of this Approach

The validity of these categories is borne out by field observations, damage assessments and the perceptions of herbalists. Examples of heavy exploitation of popular forest trees (phanerophytes) is widespread, whereas popular species with bulbs (e.g. *Scilla natalensis, Eucomis autumnalis*) or ligno-tubers (*Synaptolepis kirkii*) recorded as scarce by resource users, only show localised extinction in sites where commercial gathering is concentrated (Cunningham, 1988). Vigorous coppice production (e.g. in *Warburgia salutaris, Ocotea bullata, Curtisia dentata*) and rapid bark regrowth after debarking (e.g. in *Warburgia salutaris, Prunus africana, Nuxia floribunda*) play a very important role in the tree is survival. Conversely, sensitivity to debarking (e.g. in *Faurea macnaughtonii, Podocarpus henkelii*) and lack of coppice production (e.g. in *Faurea macnaughtonii*) greatly increase vulnerability to over-exploitation, particularly if reproductive ability is reduced by regular ring-barking of coppice so that plants rarely set seed. This is worsened if only a low percentage of seeds germinate. As a result of seed predation by insects, for example, only 0.01% of seeds of the popular medicinal plant *Ocotea bullata* will germinate (Phillips, 1931). Ring-barking ensures that few *Ocotea bullata* trees set seed in forests, further affecting population dynamics of this forest canopy species. The same situation applies to *Warburgia salutaris*. Fifty years ago, Gerstner (1938) was unable to find flowering material of *Warburgia salutaris*, recording only "poor coppices, every year cut right to the bottom, used all over and sold by native herbalists". Where mature plants occur, there is also a high percentage of seed parasitism (Nichols, pers. comm.), a factor which raises important questions about minimum critical population sizes of such species.

Commercial utilization of medicinal plants thus affects all three "vital attributes" essential for replacement of plant species (Noble & Slatyer, 1980), namely:

1. The means of dispersal or persistence at the site before and after disturbance;
2. The ability of the species to establish and grow to maturity in a developing community;
3. The time taken to reach critical life stages.

Division of plants into categories on the basis of life form, part used, distribution and demand enables selection of the most threatened species, most of which are phanerophytes. *Siphonochilus natalensis* is something of an anomaly therefore, a product of its extremely restricted distribution in Natal (no known wild populations) and low seed production.

Most rare or threatened species given in the Appendix represent the K-selected part of this spectrum. Many species are trees (phanerophytes) with a restricted distribution which take a long time to reach reproductive maturity, have highly specific habitat requirements and are popular sources of bark or roots. At the other extreme is the use of leaves/stems of forbs and grasses (chamaephytes, hemi-cryptophytes, geophytes). These plants have several advantages over large woody plants, being able to devote a greater proportion of primary production to seed formation, breed rapidly, disperse widely and can often survive as seeds during unfavourable periods. They also have less specific habitat requirements, growing in disturbed soils (e.g. *Psammotropha mucronata*, *Lippia javanica*, *Portulaca* sp.) or overgrazed sites (*Helichrysum odoratissimum* and *Ornithogalum*) (Story, 1952; Noel, 1961). Story (1952) recorded in the eastern Cape that the frequency of the poisonous *Ornithogalum thrysoides* could increase up to 60 times due to selective grazing.

Overgrazing also resulted in an increase in *Helichrysum odoratissimum* in this region (Noel, 1961) as these species are unpalatable to stock. The same could apply to other poisonous Liliaceae used for medicinal purposes (e.g. *Urginea* and *Drimea* species, *Ornithogalum* as well as to the aromatic *Helichrysum odoratissimium*).

Perceptions of Botanists and Resource Users

With few exceptions, botanical and forestry records reflecting the impact of commercial medicinal plant collecting published since 1960 concentrate almost entirely on debarking of the Afro-montane forest species *Ocotea bullata* (Cooper, 1979; Dally, 1984; Moll & Haigh, 1966; Oatley, 1979). This bias is probably due to the economic value of this species for timber, its easy recognition in the field and the lasting signs of debarking on the trunk. All reports document heavy exploitation taking place. Moll & Haigh (1966) observed that "every pole size or larger stinkwood seen had been partly or wholly ring-barked, sometimes to a height of 10 to 15 feet above the ground" in Xumeni forest. In the Karkloof forest, Oatley (1979) estimated that less than 1% of 450 trees examined were undamaged, and approximately one third were dead. Cooper (1979) similarly estimated that 95% of all stinkwoods he examined had been exploited for their bark, with 40% ring-barked and dying.

The knowledge and perceptions of herbalists and herb traders on the other hand, provides valuable insight into the scarcity of a much wider range of medicinal plant species. Such knowledge is particularly useful as it has been gathered over many years of harvesting, buying and selling medicinal plants. The validity of these observations could be gauged against current knowledge on the geographical distribution, rarity and extent of exploitation of the species that was known from personal experience or from the literature. It therefore represents a practical means of directing attention at key species for damage assessment. It also proved to be embarassingly more accurate than data on geographical distribution obtained from published records and the Botanical Research Institute. Early on in the research, for example, after careful search of all available distribution records, I disputed with herb traders that *Hydnora africana* occurred in Natal, saying that it must be transported into the province from the eastern Cape or Transvaal. A few months later material of *Hydnora africana* collected in Natal by gatherers for sale in Durban was recorded as the "first record" of this root parasite for the province!

The perceptions of resource users on species scarcity formed an important part of the study. Responses to the question "Which species are becoming scarce"? were similar for herbalists and herb traders (Cunningham, 1988). At open discussions held with an estimated 400 traditional practitioners throughout Natal/KwaZulu, *Warburgia salutaris*, *Siphonochilus* species and *Boweia volubilis* were invariably cited as becoming scarce. Nomination of *Eucomis* species, *Ocotea bullata*, *Haworthia limifolia*, *Synaptolepis kirkii*, *Curtisia dentata*, *Erythrophleum lasianthum* and the alien *Cinnamomum camphora* was also common, but less widespread than the previous three species.

Perceived scarcity can be due to either over-exploitation, limited distribution, limited access (e.g. species in private farmland or conservation areas) or a combination of these factors. Backing up (or refuting) these perceptions through the use of plant distribution records, damage assessments and field observation therefore formed a crucial part of this study. *Synaptolepis kirkii*, for example, is locally common but is perceived as being scarce by most herbalists and herb traders because of its limited distribution along the Mozambique coastal plain. The alien species *Cinnamomum camphora* was similarly perceived to be scarce due to its restriction mainly to urban gardens. Field observation shows that over-exploitation of *Warburgia salutaris*, *Curtisia dentata*, *Ocotea bullata*, and *Siphonochilus* (*S. natalensis* and *S. aethiopicus*) populations is widespread however, while localised depletion of *Boweia volubilis*, *Scilla natalensis*, *Alepidea amatymbica*, *Eucomis* species (*E. undulata* and *E. bicolor*) and *Haworthia limiifolia* populations has also resulted in increased scarcity of these species. In each case over-exploitation resulted from commercial exploitation by gatherers in order to supply the urban demand. The extent of this exploitation on "indicator" species is clearly seen from damage assessments.

Damage Assessments

Records of damage were made in forest areas throughout the region using a seven point scale for rating bark damage (Figure 6). These data were backed up by field observation of "indicator species" (i.e. those species vulnerable to over-exploitation where damage is visable in the field) and other commercially sold trees, shrubs and climbers.

Although the degree of damage varied, the level of damage at all sites where commercial gathering is taking place was high and concentrated on large diameter size classes (Figures 2 and 3).

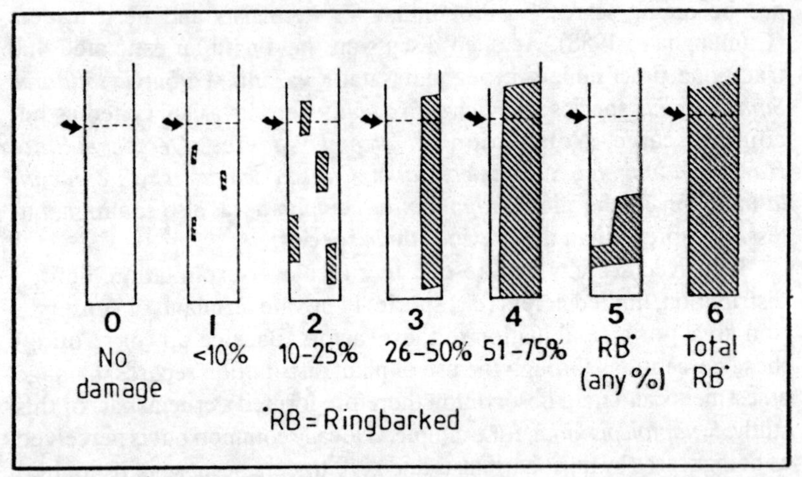

Figure 6. The seven-point scale used in field assessment of bark damage. All assessments represent the degree of bark removal below head height (2 m), which is shown by the dotted line and arrow in the figure (from Cunningham, 1988).

Field Observations

Field observations on extent of damage provide valuable supportive evidence in conjunction with damage assessments and the perceptions of resource users. With the exception of populations in three national parks, all (15) large (5 cm DBH) *Warburgia salutaris* specimens seen in the field were heavily debarked. This corresponds with the perceptions of nearly 90% of all interveiwed resource users (n = 84) that this species is becoming scarce. The same applies to other heavily utilized tree species (Cunningham, 1988).

The current level of damage to many "indicator" species shown in Figures 2 and 3 is clearly not sustainable as it is taking place at a level that is unlikely to keep pace with recruitment from either coppicing or germination. The level of exploitation of *Ocotea bullata* (Figure 2) and *Curtisia dentata* (Figure 3), for example, represents 51% and 57% respectively of stems (excluding coppice less than 2 cm) with more than 50% bark removed. Even higher levels of debarking occur in Afro-montane forest closer to large urban areas where commercial exploitation has been carried out for longer. This accounts for (i) the increasing use of thinner and thinner trees and (ii) transport of scarce species into Natal from further and further afield. The eastern Transvaal and Swaziland for example, are now the major source areas for *Warburgia salutaris* and

Siphonochilus aethiopicus which have been depleted elsewhere (Cunningham, 1988).

When favoured tree species become very scarce (*Ocotea bullata, Warburgia salutaris, Curtisia dentata*), gatherers will build ladders to maximise the quantity of bark obtained per tree. Until this stage is reached, only the first 2-3 m of trunk bark are stripped, leading to death of the tree and tremendous wastage of bark. In order to maximize income per unit time, gatherers either select forests with a high density of a few species or a high diversity, but a low abundance of medicinal plants. Extensive, species-selective exploitation, whether for many trees of one species or a few individuals of many species, can have a marked effect on forest structure.

No data are available on the impact of harvesting products besides bark. Localised impact of commercial gathering of *Gnidia kraussiana, Pentanisia prunelloides* and *Stangeria eriopus* can clearly be seen in southern Natal grasslands, with reduced densities of these species and grassland pitted with holes where plants have been removed. Commercial harvesting of *Harpagophytum procumbens* (Grapple plant) tubers in Botswana removed up to 66% of plants (Table 1) (Leloup, 1984).

Table 1 : The effect of commercial harvesting on *Harpagophytum procumbens* **(Burch.) DC. populations at two localities in Botswana (Leloup, 1984).**

LOCALITY	KHAKHEA		KGOKONG	
IMPACT	No.	%	No.	%
Not dug out	9	17	12	42
Completely dug out	36	66	16	55
Upside down	2	4	-	-
Not replanted	5	9	-	-
Destroyed	2	4	1	3

From field observation it appears that gatherers select for the largest (oldest) individuals from multiple-aged populations to obtain the thick bark, large bulbs, tubers or roots that have highest economic value. In order to obtain the most bark/bulbs/roots per unit time, gatherers also appear to select massed stands.

Walters (1986) has pointed out three conditions that are necessary requirements in the absence of management if renewable natural resources are to regenerate after a period of exploitation:

1. The resource must become more difficult to find as its abundance declines;
2. The target resource must not occur with alternative resources that can support continued harvesting;

3. The species should not experience reduced production when densities are low (e.g. low reproductive success or increased mortality due to predation).

Most threatened or vulnerable medicinal plant species do not fill these conditions. Firstly, commercial exploitation is increasing with rapid urbanisation and population growth. Secondly, from field observation, it is clear that individual species often occur in aggregated populations (e.g. *Cassipourea gerrardii, Ocotea bullata*). Thirdly, medicinal plant species often grow in association with one another, particularly in forests. Finally, and perhaps more important in the long term, is the genetic erosion that results from local extinction of species such *Warburgia salutaris* and *Siphonochilus natalensis*. Even if trees recoppice, bark-stripping continually keeps the population in a vegetative phase preventing sexual reproduction and out-breeding.

Compensatory mechanisms can, however, reduce the chance of local extinction of understorey and sub-canopy trees, but do not appear to act for canopy species (Connell *et al.*, 1984). According to Burley & Styles (1976) (quoted by Boerboom & Wiersum, 1983), at least 200 "well spaced" trees are necessary for genetic conservation and the prevention of inbreeding. While the minimum critical population size is open to debate, the populations of *Warburgia salutaris* appear to be much lower than this through much of their range, and *Siphonochilus natalensis* is no longer found in the wild at all.

Management Options

Reducing exploitation rates is clearly necessary if wild populations of species vulnerable to depletion are to recover. From the historical review, it is clear that legislation against hawking or use of medicinal plants is ineffective on its own. Alternative sources of medicine are clearly necessary to reduce the rate of exploitation of wild stocks. After detailed discussions with conservation, herbalist and herb trader groups, the following strategy is suggested:

1. Provision of alternative supplies from (i) bark obtained from areas where *Ocotea bullata* is sustainably managed for timber, (ii) co-ordinated salvage of plants from development sites (dams, new farm and forestry lands) for sale or replanting, (iii) large scale cultivation. What must be emphasised however, is that cultivation in itself is not a solution. Plants have to be grown on a large scale and

marketed at a lower cost than they can be supplied by gatherers if conservation by cultivation is to be effective. The area required for cultivation of medicinal plants depends directly on demand and growth rate. There can be little doubt, however, that cultivation will have to be on a large scale. If we assume, for example, that *Scilla natalensis* (Liliaceae) grows at the same rate and density as *Urginea maritima* cultivated by Gentry *et al.*, 1987 (harvesting after 6 years, 10,000 bulbs per acre) then the 3,000 bags (or 300,000 large bulbs) which are considered to be sold annually (Cunningham, 1988) could be grown on 70 ha. The rotational area required for tree species would be much greater due to the slow growth rates of many of the species exploited for their bark.

2. Identification and better protection of key vegetation types in conservation areas.

3. Herb traders and urban herbalists cultivating medicinal plants and encouraging the use of leaves and twigs of vulnerable species used for symbolic purposes rather than roots or bark.

4. The formulation of patent (or pharmaceutical) medicines with the same name and function as their herbal counterparts (e.g. the "bangalala" pills and "special imbiza" already in use).

5. The formation of traditional healer associations in rural areas similar to those already formed in southern Natal as (a) focal points for cultivation efforts to fill local needs, (b) conservation pressure groups to reduce ring-barking by commercial gatherers, (c) a local interest group that will encourage the maintenance of traditional resource management practices.

6. The *ex-situ* cultivation of as wide a range of genotypes as possible of rare and endangered species (Appendix).

Past Problems and Future Action

The social consequences of this over-exploitation are first felt by the poorest sector of the community as they walk further or pay more for over-exploited medicinal plants. The direct costs of such resource depletion do not only affect the rural poor however. Ultimately, the loss of potentially valuable genetic resources affects the whole of society.

Cultivation of medicinal plants as an alternative source of supply was proposed over 40 years ago by Gerstner (1946), specifying *Warburgia salutaris* as one of the tree species most in need of cultivation. Nothing was done. Today this species is considered endangered in South Africa (Cunningham, 1988) and is heavily exploited elsewhere in Africa (FAO, 1986). If such initiatives are to succeed they need the support they deserve.

Acknowledgements

I am greatly indebted to all resource users – gatherers, traditional practitioners and herb traders. Without their support and knowledge, this project would have failed. The financial support and suggestions from all organisations on the project steering commiteee (Herb Traders Association, Nyangas National Association, Natal Parks Board, CSIR, HSRC, KwaZulu Bureau of Natural Resources and DDA) is most appreciated. Financial assistance from the CSIR National Programmes Section, the Institute of Research Development of the Human Sciences Research Council and the University of Natal enabled my attendance at this conference and is gratefully acknowledged.

References

Berglund, A-I. (1976). *Zulu thought patterns and symbolism.* Uppsala: Almquist & Wiksell.

Boerboom, J.H.A. & Wiersum, K.F. (1983). Human impact on moist tropical forest. pp. 83 - 106 In *Man's Impact on Vegetation,* ed. W. Holzner, M.J.A. Werger & I. Ikusima. The Hague: W Junk.

Box, E.O. (1981). *Macroclimate and plant forms : An introduction to predictive modelling in phytogeography.* The Hague: W. Junk.

Connell, J.H., Tracey, J.G. & Wells, L.J. (1984). Compensatory recruitment, growth and mortality as factors maintaining rainforest tree diversity. *Ecological Monographs,* 54(2), 141-164.

Cooper, K.H. (1979). 'N sluikhandel in Stinkhout deur die toordokters. *African Wildlife,* 33(3), 8.

Cooper, K.H. (1985). *The conservation status of indigenous forests in Transvaal, Natal and O.F.S, South Africa.* Durban: Wildlife Society of S.A.

Cunningham, A.B. (1984). Medicinal plants and witch-doctors: are we barking up the wrong tree? *African Wildlife* ,38(6), 247-249.

Cunningham, A.B. (1988). *An investigation of the herbal medicine trade in Natal/KwaZulu.*

Investigational Report No. 29, Institute of Natural Resources, University of Natal.

Cunningham, A.B. & Milton, S.J. (1987). Effects of the basket weaving industry on the mokola palm (*Hyphaene petersiana*) and on dye plants in NW Botswana. *Economic Botany*, 42, 386-402.

Dally, K. (1984). Illegal debarking of stinkwood. *Forestry News*, 2/84, 23.

FAO (1986). *Some medicinal forest plants of Africa and Latin America.* Rome: FAO Forest Resources Development Branch, Forest Resources Division, FAO.

Gelfand, M., Mavi, S., Drummond, R.B. & Ndemera, B. (1985). *The traditional medical practitioner in Zimbabwe.* Harare: Mambo Press.

Gentry, H.S., Verbiscar, A.J. & Branigan, T.F. (1987). Red squill (*Urginea maritima*, Liliaceae). *Economic Botany*, 41(2), 267-282.

Gerstner, J. (1938). A preliminary checklist of Zulu names of plants. *Bantu Studies*, 12, 215-236.

Gerstner, J. (1946). Some factors affecting the perpetuation of our indigenous flora. *Journal of the South African Forestry Association*, 13, 4-11.

Gibbs-Russell, G.E., Welman, W.G., Retief, E, Immelman, K.L., Germishuisen, G., Pienaar, B.J., van Wyk, M. & Nicholas, A. (1987). *List of species of southern African plants*, 2nd edn, Part 2. Memoirs of the Botanical Survey of South Africa, No. 56. Pretoria: Botanical Research Institute.

Grobbelaar, J.A. (1985). *The population of Natal/KwaZulu 1904 - 2010.* Natal Town and Regional Planning Report, Vol. 68.

Hall, A.V., De Winter, M., De Winter, B. & Van Oosterhout, S.A.M. (1980). *Threatened plants of Southern Africa.* South African National Scientific Programmes Report no. 45. Pretoria: Council for Scientific and Industrial Research.

Leloup, S. (1984). *The grapple plant project: an ecophysiological approach of the influence of harvest on the population dynamics of the Grapple plant Harpagophytum procumbens* DC. National Institute for Development Research and Documentation, Gaborone.

Lucas, G. & Synge, H. (1978). *The IUCN Plant Red Data Book.* Morges: IUCN.

MacDonald, I.A.W. (1984). Witchdoctors versus wildlife in southern Africa. *African Wildlife*, 38(1), 4-9.

Medley-Wood, J. & Evans, M.S. (1898). *Natal Plants*, 1(1), 27. Durban: Bennett and Davis.

Moll, E.J. (1981). *Trees of Natal.* Cape Town: ABC Press.

Moll, E.J. & Haigh, H. (1966). A report on the Xumeni forest, Natal. *Forestry in South Africa*, 7, 99-108.

Mueller-Dombois, D. & Ellenberg, H. (1974). *Aims and methods of vegetation ecology.* New York: Wiley.

Ngubane, H. (1977). *Body and mind in Zulu medicine.* New York: Academic Press.

Noble, I.R. & Slatyer, R.O. (1980). The use of vital attributes to predict successional changes in plant communities subject to recurrent disturbances. *Vegetatio*, 43, 5-21.

Noel, A.R.A. (1961). A preliminary account of the effect of grazing upon species of *Helichrysum* in the Amatola mountains. *Journal of South African Botany*, 27, 81-86.

Oatley, T. B. (1979). *Status of Black Stinkwood in the Karkloof Forest.* Internal Report, Natal Parks Board, Pietermaritzburg, South Africa.

O'Connor, T.G. (1985). *A synthesis of field experiments concerning the grass layer in the savanna regions of southern Africa.* South African National Scientific Programmes Report No. 114.

Phillips, J.F.V. (1931). *Forest succession and ecology in the Knysna region.* Memoirs of the Botanical Survey of South Africa, No. 14.

Raunkiaer, C. (1937). *Plant life forms.*, Oxford: Clarendon Press.

Rutherford, M.C. (1978). Primary production ecology in southern Africa. In *Biogeography and ecology in southern Africa,* ed. M.J.A. Werger, pp. 621-659. The Hague: W. Junk.

Rutherford, M.C. & Westfall, R.H. (1986). Biomes of southern Africa – an objective categorization. *Memoirs of the Botanical Survey of South Africa,* 54.

Staugard, F. (1985). Traditional healers: traditional medicine in Botswana. Gaborone: Ipelegeng Publishers.

Story, R. (1952). A botanical survey of the Keiskammahoek district. *Memoirs of the Botanical Survey of South Africa,* 27.

Walters, C. (1986). *Adaptive management of renewable resources.* London: MacMillan.

Appendix

Vulnerability to over-exploitation is considered to be a function of life form, part used, distribution in Natal province, South Africa, as well as the demand for a particular species. Demand was represented by the number of bags of each species sold anually by 54 herb traders surveyed (Cunningham, 1988). The short-list of species given here represents five categories of threat or vulnerability to over-exploitation (after Hall *et al.*, 1980).

Extinct in the Wild

No wild populations in former localities although the species may exist under cultivation.

The only species falling into this category would be *Siphonochilus natalensis* which Gibbs-Russell *et al.* (1987) list separately from *S. aethiopicus* but may be synonymous with *Siphonochilus aethiopicus* (R.M. Smith, pers. comm.).

Genus	SPECIES	Zulu Name	Life Form	Part	Dist
Siphonochilus	natalensis	Phephetho, -isi	G1	r	r

Endangered

Genus	Species	Zulu Name	Life Form	Part	Dist
Warburgia	salutaris	Bhaha, -isi	P2	bk	r
Siphonochilus	aethiopicus	Phephetho, -isi	G1	r	r

Vulnerable and Declining

Genus	Species	Zulu Name	Life Form	Part	Dist
Dioscorea	sylvatica	Ngwevu, -i	VL3	wpl	ul
Bersama	tysoniana	Diyaza, -un	P3	bk	uw
Ocotea	bullata	Nukani, -u	P2	bk	ul
Curtisia	dentata	Lahleni, -um	P2 bk	uw	
Pleurostylia	capensis	Thunyelelwa, -u	P2	bk r	ul(r)
Faurea	macnaughtonii	Sefo, -isi	P2	bk	uw
Olea	woodiana	Hlwazimamba, -u	P2	bk r	uw
Loxostylis	alata	Futhu, -u	P2	bk	Ul(r)
Ocotea	kenyensis		P2	bk	r
Haworthia	limifolia	Mathithibala	C4	wpl	ul
Mystacidium	millari	Phamba, -i	E1	wpl	r

Declining

Species that were reently widespread but are likely to become vulnerable and to continue to decline if destruction of wild populations continues.

Genus	Species	Zulu Name	Life Form	Part	Dist
Mondia	shitei	Mondi, -u	VL1	r	ul(r)
Schlechterina	mitrostemmatoides	Hlalanyosi-emhlophe, -i	VL1	wpl	ul(r)
Acridocarpus	natalitius	Mabophe-omkhulu, -u	VL1	r	ul
Andrachne	ovalis	Mbesa, -u	P4	r	uw
Garcinia	gerrardii	Binda, -um	P3	bk	ul
Cryptocarya	latifolia	Khondweni, -um	P2	bk	ul
Erythrophleum	lasianthum	Khwangu, -um	P2	bk	uw
Prunus	africana	Yazangoma-elimnyama, -in	P2	bk	ul
Cassine	transvaalensis	Ngwavuma, -i	P2	bk	uw
Balanites	maughamii	Nulu, -um	P2	bk r	ul
Harpephyllum	caffrum	Gwenya, -um	P2	bk	ul
Alepidea	amatymbica	Khathazo, -i	H3	r	cw
Anemone	fanninii	Manzemnyama, -a	H2	r	ul
Knowltonia	bracteata	Vuthuza, -um	H2	wpl	ul
Anemone	caffra	Manzemnyama, -u	H2	r	ul
Cassipourea	gerrardii	Memeze-obomvu	P2	bk	ul
Cassipourea	flanaganii	Memeze-obumvu	P2	bk	ul
Pterocelastrus	tricuspidatus	Sahlulamanye, -u	P2	bk	ul
Pterocelastrus	rostratus	Sahlulamanye, -u	P2	bk	ul
Pterocelastrus	echinatus	Sahlulamanye, -u	P2	bk	ul
Cassine	papillosa	Sehlulamanye, -u	P2	bk	uw
Boweia	volubilis	Gibisila, -i	VL3	bbf	uw
Bersama	swynii	Diyaza, -un	P4	bk	ul(r)
Scilla	natalensis	Guduza, -in	G4	bb	cl
Eucomis	autumnalis	Mathunga, -u	G4	bb	cl
Eulophia	cucullata	Mabelejongosi, -u	G2	bb	cl

Rare and Vulnerable

Species with relatively small populations that are vulnerable to over-exploitation if exploitation for medicinal purposes increases.

Genus	Species	Zulu Name	Life Form	Part	Dist
Gerrardina	foliosa		P4	bk	ul
Encephalartos	species	Qikisomkovu, -i	P3	st	
Cyathea	dregei	Nkomankoma, -i	P3	st	uw
Albizia	suluensis	Ngwebunkulu, -u	P2	bk	r
Hydrostachys	polymorpha	Phophoma-lasemanzi	H4	wpl	ul(r)
Asclepias	cucullata	Delenina, -u	H3	r	ul
Peucadanum	thodii	Mpondovu, -i	H2	l/st	ul
Haemanthus	albiflos	Zaneke, -u	G4	bb	ul
Eriospermum	mackenii	Nsulansula, -i	G4	bb	cl
Eriospermum	ornithogaloides	Ncameshela, -i	G4	bb	cl
Haemanthus	deformis	Zaneke, -u	G4	bbf	ul
Schizobasis	intricata	Thondo-wemfene, -um	G4	bb	uw
Stangeria	eriopus	Fingo, -im	G3	r(lt)cl	
Eulophia	petersii	Saha, -i	G2	wpl	ul
Clivia	nobilis	Mayime, -u	G1	bb	ul
Clivia	miniata	Mayime, -u	G1	bb	ul
Ansellia	gigantea	Fenkawu, -im	E1	l/st	uw
Aloe	aristata	Mathithibala, -u	C4	wpl	ul
Euphorbia	woodii	Nhlehle, -i	C4	wop	ul
Gasteria	courcheri	Mpundi, -i	C4	wpl	uw
Euphorbia	pulvinata	Kamamasane, -in	C4	wop	ul
Justicia	capensis	Khokhela, -i	C1	wpl	ul
Myrothamnus	flabellifolia	Vukwabafile, -u	C1	l/st	ul

Indeterminate

Species whose status is uncertain, but which appear to be heavily exploited and for which more data is required.

Genus	Species	Zulu Name	Life Form	Part	Dist
Dioscorea	rupicola	Mpinyampinya, -i	VL3	wpl	uw
Discorea	dregeana	Dakwa, -isi	VL2	wpl	us
Dumasia	villosa	Khalimele, -u	VL2	l/st	ul
Adenia	gummifera	Fulwa, -im	VL1	st	ul
Helinus	integrifolius	Bhubhubhu, -u	VL1	st	cl
Embelia	ruminata	Bhinini, -in	VL1	r	uw
Tylophora	flanaganii	1Nhlanhla, -i	VL1	st	uw
Secamone	gerrardii	Phophoma, i	VL1	r	uw
Rhoicissus	tridentata	Nwazi, -isi	VL2	r	cw
Osyridicarpos	schimperianus	Malala, -u	VL1	l/st	ul
Cycnium	racemosum	Hlabahlangane, -u	PR2	r	ul
Sarcophyte	sanguinea	Mafumbuka, -u	PR1	r	ul
Hyndora	africana	Mafumbuka, -u	PR1	r	ul(r)

Indeterminate (cont.)

Genus	Species	Zulu Name	Life Form	Part	Dist
Zanthoxylum	davyi	Nungwane, -isi	P3	r	ul
Faurea	saligna	Sefo, -isi	P3	bk	ul
Ekebergia	cpensis	Mathunzini-wentaba	P2	bk	ul
Ilex	mitis	Dumo, -isi	P2	bk	uw
Protorhus	longifolia	Hlothe, -in	P2	bk	cl
Aster	bakeranus	Dlutshane, -i	H3	r	cw
Gerbera	piloselloides	Moya-wezwe, -u	H3	wpl	cw
Piminella	caffra	Bheka, -i	H2	wpl	cw
Polygala	fruitcosa	Thethe, -i	H1	wpl	uw
Tephrosia	sp.cf. marginella	Dala, -isi	H1	l/st	ul
Polygala	serpentaria	Ngqengendlela, -i	H1	wpl	uw
Polygala	ohlendorfiana	Ngqengendlela	H1	wpl	uw
Polygala	myrtifolia	Chwasha, -u	H1	wpl	uw
Polygala	gerrardii	Ngqengendlela, -i	H1	wpl	uw
Polygala	cf. natalensis	Ngqengendlela, -i	H1	wpl	uw
Polygala	confusa	Ngqengendlela, -i	H1	wpl	ul
Polygala	marensis	Ngqengendlela, -i	H1	wpl	uw
Crabbea	hirsuta	Musa, -u	H1	wpl	cl
Begonia	homonyma	Dlula, -i	H1	r	ul
Tragia	rupestris	Babazane, -im	H1	r l/st	ul
Dianthus	zeyheri	Ningizimu, -i	H1	wpl	ul
Crinum	moorei	Duze, -um	G4	bb	cl
Crinum	macowanii	Duze, -um	G4	bbf	cl
Crinum	delagoense	Duze, -um	G4	bb	ul
Drimia	elata	Klenama, -isi	G4	bb	cw
Urginea	altissima	Mahlogolosi, -u	G4	bb	cl
Eriospermum	mackenii	Nsulansula, -i	G4	bb	cl
Eriospermum	ornithogaloides	Ncamesthela, -i	G4	bb	cl
Schizobasis	intricata	Thondo-wemfene	G4	bb	uw
Boophane	disticha	Ncotho, -i	G4	bb	uw
Albuca	fastigiata	Maphipha-intelezi-u	G4	bb	cl
Drimia	robusta	Dongana-zibomvana	G4	bb	cl
Eriospermum	cooperi	Ncamashela, -i	G4	bb	cl
Eriospermum	luteo-rubrum	Mathintha, -u	G4	bb	cl
Turbina	oblongata	Bhogo, -u	G3	r	cl
Helichrysum	acutatum	Zangume, -u	G3	r	cw
Convolvulus	saggitatus	Vimbukalo, -u	G3	r	cl

Proposals for International Collaboration

Olayiwola Akerele,
Programme Manager, Traditional Medicine, Division of Diagnostic, Therapeutic and Rehabilitative Technology, World Health Organization, Geneva

If I may be allowed to play the medicinal plants farmer, the opening days of our Consultation were seeding time. Our discussions could be compared to weeding and fertilizing time. We have now reached the critical harvesting time and I am happy to announce that we have a bumper crop.

We have looked at our subject from many aspects. We have shared experiences as well as ideas as to where to go from here. We need now to summarize our discussions to capture the essence of what has gone on during the past days and to provide a framework for future activities.

I have been asked to look at the future from an international perspective with a view to identifying areas for collaboration. From the beginning, I must underscore the fact that international involvement cannot be separated from what is going on in the countries themselves. On the contrary, its whole 'raison d'etre' lies in the needs identified at national level.

Furthermore, it is only by interacting with and supporting national programmes as well as by clearly defining objectives and priorities for action that we can learn how best these needs can be met. So, I will include in my framework the main components of a national plan of action. Beginning with these, I shall outline the linkages that must be forged at both national and international levels.

The primary health care approach, amongst other things, calls for increased intersectoral planning, programming and delivery and for a common approach in solving the problems of health as part of general socioeconomic development. It also emphasizes the need to make maximum use of all available resources. This is why the Alma-Ata Conference in 1978 recommended that governments give high priority to the utilization of traditional medicine and the incorporation of proven traditional remedies into national drug policies and regulations.

The challenge that faces governments in respect of medicinal plants is how to get the many different national bodies, departments and institutions to coordinate their activities and work together.

Where it does not already exist, some form of coordinating committee may be needed to bring together the various groups involved, which include the health, agriculture, trade and industry sectors and the universities. Dr Pricha, in his background paper, has described the action being taken by the Thai Authorities in this regard.

Such an Advisory Committee on Medicinal Plants and Natural Products would assist the Ministry of Health in the formulation of a national policy in this field and would help to ensure the orderly development of work and research by the various institutions involved.

Without attempting to cover all aspects of the subject, a national programme concerning medicinal plants should include the following institutions and activities:

Ministry of Health

- Formulation of national policy
- Legislation and licensing
- Collection, analysis and dissemination of information on medicinal plants
- Approval of selected plant remedies for use by health services
- Banning the use of dangerous plants and products

Ministry of Agriculture

- Cultivation of medicinal plants (small and large-scale production)
- Protection of endangered species

Universities

Pharmacy

- Inventory of indigenous medicinal plants and natural products
- Identification of constituents of traditional remedies
- Pharmacological evaluation of medicinal plants and natural products
- Identification of active substances, their extraction and toxicity testing
- Dosage and formulation

Clinical Medicine

- Clinical trials and field testing

Public Health

- Studies on indigenous remedies and their uses (ethno-medicine)
- Training of health personnel (manuals)

Botany

- Cultivation of medicinal plants
- Cloning and cell culture
- Taxonomic identification

	• Studies on ethnobotany
Pharmaceutical Industry (government or private)	• Pharmaceutical development • Processing and pilot production • Trial marketing • Full-scale production
Ministry of Trade	• Assessment of local trade in medicinal plants • Exports and imports

Certain functions and responsibilities might be allocated in different ways, depending on the national situation. For example, the creation of a Medicinal Plant Information System, with access to NAPRALERT and other data bases, might be undertaken by a Ministry of Health or by a University Faculty of Medicine or Science.

It may of course be desirable to involve many other groups but not necessarily on a permanent basis.

One way of getting the various individuals and institutions concerned to work together would be to hold a national workshop to focus attention on the need for rational utilization of medicinal plants. This should bring together national experts on different aspects of the subject to make a general assessment of the situation, define objectives, set priorities, and draw up a national plan of action.

The role of international organizations, as I see it, would now be to assist national authorities to design and implement their own programmes. Within their different fields of expertise, the international organizations would collaborate with their national counterparts in applying the guidelines you have prepared and adapting them to local circumstances.

During the week we have been able to provide some guidance on the ways in which national authorities can become involved in the rational utilization of medicinal plants in their health systems. There are, of course, still some topics which have not been covered but, in spite of these gaps, we are now in a position to encourage national authorities to take the initiative in promoting the rational utilization of medicinal plants.

In the first instance, what we need is a few countries in the African, American, Eastern Mediterranean, South-East Asian and the Western Pacific Regions which are willing to establish at national level the kind of coordination mechanism that I have outlined. If these countries could test the guidelines that have been developed during the past few days, they could then be modified in the light of experience and subsequently used as a basis for intercountry or regional workshops. This would be a prelude to the introduction of simiiar programmes in other countries.

In summary, international collaboration in the rational utilization and conservation of medicinal plants should, from WHO's point of view, include:

- Promotion of national and international commitment
- International technical support
- Assisting in the development of national strategies
- Increasing public awareness
- Mobilizing resources
- Involving non-governmental organizations
- Technical cooperation among developing countries
- International coordination

This framework provides a convenient way of classifying the various points of discussion. It does not, however, tell us what we must do next. This requires further consideration. No doubt we will have other consultations, probably at the regional level, in order to reach more countries and to be better in tune with regional aspirations and activities. But perhaps the fundamental question is what mechanism should we adopt to ensure that our recommendations are followed up. I would include this under the last item – international coordination.

As I mentioned in my opening remarks, new relationships have to be created together with new mandates and policies supporting such relationships. We need to see that our efforts are backed up by precise mandates from our respective governing bodies so that funds may be made available for the development and implementation of this exciting joint programme.